Learning in Energy-Efficient Neuromorphic Computing

Learning in Energy-Efficient Neuromorphic Computing

Algorithm and Architecture Co-Design

Nan Zheng

The University of Michigan
Department of Electrical Engineering and Computer Science
Michigan, USA

Pinaki Mazumder

The University of Michigan
Department of Electrical Engineering and Computer Science
Michigan, USA

Registered Offices
John Wiley & Sons, Inc., 111 River Street, Hoboken, NJ 07030, USA
John Wiley & Sons Ltd, The Atrium, Southern Gate, Chichester, West Sussex, PO19 8SQ, UK

Editorial Office
The Atrium, Southern Gate, Chichester, West Sussex, PO19 8SQ, UK

For details of our global editorial offices, customer services, and more information about Wiley products visit us at www.wiley.com.

Wiley also publishes its books in a variety of electronic formats and by print-on-demand. Some content that appears in standard print versions of this book may not be available in other formats.

Library of Congress Cataloging-in-Publication Data

Names: Zheng, Nan, 1989- author. | Mazumder, Pinaki, author.
Title: Learning in energy-efficient neuromorphic computing : algorithm and
    architecture co-design / Nan Zheng, Pinaki Mazumder.
Description: Hoboken, NJ : Wiley-IEEE Press, [2020] | Includes
    bibliographical references and index.
Identifiers: LCCN 2019029946 (print) | LCCN 2019029947 (ebook) | ISBN
    9781119507383 (cloth) | ISBN 9781119507390 (adobe pdf) | ISBN
    9781119507406 (epub)
Subjects: LCSH: Neural networks (Computer science)
Classification: LCC QA76.87 .Z4757 2019 (print) | LCC QA76.87 (ebook) |
    DDC 006.3/2–dc23
LC record available at https://lccn.loc.gov/2019029946
LC ebook record available at https://lccn.loc.gov/2019029947

Cover design: Wiley
Cover image: © monsitj/Getty Images

Set in 10/12pt WarnockPro by SPi Global, Chennai, India
Printed and bound in Singapore by Markono Print Media Pte Ltd

10 9 8 7 6 5 4 3 2 1

Learning is the beginning of wealth.
Learning is the beginning of health.
Learning is the beginning of spirituality.
Searching and learning is where the miracle process all begins.
–Emanuel James Rohn

In loving memory of my father Animesh Chandra Mazumder (1918–2011).

Contents

Preface

In 1987, when I was wrapping up my doctoral thesis at the University of Illinois, I had a rare opportunity to listen to Prof. John Hopfield of the California Institute of Technology describing his groundbreaking research in neural networks to spellbound students in the Loomis Laboratory of Physics at Urbana-Champaign. He didactically described how to design and fabricate a recurrent neural network chip to rapidly solve the benchmark Traveling Salesman Problem (TSP), which is provably NP-complete in the sense that no physical computer could solve the problem in asymptotically bounded polynomial time as the number of cities in the TSP increases to a very large number.

This discovery of algorithmic hardware to solve intractable combinatorics problems was a major milestone in the field of neural networks as the prior art of perceptron-type feedforward neural networks could merely classify a limited set of simple patterns. Though, the founder of neural computing, Prof. Frank Rosenblatt of Cornel University had built a Mark 1 Perceptron computer in the late 1950s when the first waves of digital computers such as IBM 650 were just commercialized. Subsequent advancements in neural hardware designs were stymied mainly because of lack of integration capability of large synaptic networks by using the then technology, comprising vacuum tubes, relays, and passive components such as resistors, capacitors, and inductors. Therefore, in 1985, when AT&T Bell Labs fabricated the first solid-state proof-of-concept TSP chip by using MOS technology to verify Prof. John Hopfield's neural net architecture, it opened the vista for solving non-Boolean and brain-like computing on silicon.

Prof. John Hopfield's seminal work established that if the "objective function" of a combinatorial algorithm can be expressed in quadratic form, the synaptic links in a recurrent artificial neural network could be accordingly programmed to reduce (i.e. locally minimize) the value of the objective function through massive interactions between the constituent neurons. Hopfield's neural network consists of laterally connected neurons that can be randomly initialized and then the network can iteratively reduce the intrinsic Lyapunov energy function of the network to reach a local minima state. Notably, the Lyapunov function decreases in a monotone fashion under the dynamics of the recurrent neural networks, where neurons are not provided with self-feedback.[1]

1 In Mazumder and Yih [1], we demonstrated that the quality of solutions achieved by the Hopfield network could be improved considerably by selectively providing self-feedback to allow neurons to escape the local minima. Such a method is akin to hill climbing in a gradient descent search that often is trapped in a local minimum point. As self-feedback to neurons impair the stability of a Hopfield neural network, we did not apply any self-feedback to neurons until the network converged to a local minimum state. Then we pulled the network out of the local minimum by elevating the energy of the network through the hill climbing

Prof. Hopfield used a combination of four separate quadratic functions to represent the objective function of the TSP. The first part of the objective function ensures that the energy function minimizes if the traveling salesman traverses cities exactly once, the second part ensures that the traveling salesman visits all cities in the itinerary, the third part ensures that no two cities are visited simultaneously, and the fourth part of the quadratic function is designed to determine the shortest route connecting all cities in the TSP. Because of massive simultaneous interactions between neurons through the connecting synapses that are precisely adjusted to meet the constraints in the above quadratic functions, a simple recurrent neural network could rapidly generate a very good quality solution. However, unlike well-tested software procedures such as simulated annealing, dynamic programming, and the branch-and-bound algorithm, neural networks generally fail to find the best solution because of their simplistic connectionist structures.

Therefore, after listening to Prof. Hopfield's fascinating talk I harbored a mixed feeling about the potential benefit of his innovation. On the one hand, I was thrilled to learn from his lecture how computationally hard algorithmic problems could be solved very quickly by using simple neuromorphic CMOS circuits having very small hardware overheads. On the other hand, I thought that the TSP application that Prof. Hopfield selected to demonstrate the ability of neural networks to solve combinatorial optimization problems was not the right candidate, as software algorithms are well crafted to obtain nearly the best solution that the neural networks can hardly match. I started contemplating developing self-healing VLSI chips where the power of neural-inspired self-repair algorithms could be used to automatically restructure faulty VLSI chips. Low overheads and the ability to solve a problem concurrently through parallel interactions between neurons are two salient features that I thought could be elegantly deployed for automatically repairing VLSI chips by built-in neural net circuitry.

Soon after I joined the University of Michigan as an assistant professor, working with one of my doctoral students [2], and, at first, I developed a CMOS analog neural net circuitry with asynchronous state updates, which lacked robustness due to process variation within a die. In order to improve the reliability of the self-repair circuitry, an MS student [3] and I designed a digital neural net circuitry with synchronous state updates. These neural circuits were designed to repair VLSI chips by formulating the repair problem in terms of finding the node cover, edge cover, or node pair matching in a bipartite graph. In our graph formalism, one set of vertices in the bipartite graph represented the faulty circuit elements, and the other set of vertices represented the spare circuit elements. In order to restructure a faulty VLSI chip into a fault-free operational chip, the spare circuit elements were automatically invoked through programmable switching elements after identifying the faulty elements through embedded built-in self-testing circuitry.

Most importantly, like the TSP problem, the two-dimensional array repair can be shown to be an NP-complete problem because the repair algorithm seeks the

mechanism. By using this innovative technique, we showed in the above paper that yield of VLSI memories could be ameliorated by increasing the repair by about 25%.

optimal number of spare rows and spare columns that can be assigned to bypass faulty components such as memory cells, word-line and bit-line drivers, and sense amplifier bands located inside the memory array. Therefore, simple digital circuits comprising counters and other blocks woefully fail to solve such intractable self-repair problems. Notably, one cannot use external digital computers to determine how to repair embedded arrays, as input and output pins of the VLSI chip cannot be deployed to access the fault patterns in the deeply embedded arrays.

In 1989 and 1992, I received two NSF grants to expand the neuromorphic self-healing design styles to a wider class of embedded VLSI modules such as memory array [4], processors array [5], programmable logic array, and so on [6]. However, this approach to improving VLSI chip yield by built-in self-testing and self-repair was a bit ahead of its time as the state-of-the-art microprocessors in the early 1990s contained only a few hundred thousands of transistors as well as the submicron CMOS technology that was relatively robust. Therefore, after developing the neural-net based self-healing VLSI chip design methodology for various types of embedded circuit blocks, I stopped working on CMOS neural networks. I was not particularly interested in pursuing applications of neural networks for other types of engineering problems, as I wanted to remain focused on solving emerging problems in VLSI research.

On the other hand, in the late 1980s there were mounting concerns among CMOS technology prognosticators about the impending red brick wall heralding the end of the shrinking era in CMOS. Therefore, to promote several types of emerging technologies that might push the frontier of VLSI technology, the Defense Advanced Research Projects Agency (DARPA) in the USA had initiated (around 1990) the Ultra Electronics: Ultra Dense, Ultra Fast Computing Components Research Program. Concurrently, the Ministry of International Trade & Industry (MITI) in Japan had launched the Quantum Functional Devices (QFD) Project. Early successes with a plethora of innovative non-CMOS technologies in both research programs led to the launching of the National Nanotechnology Initiative (NNI), which is a U.S. Government research and development (R&D) initiative, involving 20 departments and independent agencies to bring about revolution in nanotechnology to impact the industry and society at large.

During the period of 1995 and 2010, my research group had at first focused on a quantum physics based device and circuit modeling for quantum tunneling devices, and then we extensively worked on cellular neural network (CNN) circuits for image and video processing by using one-dimensional (double barrier resonant tunneling device), two-dimensional (self-assembled nanowire), and three-dimensional (quantum dot array) constrained quantum devices. Subsequently, we developed learning-based neural network circuits by using resistive synaptic devices (commonly known as memristors) and CMOS neurons. We also developed analog voltage programmable nanocomputing architectures by hybridizing quantum tunneling and memristive devices in computing nodes of a two-dimensional processing element (PE) ensemble. Our research on nanoscale neuromorphic circuits will soon be published in our new book, titled: *Neuromorphic Circuits for Nanoscale Devices*, River Publishing, U.K., 2019.

After spending a little over a decade developing neuromorphic circuits with various types of emerging nanoscale electronic and spintronic devices, I decided to embark on research on learning-based digital VLSI neuromorphic chips using nanoscale CMOS technology in both subthreshold and superthreshold modes of operation. My student and coauthor of this book, Dr. Nan Zheng, conducted his doctoral dissertation work on architectures and algorithms for digital neural networks. We started our design from both machine learning and biological learning perspectives to design and fabricate energy-efficient VLSI chips using TSMC 65 nm CMOS technology.

We captured the actor-critic type reinforcement learning (RL) [7] and an example of temporal difference (TD) learning with off-line policy updating, called Q-learning [8] on VLSI chips from the machine learning perspectives. Further, we captured spiking correlation-based synaptic plasticity commonly used in biological unsupervised learning applications. We also formulated hardware-friendly spike-timing-dependent plasticity (STDP) learning rules [9], which achieved classification rates of 97.2% and 97.8% for the one-hidden-layer and two-hidden-layer neural networks, respectively, on the Modified National Institute of Standards and Technology (MNIST) database benchmark. The hardware-friendly learning rule enabled both energy-efficient hardware design [10] as well as implementations that were robust to the process-voltage-temperature (PVT) variations associated with chip manufacturing [11]. We demonstrated that the hardware accelerator VLSI chip for the actor-critic network solved some control-theoretic benchmark problems by emulating the adaptive dynamic programming (ADP), which is at the heart of the RL software program. However, compared with traditional software RL running on a general-purpose processor, the VLSI chip accelerator operating at 175 MHz achieves two orders of magnitude improvement in computational time while consuming merely 25 mW [12].

The chip layout diagrams included in the Preface contain a sample of numerous digital neural network chips using CMOS technology that my research group has designed over the course of the last 35 years. On the left column: a self-healing chip was designed in 1991 to repair faulty VLSI memory arrays automatically by running node-covering algorithms on a bipartite graph representing the set of faulty components and the available spare circuit elements. The STDP chip was designed in 2013 for controlling the motion of a virtual insect from an initial source to the selected destination by avoiding collisions while navigating through a set of arbitrarily shaped blocked spaces. A deep learning chip described in the previous paragraph was designed in 2016.

On the right column is shown the RL chip described in the above paragraph and designed in 2016. Also included on the right column are two ultra-low-power (ULP) CMOS chips, which were biased in the subthreshold mode for wearable applications in health care. In one application, Kohonen's self-organizing map (SOM) of neural networks was implemented to classify electrocardiogram (ECG) waveforms, while the body-sensing network with a wireless transceiver was designed to sense analog neuronal signals by using an implantable multielectrode sensor and to provide the digitized data through a built-in wake-up transceiver to doctors who could monitor the efficacy of drugs at the neuronal and synaptic levels in brain-related diseases such as schizophrenia, chronic depression, Alzheimer disease, and so on.

Initially, when we decided to publish a monograph highlighting our work in the form of CMOS neuromorphic chips for brain-like computing, we wanted to aggregate various results of the cited papers in the Preface to compose the contents of the book. However, in the course of preparation of the manuscript, we modified our initial narrow goal, as it would be rather limiting to adopt the book in a regular course for teaching undergraduate and graduate students about the latest generation neural networks with learning capabilities.

Instead, we decided to write a comprehensive book on energy-efficient hardware design for neural networks with various types of learning capability by discussing expansive ongoing research in neural hardware. This is evidently a Herculean task warranting mulling through hundreds of archival sources of references and describing co-design and co-optimization methodologies for building hardware neural networks that can learn to perform various tasks. We attempted to provide a comprehensive perspective, from high-level algorithms to low-level implementation details by covering many fundamentals and essentials in neural networks (e.g. deep learning), as well as hardware implementation of neural networks. In a nutshell, the present version of the book has the following salient features:

- It includes a cross-layer survey of hardware accelerators for neuromorphic algorithms;
- It covers the co-design of architecture and algorithms with emerging devices for much-improved computing efficiency;
- It focuses on the co-design of algorithms and hardware, which is paramount for deploying emerging devices such as traditional memristors or diffusive memristors for neuromorphic computing.

Finally, due to the stringent time constraint to complete this book in conjunction with the commitment to concurrently finish the complementary book (*Neuromorphic Circuits for Nanoscale Devices*, River Publishing, U.K., 2019), the present version of the book has been completed without describing the didactic materials pedagogically as expected in a textbook along with exercise problems at the end of each chapter. Hopefully, those goals will be achieved in the next edition of the book after gathering valuable feedback from students, instructors, practicing engineers, and other readers. I shall truly appreciate it if you give me such guiding feedback, both positive and negative, which will enable me to prepare the Second Edition of the book. My contact information is included below for your convenience.

Pinaki Mazumder   February 14, 2019
Address:
4765 BBB Building
Division of Computer Science and Engineering Department of Electrical Engineering and Computer Science University of Michigan, Ann Arbor, MI 48109–2122; Ph: 734-763-2107; E-mail: mazum@eecs.umich.edu, pinakimazum@gmail.com
Website: http://www.eecs.umich.edu/~mazum

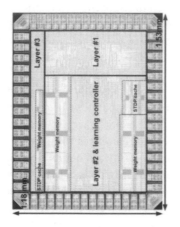

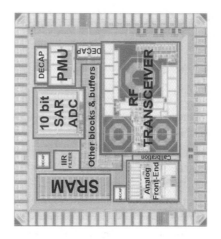

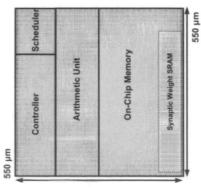

References

1 Mazumder, P. and Yih, J. (1993). A new built-in self-repair approach to VLSI memory yield enhancement by using neural-type circuits. *IEEE Trans. Comput. Aided Des. Integr. Circuits Syst.* 12 (1): 124–136.

2 Mazumder, P. and Yih, J. (1989). Fault-diagnosis and self-repairing of embedded memories by using electronic neural network. In: *Proc. of IEEE 19th Fault-Tolerant Computing Symposium*, 270–277. Chicago.

3 Smith, M.D. and Mazumder, P. (1996). Analysis and design of Hopfield-type network for built-in self-repair of memories. *IEEE Trans. Comput.* 45 (1): 109–115.

4 Mazumder, P. and Yih, J. (1990). Built-in self-repair techniques for yield enhancement of embedded memories. In: *Proceedings of IEEE International Test Conference*, 833–841.

5 Mazumder, P. and Yih, J. (1993). Restructuring of square processor arrays by built-in self-repair circuit. *IEEE Trans. Comput. Aided Des. Integr. Circuits Syst.* 12 (9): 1255–1265.

6 Mazumder, P. (1992). An integrated built-in self-testing and self-repair of VLSI/WSI hexagonal arrays. In: *Proceedings of IEEE International Test Conference*, 968–977.

7 Zheng, N. and Mazumder, P. (2017). Hardware-friendly actor-critic reinforcement learning through modulation of spiking-timing dependent plasticity. *IEEE Trans. Comput.* 66 (2).

8 Ebong, I. and Mazumder, P. (2014). Iterative architecture for value iteration using memristors. In: *IEEE Conference on Nanotechnology, Toronto*, 967–970. Canada.

9 Zheng, N. and Mazumder, P. (2018). Online supervised learning for hardware-based multilayer spiking neural networks through the modulation of weight-dependent spike-timing-dependent plasticity. *IEEE Trans. Neural Netw. Learn. Syst.* 29 (9): 4287–4302.

10 Zheng, N. and Mazumder, P. (2018). A low-power hardware architecture for on-line supervised learning in multi-layer spiking neural networks. In: *2018 IEEE International Symposium on Circuits and Systems (ISCAS)*, 1–5. Florence.

11 Zheng, N. and Mazumder, P. (2018). Learning in Memristor crossbar-based spiking neural networks through modulation of weight dependent spike-timing-dependent plasticity. *IEEE Trans. Nanotechnol.* 17 (3): 520–532.

12 Zheng, N. and Mazumder, P. (2018). A scalable low-power reconfigurable accelerator for action-dependent heuristic dynamic programming. *IEEE Trans. Circuits Syst. Regul. Pap.* 65, 6: 1897–1908.

Acknowledgment

First, I would like to thank several of my senior colleagues who encouraged me to carry on my research in neural computing during the past three decades after I published my first paper on self-healing of VLSI memories in 1989 by adopting the concept of the Hopfield network. Specifically, I would like to thank Prof. Leon O. Chua and Prof. Ernest S. Kuh of the University of California at Berkeley, Prof. Steve M. Kang, Prof. Kent W. Fuchs, and Prof. Janak H. Patel of the University of Illinois at Urbana-Champaign, Jacob A. Abraham of the University of Texas at Austin, Prof. Supriyo Bandyopadhyay of Virginia Commonwealth University, Prof. Sudhakar M. Reddy of the University of Iowa, and Prof. Tamas Roska and Prof. Csurgay Arpad of the Technical University of Budapest, Hungary.

Second, I would like to thank several of my colleagues at the National Science Foundation where I served in the Directorate of Computer and Information Science and Engineering (CISE) as a Program Director of the Emerging Models and Technologies program from January 2007 to December 2008, and then served in the Engineering Directorate (ED) as a Program Director of Adaptive Intelligent Systems program from January 2008 to December 2009. Specifically, I would like to thank Dr. Robert Grafton and Dr. Sankar Basu of the Computing and Communications Foundation Division of CISE, and Dr. Radhakrisnan Baheti, Dr. Paul Werbos, and Dr. Jenshan Lin of the Electrical Communications and Cyber Systems (ECCS) Division of ED for providing me with research funds over the past so many years to conduct research on learning-based systems that enabled me to delve deep into CMOS chip design for brain-like computing. I had the distinct pleasure of interacting with Dr. Michael Roco during my stint at NSF, and subsequently, when I was invited to present our group's brain-like computing research in US-Korea Forums in Nanotechnology in 2016 at Seoul, Korea, and in 2017 at Falls Church, Virginia, USA.

Third, I would like to thank Dr. Jih Shyr Yih, who was my first doctoral student and started working with me soon after I joined the University of Michigan. After taking a course with me, where I taught the memory repair algorithms, he had enthusiastically implemented the first self-healing VLSI chip using the analog Hopfield network. Next, I would like to acknowledge the contributions of Mr. Michael Smith, who implemented the digital self-healing chip as indicated above. My other doctoral students, Dr. W. H. Lee, Dr. I. Ebong, Dr J. Kim, Dr. Y. Yalcin, and Ms. S. R. Li, worked on RTD, quantum dots, and memristors to build neuromorphic circuits. Their research works were included in a separate book. The bulk of this book is drawn from the doctoral dissertation manuscript of my student and coauthor of the book, Dr. N. Zheng. It was a joy to work with an assiduous student like him. I would like to also thank Dr. M. Erementchouk for scrutinizing the manuscript and providing some suggestions.

Finally, I would like to thank my wife, Sadhana, my son Bhaskar, my daughter Monika and their spouses Pankti Pathak and Thomas Parker, respectively, for their understanding and support despite the fact that I spent most of my time with my research group.

1

Overview

Learning never exhausts the mind.

Leonardo da Vinci

1.1   History of Neural Networks

Even though the modern von Neumann architecture-based processors are able to conduct logic and scientific computations at an extremely fast speed, they perform poorly on many tasks that are common to human beings, such as image recognition, video motion detection, and natural language processing. Aiming to emulate the capability of human brains, a non-Boolean paradigm of computation, called the neural network, was developed since the early 1950s and evolved slowly over many decades. So far, at least three major forms of neural networks have been presented in the literature, as shown in Figure 1.1.

The simplest neural network is the perceptron, where hand-crafted features are employed as input to the network. Outputs of the perceptron are binary numbers obtained through hard thresholding. Therefore, the perceptron can be conveniently used for classification problems where inputs are linearly separable. The second type of neural network is sometimes called a multilayer perceptron (MLP). Nevertheless, the "perceptrons" in an MLP are different from the simple perceptrons in the earlier neural network. In an MLP, a non-linear activation function is associated with each neuron. Popular choices for the non-linear activation function are the sigmoid function, the hyperbolic tangent function, and the rectifier function. The output of each neuron is a continuous variable instead of a binary state. The MLP is widely adopted by the machine learning community, as it can be easily implemented on general-purpose processors. This type of neural network is so popular that the phrase "artificial neural network" (ANN) is often used to specify it exclusively, even though the word ANN should have been referred to any other neural network besides biological neural networks. ANN is the backbone for the concept of a widely popular mode of learning, called deep learning. A less well-known type of neural network is called a spiking neural network (SNN). Compared to the previous two types of neural networks, SNN resembles more to a biological neural network in the sense that spikes are used to transport information. It is believed that SNNs are more powerful and advanced than ANNs, as the dynamics of an SNN is much more complicated and the information carried by an SNN could be much richer.

Learning in Energy-Efficient Neuromorphic Computing: Algorithm and Architecture Co-Design,
First Edition. Nan Zheng and Pinaki Mazumder.

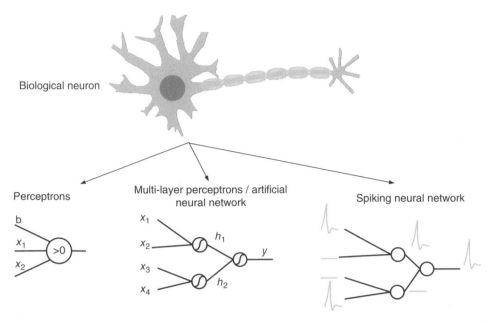

Figure 1.1 The development of neural networks over time. One of the earliest neural networks, called perceptron, is similar to a linear classifier. The type of neural network that is widely used nowadays is referred as an artificial neural network in this book. This kind of neural network uses real numbers to carry information. The spiking neural network is another type of neural network that has been gaining popularity in recent years. A spiking neural network uses spikes to represent information.

1.2 Neural Networks in Software

1.2.1 Artificial Neural Network

Tremendous advancements have occurred in the late 1980s and early 1990s for neural networks constructed in software. One powerful technique that significantly propelled the development of ANNs was the invention of backpropagation [1]. It turned out that backpropagation was very efficient and effective in training multilayer neural networks. It was the backpropagation algorithm that enabled neural networks to solve numerous real-life problems, such as image recognition [2, 3], control [4, 5], and prediction [6, 7].

In the late 1990s, it was found that other machine-learning tools, such as support vector machines (SVMs) and even much simpler linear classifiers, were able to achieve comparable and even better performances in classification tasks, which was one of the most important applications of neural networks at that time. In addition, it was observed that training of neural networks was often stuck at local minima, and consequently failed to converge to the global minimum point. Furthermore, it was generally believed that one hidden layer was enough for neural networks, as more hidden layers did not improve the performance remarkably. Since then, research interest in neural networks started to decline in the computational intelligence community.

Interest in neural networks was revived around 2006 as researchers demonstrated that a deep feedforward neural network was able to achieve outstanding classification accuracy with proper unsupervised pretraining [8, 9]. Despite its success, the deep neural network was not fully recognized by the computer vision and machine learning community until 2012 when astonishing results were achieved by AlexNet, a deep convolutional neural network (CNN) [10]. Since then, deep learning has emerged as the mainstream method in various tasks such as image recognition and audio recognition.

1.2.2 Spiking Neural Network

As another type of important neural network, SNNs did not receive much attention in comparison to the widely used ANNs. Interest in SNNs mainly came from the neuroscience community. Despite being less popular, many researchers believe that SNNs have a more powerful computational capability compared to their ANN counterparts, thanks to the spatiotemporal patterns used to carry information in SNNs. Even though SNNs are potentially more advanced, there are difficulties in harnessing the power of SNNs. Dynamics of an SNN is much more complicated in comparison to that of an ANN, which makes the purely analytical approach intractable. Furthermore, it is considerably harder to implement event-driven SNNs efficiently on a conventional general-purpose processor. This is also one of the main reasons that SNNs are not as popular as ANNs in the computational intelligence community.

Over the past decades, there were numerous efforts from both the computational intelligence community and the neuroscience community to develop learning algorithms for SNNs. Spike-timing-dependent plasticity (STDP), which was first observed in biological experiments, was proposed as an empirically successful learning rule for unsupervised learning [11–14]. In a typical STDP protocol, synaptic weight updates according to the relative order and the difference between the presynaptic and postsynaptic spike timings. Unsupervised learning is useful in discovering the underlying structure of data, yet it is not as powerful as supervised learning in many real-life applications, at least at the current stage.

1.3 Need for Neuromorphic Hardware

The development of hardware-based neural network or neuromorphic hardware started along with their software counterpart. There was a period of time (late 1980s to early 1990s) when many neuromorphic chips and hardware systems were introduced [15–18]. Later on, after finding out that the performance of neural networks was hard to keep apace with digital computers due to the inadequate level of integration of synapses and neurons, hardware research in neural networks took a back seat while Boolean computing advanced by leaps and bounds, leveraging the scaling and Moore's Law. Around 2006 when the breakthrough was made in the field of deep learning, research interests in hardware implementation of neural networks also revived. The possibilities of deploying neuromorphic computing in real-life applications were explored, as the development of the conventional von Neumann architecture-based computing slowed down because of the looming end of Moore's law.

Electronic computing devices have evolved for several decades, as shown in Figure 1.2. Two trends can be observed from the figure. The first trend is that computing devices are becoming smaller and cheaper. Indeed, partially driven by Moore's law, the sizes and

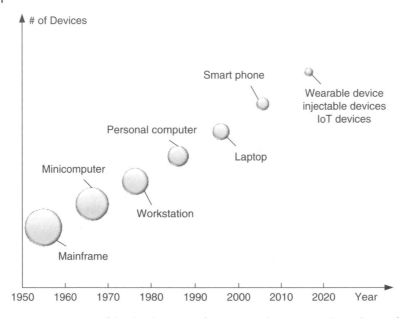

Figure 1.2 History of the development of computing devices. From the early mainframe computer that occupied the entire laboratory to nowadays ubiquitous IoT devices, the sizes of the computing devices have been shrinking over the past few decades, partially driven by Moore's law. The number of devices, on the other hand, has kept growing. As a consequence of increasing portability, more and more devices are powered by batteries today.

prices of consumer electronics are decreasing continuously. The second trend is that both the variety and the amount of information processed by the computing devices are increasing. Nowadays there are many types of sensors, such as motion, temperature, and pressure sensors, in our smart phones and wearable devices that keep gathering data for further processing. Therefore, we are experiencing a transition from the conventional rule-based computing to a data-driven computing.

With more and more low-power sensor devices and platforms being deployed in our everyday life, an enormous amount of data is being collected continuously from these ubiquitous sensors. One dilemma we often encounter is that despite the amount of data we collate, we lack the capability to fully exploit the gathered information. There is a strong need to provide these sensor platforms with built-in intelligence so that they can sense, organize, and utilize the data more wisely. Fortuitously, deep learning has emerged as a powerful tool for solving this problem [8, 19–24]. In fact, machine learning, especially deep learning, has become such a hot technology recently that it has a huge impact on how the commercial world operates. With more and more startups and big companies investigating this field, it is expected that use of artificial intelligence (AI) and machine learning will grow much faster in the near future.

Despite its successes in smaller applications, a deep neural network can only be employed in a real-life application if millions or even billions of synapses can be integrated in the system. Training of such a huge neural network usually takes weeks and burns an excessive amount of energy, even when highly optimized hardware such as graphics processing units (GPUs) are employed and matrix solving are being largely parallelized [20]. In the near future, we will have more and more ultra-low-power sensor systems for health and environment monitoring [25–27], mobile microrobots that chiefly rely on energy scavenging from the environment [28–31], and over 10 billion internet-of-things (IoT) devices [32]. For all these applications where power consumption is an important consideration, neither power-hungry GPUs nor sending raw data to the cloud computer for further analysis is a viable option. To tackle this difficulty, sustained endeavors from both industry and academia have culminated in developing low-power deep learning accelerators.

Google has built a customized application-specific integrated circuit (ASIC) called a tensor processing unit (TPU) in order to accelerate deep-learning applications in their datacenter [33], while Microsoft has utilized field-programmable gate arrays (FPGAs) in their datacenter for deep learning [34]. The realization of deep learning with FPGAs provides a cost-effective and energy-efficient solution compared to the more conventional GPU-based approach. Intel has unveiled its Nervana chip, which is the neural network processor Intel has developed with the aim to revolutionize AI computing across many different fields. In addition to the efforts made by the industry, an increasing number of papers have been published in recent years to discuss various architectures and design techniques for building energy-efficient ANN accelerators. With the growing popularity of the deep neural network, more innovations are expected in the near future.

Despite its mathematical simplicity, ANN-based learning faces challenges in scalability and power efficiency. To tackle these issues, more and more researchers in the hardware community started working on SNN-based hardware accelerators. This trend is attributed to many unique advantages that SNNs have. The event-triggered nature of an SNN can lead to a very power-efficient computation. The spike-based data encoding also facilitates the communication between neurons, providing good scalability. Nevertheless, building and utilizing specialized spike-based neuromorphic hardware is still in its early stage and there are many difficulties that need to be addressed before the hardware can become meaningfully useful. One main challenge we are encountering is how to train a spike-based neural network properly. After all, it is the learning capability of neural networks that empowers the neuromorphic system with the intelligence that can be exploited by many applications.

1.4 Objectives and Outlines of the Book

Machine learning, especially deep learning, has emerged as an important discipline through which many conventionally difficult problems, such as pattern recognition, decision making, and natural language processing, can be addressed. Nowadays, millions and even billions of neural networks are running in data centers, personal computers and portable devices to perform various tasks. In the future, it is expected that more complex neural networks with larger sizes will be needed. Such a trend demands specialized hardware to accommodate the ever-increasing requirements on power consumption and response time.

In this book, we focus on the topic of how to build energy-efficient hardware for neural networks with a learning capability. This book strives to provide co-design and co-optimization methodologies for building hardware neural networks that can learn to perform various tasks. The book provides a complete picture from high-level algorithms to low-level implementation details. Hardware-friendly algorithms are developed with the objective to ease implementation in hardware, whereas special hardware architectures are proposed to exploit the unique features of the algorithms. In the following chapters, algorithms and hardware architectures for energy-efficient neural network accelerators are discussed. An overview of the organization of this book is illustrated in Figure 1.3.

In Chapter 2, algorithms for utilizing and training rate-based ANNs are discussed as well as several basic concepts involved in the learning and inference of an ANN. Popular network structures, such as a fully connected neural network, a CNN, and a recurrent neural network, are introduced, and their advantages are discussed. Different types of learning schemes, such as supervised learning, unsupervised learning, and reinforcement learning, are demonstrated, and a concrete case study is provided to show how to employ an ANN in a reinforcement-learning task. Considering many astonishing results achieved by deep learning recently, emphasis is placed on concepts and techniques commonly used in deep learning in this chapter.

In Chapter 3, various options of executing neural networks are introduced, ranging from general-purpose processors to specialized hardware and from digital accelerators to analog accelerators. Hardware realizations of many neural network structures and deep-learning techniques presented in Chapter 2 are discussed in this chapter. Various architecture- and circuit-level techniques and innovations that can help build energy-efficient accelerators are presented for both digital and analog accelerators. A case study of building a low-power accelerator for adaptive dynamic programming with neural networks is discussed in detail to provide a concrete example.

In Chapter 4, fundamental concepts and popular learning algorithms for SNNs are discussed, starting with the basic operational principle of typical SNNs. The similarities and key differences between SNNs and ANNs are identified. Many classic learning algorithms that are capable of training shallow neural networks are first discussed. Inspired by the recent success achieved by deep ANNs, how to extend learning into deep SNNs is also explored. Popular ways of training multilayer SNNs are examined. To demonstrate the feasibility of training deep SNNs, a supervised learning algorithm that exploits spike timings for estimating gradient information needed by backpropagation is presented in great detail.

In Chapter 5, hardware implementations for SNNs are discussed. Several advantages of SNN hardware are highlighted, which serve as the motivation for implementing SNN hardware. A few general-purpose large-scale spiking systems that target simulating biological neural networks or performing cognitive tasks are presented, including both digital systems such as TrueNorth and SpiNNaker and analog systems such as Neurogrid and BrainScaleS. In addition to these large neuromorphic systems, compact customized SNN hardware aiming at accelerating specific tasks with a high energy efficiency is also discussed. To implement the learning algorithm presented in Chapter 4 efficiently in hardware, three design examples are presented. Two of the designs are digital accelerators based on conventional CMOS technology, whereas the third design is an analog system based on an emerging nanotechnology. Through these three design examples, many important aspects of designing SNN hardware are covered.

Chapter 6 concludes this book and provides some thoughts and outlooks concerning the future research directions in the field of neural network hardware.

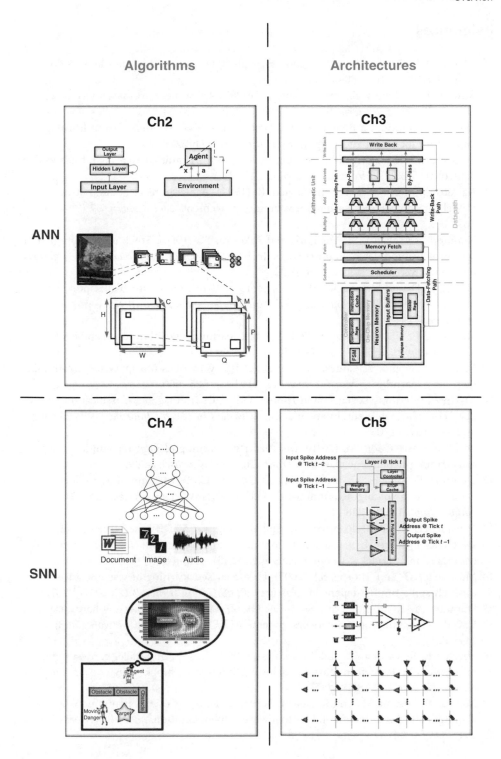

Figure 1.3 Overview of the organization of the book.

References

1 Werbos, P.J. (1990). Backpropagation through time: what it does and how to do it. *Proc. IEEE* 78 (10): 1550–1560.

2 Rowley, H.A., Baluja, S., and Kanade, T. (1998). Neural network-based face detection. *IEEE Trans. Pattern Anal. Mach. Intell.* 20 (1): 23–38.

3 LeCun, Y., Bottou, L., Bengio, Y., and Haffner, P. (1998). Gradient-based learning applied to document recognition. *Proc. IEEE* 86 (11): 2278–2323.

4 Psaltis, D., Sideris, A., and Yamamura, A.A. (1988). A multilayered neural network controller. *IEEE Control Syst. Mag.* 8 (2): 17–21.

5 Kawato, M., Furukawa, K., and Suzuki, R. (1987). A hierarchical neural network model for control and learning of voluntary movement. *Biol. Cybern.* 57 (3): 169–185.

6 Kimoto, T., Asakawa, K., Yoda, M., and Takeoka, M. (1990). Stock market prediction system with modular neural networks. In: *1990 IJCNN International Joint Conference on Neural Networks*, vol. 1, 1–6. IEEE.

7 Odom, M.D. and Sharda, R. (1990). A neural network model for bankruptcy prediction. In: *1990 IJCNN International Joint Conference on Neural Networks*, vol. 2, 163–168. IEEE.

8 Hinton, G.E. and Osindero, S. (2006). A fast learning algorithm for deep belief nets. *Neural Comput.* 1554 (7): 1527–1554.

9 Erhan, D., Bengio, Y., Courville, A. et al. (2010). Why does unsupervised pre-training help deep learning? *J. Mach. Learn. Res.* 11 (Feb): 625–660.

10 Krizhevsky, A., Sutskever, I., and Hinton, G.E. (2012). Imagenet classification with deep convolutional neural networks. In: *Advances in Neural Information Processing Systems*, 1097–1105.

11 Diehl, P.U. and Cook, M. (2015). Unsupervised learning of digit recognition using spike-timing-dependent plasticity. *Front. Comput. Neurosci.* 9: 99.

12 Querlioz, D., Bichler, O., Dollfus, P., and Gamrat, C. (2013). Immunity to device variations in a spiking neural network with memristive nanodevices. *IEEE Trans. Nanotechnol.* 12 (3): 288–295.

13 Masquelier, T. (2012). Relative spike time coding and STDP-based orientation selectivity in the early visual system in natural continuous and saccadic vision: a computational model. *J. Comput. Neurosci.* 32 (3): 425–441.

14 Masquelier, T. and Thorpe, S.J. (2007). Unsupervised learning of visual features through spike timing dependent plasticity. *PLoS Comput. Biol.* 3 (2): 0247–0257.

15 Duranton, M., Gobert, J., and Sirat, J.A. (1992). Lneuro 1.0: a piece of hardware LEGO for building neural network systems. *IEEE Trans. Neural Networks* 3 (3): 414–422.

16 Eberhardt, S., Duong, T., and Thakoor, A. (1989). Design of parallel hardware neural network systems from custom analog VLSI 'building block' chips. *Int. Jt. Conf. Neural Networks* 3: 183–190.

17 Maeda, Y., Hirano, H., and Kanata, Y. (1995). A learning rule of neural networks via simultaneous perturbation and its hardware implementation. *Neural Networks* 8 (2): 251–259.

18 Mazumder, P. and Jih, Y.-S. (1993). A new built-in self-repair approach to VLSI memory yield enhancement by using neural-type circuits. *IEEE Trans. Comput. Aided Des. Integr. Circuits Syst.* 12 (1): 124–136.

19 Bengio, Y., Lamblin, P., Popovici, D., and Larochelle, H. (2007). Greedy layer-wise training of deep networks. In: *Advances in Neural Information Processing Systems*, 153–160.

20 Le, Q.V., Ranzato, M.A., Monga, R. et al. (2011). Building high-level features using large scale unsupervised learning. In: *Acoustics, Speech and Signal Processing (ICASSP), 2013 IEEE International Conference on*, 8595–8598.

21 LeCun, Y., Bengio, Y., and Hinton, G. (2015). Deep learning. *Nature* 521 (7553): 436–444.

22 Mnih, V., Kavukcuoglu, K., Silver, D. et al. (2015). Human-level control through deep reinforcement learning. *Nature* 518 (7540): 529–533.

23 Schmidhuber, J. (2015). Deep learning in neural networks: an overview. *Neural Networks* 61: 85–117.

24 Silver, D., Huang, A., Maddison, C.J. et al. (2016). Mastering the game of Go with deep neural networks and tree search. *Nature* 529 (7587): 484–489.

25 Lee, Y., Bang, S., Lee, I. et al. (2013). A modular $1 \, mm^3$ die-stacked sensing platform with low power I^2C inter-die communication and multi-modal energy harvesting. *IEEE J. Solid-State Circuits* 48 (1): 229–243.

26 Chen, Y.P., Jeon, D., Lee, Y. et al. (2015). An injectable 64 nW ECG mixed-signal SoC in 65 nm for arrhythmia monitoring. *IEEE J. Solid-State Circuits* 50 (1): 375–390.

27 Lee, I., Kim, G., Bang, S. et al. (2015). System-on-mud: ultra-low power oceanic sensing platform powered by small-scale benthic microbial fuel cells. *IEEE Trans. Circuits Syst. I Regul. Pap.* 62 (4): 1126–1135.

28 Pérez-Arancibia, N.O., Ma, K.Y., Galloway, K.C. et al. (2011). First controlled vertical flight of a biologically inspired microrobot. *Bioinspiration Biomimetics* 6 (3): 036009.

29 Mazumder, P., Hu, D., Ebong, I. et al. (2016). Digital implementation of a virtual insect trained by spike-timing dependent plasticity. *Integr. VLSI J.* 54: 109–117.

30 Wood, R.J. (2008). The first takeoff of a biologically inspired at-scale robotic insect. *IEEE Trans. Rob.* 24 (2): 341–347.

31 Hu, D., Zhang, X., Xu, Z. et al. (2014). Digital implementation of a spiking neural network (SNN) capable of spike-timing-dependent plasticity (STDP) learning. In: *14th IEEE International Conference on Nanotechnology, IEEE-NANO 2014*, 873–876. IEEE.

32 Friess, P. (2011). *Internet of Things-Global Technological and Societal Trends from Smart Environments and Spaces to Green ICT*. River Publishers.

33 Jouppi, N.P., Young, C., Patil, N. et al. (2017). In-datacenter performance analysis of a tensor processing unit. In: *Proceedings of the 44th Annual International Symposium on Computer Architecture*, 1–12. IEEE.

34 Chung, E., Fowers, J., Ovtcharov, K. et al. (2018). Serving DNNs in real time at data-center scale with project brainwave. *IEEE Micro* 38 (2): 8–20.

2

Fundamentals and Learning of Artificial Neural Networks

Live as if you were to die tomorrow. Learn as if you were to live forever.

Mahatma Gandhi

In this chapter, basic concepts of the rate-based artificial neural networks (ANNs) are presented with the emphasis on how learning is conducted. Indeed, the usefulness of a neural network is mainly due to its intrinsic ability to incorporate learning by updating the synaptic weights between pairs of neurons. Starting from the popular backpropagation-based gradient descent learning, different network architectures are covered, including fully connected neural networks, convolutional neural networks, and recurrent neural networks. Additionally, important concepts and techniques widely used in deep learning are also discussed.

2.1 Operational Principles of Artificial Neural Networks

An ANN can be considered as an abstract version of a biological spiking neural network. The spiking rate of a neuron is abstracted as real numbers in an ANN. The synapse in a biological neural network is treated as a multiplicative edge in the neural network. For an ANN, there are two basic modes: inference and learning. Inference is a process of computing output values based on current inputs as well as parameters of the neural network. For example, in a classification application, input can be a song and the inference result outputted by the neural network can be the name of that song. A correct inference means that the name of the song outputted by the neural network matches the actual name of that song, or the ground truth. Learning, on the other hand, is a process of acquiring the parameters of the neural network in order to produce the correct inference results.

2.1.1 Inference

Figure 2.1 illustrates an example with two presynaptic neurons and one postsynaptic neuron. In the neural network literature, especially when discussing neural network hardware, the edge that carries the outputted activation value of a neuron is often referred as an axon and the edge that carries an input to a neuron is often called a dendrite. During the inference operation of the neural network, activation levels from

Learning in Energy-Efficient Neuromorphic Computing: Algorithm and Architecture Co-Design,
First Edition. Nan Zheng and Pinaki Mazumder.
© 2020 John Wiley & Sons Ltd. Published 2020 by John Wiley & Sons Ltd.

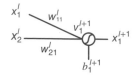

Figure 2.1 Illustration of the basic operations in an ANN. Activations from the presynaptic neurons, x_1^l and x_2^l, are first scaled by the synaptic weights, w_{11}^l and w_{21}^l, and are added with the bias b_1^{l+1}. The obtained sum is fed to a non-linear activation function to obtain x_1^{l+1}, the activation level of the postsynaptic neuron.

the previous layer (or input levels if the current layer is the first layer), x_i^l, are multiplied by the corresponding synaptic weights, w_{ij}^l, and are summed up at the location of each neuron in the layer, as shown in Eq. (2.1). In the equation, i and j denote the indexes of the neurons, whereas l denotes the index of the layer. The bias term b_j^{l+1} is included to provide more design flexibility:

$$v_j^{l+1} = \sum_i w_{ij}^l x_i^l + b_j^{l+1} \tag{2.1}$$

For ease of notation, we use x_i^l to refer to both the neuron and its activation level. The meaning, nevertheless, should be clear based on the context. The activation outputted by each neuron in the layer can then be obtained by passing the sum through a non-linear activation stage, as shown in Eq. (2.2)

$$x_j^{l+1} = f(v_j^{l+1}) \tag{2.2}$$

In Eq. (2.2), $f(\cdot)$ is a non-linear activation function. Popular activation functions are sigmoid function, hyperbolic tangent function, and rectified linear unit (ReLU), as shown in Figure 2.2. It is important to bear in mind that a non-linear activation function is critical to a neural network. It is the non-linearity of the activation function that empowers neural networks with the capability to approximate any function with a moderate number of parameters. Indeed, without the non-linear activation, a multi-layer neural network essentially degenerates into a one-layer neural network with a weight matrix being the cascaded product of all the weight matrices in the network. Another requirement that is usually imposed on the activation function is that it should

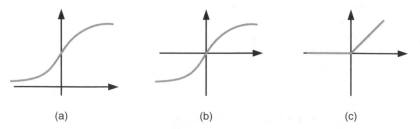

      (a)   (b)   (c)

Figure 2.2 Popular activation functions. (a) A sigmoid function. (b) A hyperbolic tangent function. (c) A rectified linear unit function. One of the important roles of an activation function is to introduce non-linearity into the neural network. Without the non-linearity, a multilayer neural network is degenerated to a simple linear matrix–vector multiplication.

Figure 2.3 Illustration of a feedforward neural network. The layers connected to the input and the output of the neural network are called input layer and output layer, respectively. The layers that are sandwiched in between the input and output layers are called hidden layers.

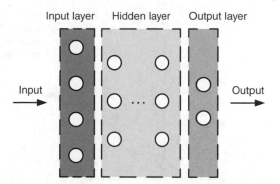

be differentiable (at least piece-wisely). This requirement is essential for many forms of learning, as demonstrated in the next section.

Neurons and synapses are basic building blocks for a neural network. With these components, a neural network can be formed. A simple feedforward neural network (FNN) is illustrated in Figure 2.3. The first layer and the last layer are called the input layer and the output layer, respectively. They take the input to the neural network and output the computed results. In between these two layers, there are layers called hidden layers. Most FNNs have at least one hidden layer even though many deep networks can have tens and even hundreds of hidden layers. The forward operation of a neural network is a process of passing the input to the network and evaluating the activations of each neuron, layer by layer, and reaching the final output. In the next section, the operation in the opposite direction is introduced, which runs a backward operation for learning.

2.1.2 Learning

Learning of a neural network is normally performed in the form of minimizing certain loss (cost) functions $L(\mathbf{w})$, where $\mathbf{w}$ represents all the weights in the neural network. They are the parameters to learn in a training process. It is worth noting that a bias in a neural network can be treated as a synapse with its input always being one, as can be seen from Eq. (2.1). Such an arrangement can help simplify the notation. Therefore, we treat biases in the network as synaptic weights unless otherwise stated.

Given a set of inputs and outputs, the loss function is essentially a function of the parameters of the network, and one can minimize the loss function through adjusting the value of $\mathbf{w}$. The detailed definition of $L(\mathbf{w})$ depends on the type of learning involved, which is elaborated in Section 2.2. The minimization of the loss function is mostly done through a gradient descent-based optimization or its derivatives, even though some other standard optimization methods, such as the genetic algorithm [1], can be used as well. In gradient descent, the most important step is to find out the gradients of the loss function with respect to each tunable parameter, i.e. $\partial L(\mathbf{w})/\partial w_{ij}^l$.

With the gradient information, one can change the weights and biases toward the descent direction in order to minimize the loss function. This process is conceptually illustrated in Figure 2.4. In the figure, a one-dimensional example is shown for visual clarity, yet it is straightforward to extend this example to higher-dimensional ones. The loss function is a function of the synaptic weight w in the figure. The target of the learning is to find out the correct weight so that the loss function can be minimized. Circles in the

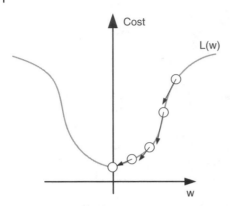

Figure 2.4 Illustration of the gradient descent process. The parameter *w* is adjusted such that the loss function is moving toward the direction where its value decreases.

figure represent the weight and its corresponding loss function in the learning process. Based on the locally linearized gradient of the cost function, the parameters are adjusted. Starting from an initial state, which is usually selected randomly, the parameters of the network are adjusted slightly toward the descent direction during every learning iteration. Mathematically, this can be described as

$$w_{ij}^l[n+1] = w_{ij}^l[n] - \alpha \frac{\partial L(\mathbf{w})}{\partial w_{ij}^l} \tag{2.3}$$

where $w_{ij}^l[n]$ represents the synaptic weight at the *n*th learning iteration; α is called the learning rate, which is a hyperparameter used to control the learning speed. The name hyperparameter is used to refer to those parameters that are not part of the neural network model but are needed in order to control the learning process.

The implication of α is conceptually illustrated in Figure 2.5. When α is too small, the updates in the parameters are marginal. Consequently, it takes more learning iterations to reach the minimum, as shown in Figure 2.5a. Intuitively, when the change in the parameters are small, the localized gradients for iteration *n* and iteration *n* − 1 are close in value. Therefore, one could have just combined these two movements into one to save one learning iteration. Another extreme, on the other hand, is one when the

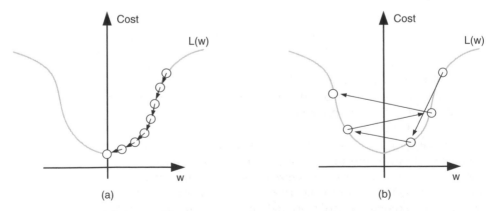

Figure 2.5 Illustration of the learning process when (a) the learning rate is too small and (b) the learning rate is too large. When the learning rate is small, the loss function can be minimized, yet it takes too many iterations. When the learning rate is large, the learning process may fail to converge.

learning rate is too large. Under this circumstance, the learning might diverge, as shown in Figure 2.5b. With gradient descent learning, we use a linearized model to approximate the true loss function locally. When the changes in the parameters are small, the linearized model matches well with the true loss function. When the changes in the parameters are too large, however, the true loss function behaves significantly differently from the assumed linearized model, which results in divergence in the learning. Furthermore, even when the learning rate is not large enough for the learning to diverge, the learning might still oscillate around minima instead of converging. Clearly, the choice of the hyperparameter α is very critical in achieving a fast and effective learning. Conventionally, the pick of this learning rate is more or less a process of trial and error. A common practice is to choose a reasonable number to start with and reduce the learning rate gradually over time in order for the learning to converge smoothly. Recently, many advanced techniques have been proposed to avoid the manual setting of the learning rate. This is detailed in Section 2.5.3.1.4.

Clearly, the gradient descent method cannot guarantee the convergence to the global minimum unless the problem is a convex one. This is illustrated in Figure 2.6. A good analogy to this problem is a ball rolling from the peak to the valley. Even though the ball is able to move toward the direction where the gravitational potential energy reduces, it may be stuck in some local valleys. In the early days, it was often argued that the local minima are the main reasons that many neural networks could not be trained properly. In order to tackle this issue, researchers have proposed some deliberately designed neural networks that were local-minima-free [2, 3]. Alternatively, global optimization methods, such as a simulated annealing and genetic algorithm, have also been applied to training neural networks in order to obtain a global optimum solution [1, 4]. Nevertheless, recent developments in deep neural networks suggest that the local minimum might not be as problematic as one would think. A deep neural network is almost always able to reach a solution that is close to the global minimum [5].

The process of finding out the gradient components $\partial L(\mathbf{w})/\partial w_{ij}^l$ is performed through a well-formulated method called *backpropagation*. The method of backpropagation is widely used as a definitive method for training neural networks. Even though ideas similar to backpropagation have existed for a while and have been utilized to solve certain

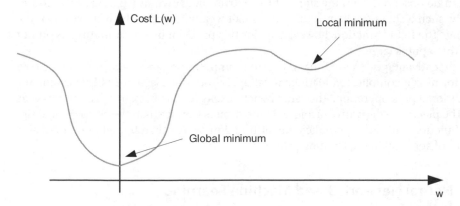

Figure 2.6 Illustration of the local minimum and the global minimum of a cost function. Depending on the initial point, the learning might be stuck at the local minimum instead of converging to the global minimum.

optimization problems, backpropagation was not introduced to the field of the neural network until the 1970s and 1980s. The concept of backpropagation in neural networks was studied by many researchers independently, such as Werbos [6, 7] and Rumelhart et al. [8]. The algorithm relies on a linearized model of the neural network. It utilizes the chain rule in calculus to obtain the gradient associated with each synaptic weight that is not directly connected to the output neurons. For example, if the error at neuron x_j^{l+1}, e_j^{l+1}, is known, then we are able to propagate the error to neuron x_i^l through

$$\frac{\partial e_j^{l+1}}{\partial x_i^l} = \frac{\partial e_j^{l+1}}{\partial v_j^{l+1}} \cdot \frac{\partial v_j^{l+1}}{\partial x_i^l} \tag{2.4}$$

This is the simple chain rule widely used in calculus. The term $\partial e_j^{l+1}/\partial v_j^{l+1}$ is the gradient of the activation function at the bias point x_i^l, or mathematically $f'(x_i^l)$. The term $\partial v_j^{l+1}/\partial x_i^l$ can be easily obtained as w_{ij}^l from the relationship shown in Eq. (2.1). Therefore, Eq. (2.4) can be rewritten as

$$\frac{\partial e_j^{l+1}}{\partial x_i^l} = f'(x_i^l) \cdot w_{ij}^l \tag{2.5}$$

With Eq. (2.5), the errors computed at output neurons can be propagated back to each neuron layer-by-layer. Similarly, the gradient with respect to each weight can be expressed as

$$\frac{\partial e_j^{l+1}}{\partial w_{ij}^l} = f'(x_i^l) \cdot x_i^l \tag{2.6}$$

Historically, the sigmoid function and hyperbolic tangent function were used extensively in neural networks. Nowadays, the ReLU function and its derivatives are predominantly used in deep neural networks. The introduction of ReLU is mainly because of two reasons. The first reason is to counteract the diminishing gradients that exist when the neural network structure is deep. When a conventional sigmoid function or a hyperbolic tangent function is used, $f'(x_i^l)$ becomes very small if x_i^l is too small or too large. This can be observed from the slope of the activation shown in Figure 2.2a, b. Consequently, such a small gradient results in a small weight update, which slows down the learning. The ReLU function, however, avoids this problem by not saturating its output when the input is large.

Another advantage that ReLU has is that the computation can be very efficient. Instead of performing a complicated mathematical operation, the usage of ReLU is essentially a conditional pass operation. The same benefit exists for the backpropagation phase as well. The piecewise derivative of an ReLU function is simpler than the conventional sigmoid function and the hyperbolic tangent function, which thereby yields a remarkable amount of speedup in the learning process.

2.2 Neural Network Based Machine Learning

There are mainly three common types of machine learning methods: (i) supervised learning, (ii) reinforcement learning, and (iii) unsupervised learning. In the past, neural networks have been successfully employed in all these three types of machine learning methods. Despite the striking differences that exist in these three learning methods, the learning for a neural network can often be formulated as a problem of approximating some functions. This problem can then be solved through minimizing certain loss functions associated with the neural networks, which has been discussed in the previous section. In this section, the concepts of using neural networks for these three types of learning schemes are briefly reviewed. How the loss functions are constructed for different types of learning mechanisms are discussed. A concrete case study is also provided in this section to illustrate how a neural network can be employed in a reinforcement-learning application.

2.2.1 Supervised Learning

Supervised learning is the most developed type of machine-learning task using neural networks. One of the most popular applications of supervised learning is classification. In a supervised-learning task, the objective is to learn certain input-output mapping based on available data. For example, in a classification task where we want to classify images into different categories, the input data should contain two parts: images and labels. Each image has an associated label that helps to categorize the image. The labels may be generated by an expert. The objective of supervised learning is that after learning, the neural network should be able to associate the correct labels with the images presented.

In supervised learning, a popular definition for the loss function is

$$L(\mathbf{w}) = \frac{1}{2}(\mathbf{x_o} - \mathbf{t})^{\mathrm{T}}(\mathbf{x_o} - \mathbf{t}) \tag{2.7}$$

where $\mathbf{x_o}$ is the column vector comprising the output of the neural network and $\mathbf{t}$ is the desired or target output vector. The name "supervised learning" comes from the fact that a supervisory signal, $\mathbf{t}$, is explicitly provided in the learning phase. The need of the label makes the classification problem belong to this category. However, in practice it may not be possible to provide explicit supervisory signals for many problems that we are interested in. In that case, we have to resort to other learning schemes. It is worth noting that the divide-by-two operation shown in Eq. (2.7) is not critical. It is mostly a convention used by many researchers. The introduction of a factor of 1/2 leads to a clearer notation when calculating the gradient.

In Eq. (2.7), $\mathbf{x_o}$ can be written explicitly as $\mathbf{x_o}(\mathbf{w}, \mathbf{x_i})$, where $\mathbf{x_i}$ is the input vector. From the perspective of learning, the neural network is essentially a function with a tunable parameter $\mathbf{w}$ and a fixed parameter $\mathbf{x_i}$. The learning is a process of setting the parameter $\mathbf{w}$ to a proper value such that the output from the neural network matches the supervisory signal when the given data is presented at the input of the neural network. The tool to achieve the above goal is the gradient decent learning described in Section 2.1.2.

Clearly, just to remember the correct answer is not always very useful. The hope in supervised learning is that by seeing several examples with the correct answers, the neural network can make correct inference on previously unseen data. Such an expectation seems reasonable, although, in practice, it may not be trivial to achieve. Mathematically, in supervised learning, the neural network attempts to learn a mapping function that can map the given input to the correct output. By providing the labeled data, we are

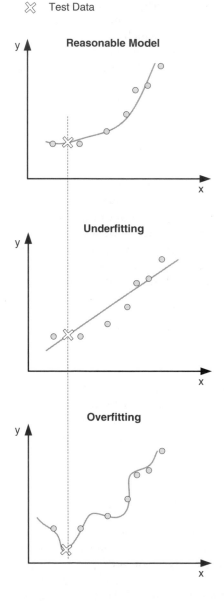

⊙ Training Data
— Learned Model
✕ Test Data

Figure 2.7 Illustration of proper-fitting, underfitting and overfitting.

essentially providing samples of the input–output mapping relationship to the neural network. This can be illustrated with the help of Figure 2.7. The samples are represented as green circles in the figure. The neural network is expected to figure out the correct function mapping through minimizing the difference between its output and the sample data. Obviously, when the network does not have enough free parameters or, in other words, the network does not have enough capacity, it cannot properly fit the data. This case corresponds to the underfitting case shown in Figure 2.7. Another extreme is overfitting, which typically occurs when the neural network has too many degrees of freedom, yet the training data is not enough. An analogy that might be helpful in understanding the concept of underfitting and overfitting is that say we have a set of equations. In one extreme, we may have many less unknowns than the number of equations. Consequently, it is most likely that we cannot solve this set of equations perfectly. The best thing we can do is to achieve an approximate solution that can minimize the error we obtain. This corresponds to the underfitting case. In another extreme, we have far more unknowns than the number of equations. Under this circumstance, we have an infinite number of solutions and we do not know which one is the correct one. This corresponds to the overfitting case. A good way to distinguish underfitting from overfitting is to observe the test error and training error. Typically, an underfitted neural network performs poorly on both the training data and the test data. On the other hand, an overfitted network generally performs well on the training data yet is not so good on the unseen test data.

Apparently, one can try to avoid underfitting by simply increasing the capacity of the network. This can be done through, for example, increasing the size of the network. Overfitting, on the other hand, can be avoided through posing more constraints in the learning. One good practice that many researchers often use to avoid both underfitting and overfitting is to start with a small network and then gradually increase the size of the network. A reasonable network size can then be chosen by monitoring the performance of learning in this process. In addition to adjusting the network size, a very popular alternative to counteract overfitting is through regularization. The intuition behind a regularization method is that since the reason for overfitting is that we have too much freedom in the learning, we can pose more constraints on the parameters of the neural network in order to reduce the degrees of freedom. Mathematically, the loss function with regularization can be written as

$$L(\mathbf{w}) = \frac{1}{2}(\mathbf{x_o} - \mathbf{t})^T(\mathbf{x_o} - \mathbf{t}) + \lambda R(\mathbf{w}) \tag{2.8}$$

Comparing Eq. (2.8) with Eq. (2.7), one may notice that a new term $\lambda R(\mathbf{w})$ has been added. This term is called a regularization term and λ is called the regularization coefficient. In principle, the function $R(\mathbf{w})$ can take any form as long as it poses some reasonable constraints on $\mathbf{w}$. In practice, L_2 regularization is most common and L_1 regularization is also very popular. Sometimes, L_0 regularization is also used to encourage sparsity in the network. Mathematically, $R(\mathbf{w})$ in these three regularizations can be written as $\|\mathbf{w}\|_2$, $\|\mathbf{w}\|_1$, and $\|\mathbf{w}\|_0$, respectively. In plain English, L_2 norm is the sum of the squares of all the synaptic weights, L_1 norm is the sum of the absolute values of all the synaptic weights, and L_0 norm is the number of non-zero synaptic weights.

With a regularization term, a new hyperparameter λ is introduced. A large λ encourages a simple model, which might cause underfitting. A small λ, on the other hand, puts less emphasis on the regularization, which might lead to overfitting. The selection of λ is very critical in training neural networks. A common practice is to pick the correct regularization coefficient by sweeping the regularization term and monitoring the learning progress.

In general, it is non-trivial to train a neural network that can properly fit to the problem at hand. A typical procedure for supervised learning is as follows:

1. Prepare dataset by collecting labeled data.

2. Conduct preprocessing on the data, e.g. data normalization.
3. Divide the data into a training set and a test set, and often also a validation set.
4. Use the training data to train the neural network model. Training data are typically presented to the neural networks multiple times. It is often advantageous to shuffle the order of the training data.
5. Tune the hyperparameters such as the learning rate, regularization coefficient, and the number of layers with the validation set.
6. Test the performance of the neural network by applying the trained neural network on the test dataset.

In the learning process, the dataset is divided into three parts, which serve different purposes.

Training set. This set of data is used to train the model.
Validation set. This set of data is to evaluate the performance of the neural network with a certain set of hyperparameters.
Test set. This set of data is used to evaluate the final performance of the trained neural network. The measured performance serves as an estimation on how well the trained neural network will perform on unseen data.

It is worth noting that a common pitfall is to use the test set for hyperparameter tuning. The catch is that there exist biases when the test set is used for both the test and the validation. The obtained test accuracy is biased, as the model that performs the best on the test set has been picked. In other words, the model is likely to be overfitted to the test set data. Therefore, although the obtained model may perform well on the test set, this does not mean that the selected model actually performs well on unseen data.

2.2.2 Reinforcement Learning

Different from supervised learning, there is no explicit supervisory signal that instructs what the output of the neural network should be in a reinforcement-learning task. Compared to supervised learning and unsupervised learning, reinforcement learning is a relatively well-developed and standalone discipline. Only a few aspects of reinforcement learning that are tightly related to neural networks are covered in this section. Interested readers are referred to the textbook on reinforcement learning for more details [9].

Figure 2.8 illustrates the problem that many reinforcement-learning algorithms attempt to solve. An agent is interacting with its surrounding environment. At every time step t, the agent is able to observe the states of the environment $\mathbf{x}(t)$ and a reward signal $r(t)$. The objective of the agent is to pick the correct action $\mathbf{a}(t)$ such that the accumulated reward it receives in the future is maximized. To give an example, suppose that the reinforcement-learning problem at hand is to teach an agent how to play a

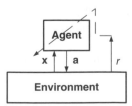

Figure 2.8 Illustration of the configuration of a reinforcement-learning task. The agent is interacting with its surrounding environment by taking the action **a** based on its observation of the state **x** and reward *r*. The target of the agent is often to maximize the total reward it receives from the environment.

basketball game. In this case, the state $\mathbf{x}(t)$ could be the location of the player who possesses the ball, the location of the defensive player, the distance between the player and the basket, and so on. The action signal $\mathbf{a}(t)$ could be to move forward, move left, shoot, and so on. The reward signal $r(t)$ could be set as the current score of the team that the agent is controlling. Under this circumstance, the agent should learn how to maximize its score by controlling the basketball players according to the information it possesses.

Suppose that the discrete-time system the agent is interacting with can be modeled by

$$\mathbf{x}(t+1) = f[\mathbf{x}(t), \mathbf{a}(t)] \tag{2.9}$$

where $\mathbf{x}(t)$ is the n-dimensional state vector at time t, $\mathbf{a}(t)$ is the m-dimensional action vector, and $f(\cdot)$ is the model of the system. The target of the algorithm is to maximize the reward-to-go J, which can be expressed as follows:

$$J[\mathbf{x}(t)] = \sum_{k=1}^{\infty} \gamma^{k-1} r[\mathbf{x}(t+k)] \tag{2.10}$$

where γ is the discount factor used to promote the reward received soon over a long-term reward and $r[\mathbf{x}(t)]$ is the reward received at state $\mathbf{x}(t)$.

The target of the reinforcement learning is to maximize the reward-to-go in Eq. (2.10). This can be accomplished through solving the Bellman equation

$$J^*[\mathbf{x}(t)] = \max_{\mathbf{a}(t)}(r[\mathbf{x}(t+1)] + \gamma J^*[\mathbf{x}(t+1)]) \tag{2.11}$$

Solving the Bellman equation can be a hard problem. It can be directly solved through dynamic programming [10] or it can be solved approximately with adaptive dynamic programming (ADP) [11, 12], Q-learning [13, 14], Sarsa [15], etc. One difficulty in solving a reinforcement-learning problem is how to assign credits for the ultimate reward. For example, in a task of playing chess, the reward is revealed at the end of the game; that is, the agent is rewarded only if it wins the game. How to assign credits of this delayed reward to each state is not straightforward. It requires the agent to have an internal evaluating system that can estimate the potential reward that it can receive, starting from specific states. There are, in general, two types of methods to acquire or learn this information in order to evaluate each state internally: the Monte Carlo (MC) method and temporal difference (TD) learning. With the MC method, the agent simply averages all the rewards received starting from one state and uses that average to approximate the expected reward. For each update, the agent needs to run a complete episode of tests. On the other hand, TD learning is a type of incremental learning. The agent is able to learn before knowing the final result. It uses a newer estimated value to update an old estimation.

Reinforcement learning itself is a standalone discipline that could be totally independent of neural networks. Nevertheless, neural networks are often used as universal function approximators in a reinforcement learning framework. In problems with a discrete state space, a look-up table can be used to store the value functions that are used to predict the reward-to-go. However, such a method soon becomes intractable for a large state space or a continuous-state problem, which is known as the curse of dimensionality [16]. To tackle this problem, a function approximator can be used to represent

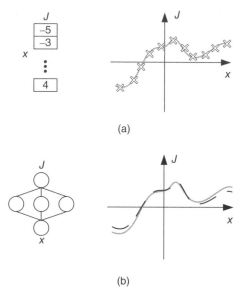

(a)

(b)

Figure 2.9 Illustration of two ways of storing the state-value functions needed in a reinforcement-learning task. (a) A tabular approach. (b) A function-approximator approach. In the tabular approach, estimated reward-to-go J is stored and maintained in a table. The size of the table grows rapidly with the size and the dimension of the problem. In a function-approximator approach, function approximators, e.g. a neural network, is employed to fit the function J. This approach can significantly reduce the memory requirement for most practical problems.

the state-value function. Figure 2.9 illustrates this idea. In a tabular approach, the agent needs to update entries in the look-up table to update the state-value function. However, since there is one entry for each state in the look-up table, the size of the table grows rapidly as the number of states or the dimensions of the states increase. On the other hand, in a function-approximator approach, the agent changes the coefficients in the parameterized function, which are the weights associated with the neural network in our example, to update the state-value function. In a sense, the function-approximator approach utilizes the correlation between states to compress the model in order to avoid the curse of dimensionality. To help understand neural network-based reinforcement learning, a concrete example is presented in Section 2.2.4.

2.2.3 Unsupervised Learning

Unsupervised learning is the type of learning that requires the least amount of supervision, as its name implies. Typically, the target of the learning is to find the underlying structure in the data. This normally appears in the form of dimensionality reduction, as most natural signals we are interested in have certain built-in structures. This idea is conceptually illustrated in Figure 2.10. In this example, even though the input data span a two-dimensional space, the inherent structure of the data is a one-dimensional hyperplane. Such a reduction in dimension normally occurs from the fact that information from different dimensions is correlated or dependent on each other. The dimensionality reduction can be quite useful in many real-life applications. The target of unsupervised learning is to find such a built-in structure through going over many sample data.

One of the most famous examples of using neural networks for unsupervised learning is the self-organizing map (SOM) or the Kohonen map, which is named after Teuvo Kohonen [17, 18]. An SOM quantizes and projects high-dimensional raw input data into low-dimensional (usually two-dimensional) spaces. The SOM is able to preserve the topological structure of the data while conducting the dimensionality reduction. Such a property makes the SOM suitable for visualizing high dimensional data [18] as well as performing data clustering [19].

Figure 2.10 Illustration of the dimensionality reduction in unsupervised learning. Even though the raw data are two-dimensional, they can be approximated by a one-dimensional hyperplane.

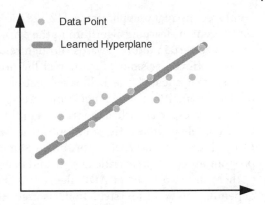

Figure 2.11 Configuration of a typical autoencoder. The autoencoder encodes the input into hidden-layer activations and then decodes the information. The objective of the autoencoder is to minimize the reconstruction error. Through minimizing this error, the autoencoder can find an alternative and often more efficient way of representing the input data.

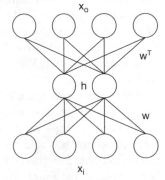

Another emerging application of unsupervised learning in neural networks is conducting pretraining for supervised learning, such as the Restricted Boltzmann Machine (RBM) [20–22] and autoencoder [23–26]. An example of an autoencoder is shown in Figure 2.11. The input data is fed into the neural network as x_i. Output x_o is read out from the output layer. There is one hidden layer in between the input layer and the output layer. In an autoencoder, the loss function is often defined as the reconstruction error, as shown below:

$$L(\mathbf{w}) = \frac{1}{2}(\mathbf{x_o} - \mathbf{x_i})^T(\mathbf{x_o} - \mathbf{x_i}) \qquad (2.12)$$

Conceptually, training an autoencoder can be thought of as a supervised-learning problem where the supervisory signal is the input signal. In an autoencoder, the number of hidden-layer neurons is typically less than the number of neurons in input/output layer. Therefore, the neural network is forced to resort to a more compact representation of the data in order to achieve a low reconstruction error while having fewer hidden-layer units. Such an unsupervised-learning approach is one of the earliest techniques that helps to realize deep learning.

2.2.4 Case Study: Action-Dependent Heuristic Dynamic Programming

In this section, a concrete example of learning with neural networks is presented to help understand how to use and train an ANN for real applications. The material in this section is partially based on one of our previous works on building customized ADP accelerators [27]. ADP is a popular and powerful algorithm that has been employed in many applications [12, 28–30]. It can be considered as one type of reinforcement-learning algorithm in the sense that the objective of the algorithm is to maximize certain future rewards through reinforcement. The optimal control policies or the optimal decisions of many real-life problems can be obtained through solving the optimal equality, the Bellman equation, as shown in Eq. (2.11). As mentioned earlier, solving the Bellman equation directly through dynamic programming can be very costly, especially when the size of a problem is large. Therefore, ADP was invented to adaptively approach the solution to the Bellman equation through reinforcement. In the literature, the ADP algorithm is sometimes also called approximate dynamic programming, adaptive critic design, and neurodynamic programming.

There are two types of ADP algorithms: model-based ADP and model-free ADP, depending on how the system model is used. The model-based ADP algorithm requires the dynamics of the system, such as the one shown in Eq. (2.9), to be explicitly known in order to conduct the learning. The model information is utilized in the process of solving the Bellman equation. On the other hand, the model of the system is not necessary in a model-free ADP algorithm. The model-free ADP algorithm learns the information associated with the model of the system during the process of interacting with the system.

In the case where the model of the system is simple and available, the model-based ADP algorithm tends to be more convenient because the readily available model information can be employed to accelerate the learning directly. Unfortunately, the models for many complex systems are hard to build and are possibly time-varying, which makes the model-based method less attractive. In this sense, the model-free ADP algorithm is more general and more powerful. The action-dependent heuristic dynamic programming (ADHDP) presented in this case study is one type of model-free ADP algorithm. It is one of the most popular and powerful forms of the ADP algorithms [11, 31–35]. By proceeding through this algorithm, the usage and training of neural networks are illustrated in great detail.

2.2.4.1 Actor-Critic Networks

The ADHDP algorithm utilizes two function approximators to learn the target functions. Figure 2.12 illustrates the basic configuration of an ADHDP algorithm. The agent consists of two networks: an actor network as well as a critic network. The function of the actor network is to choose the best policy that it believes can maximize all the future rewards, whereas the job of the critic network is to estimate the reward-to-go $J[\mathbf{x}(t)]$, shown in Eq. (2.10) with its output $\hat{J}[\mathbf{x}(t)]$. Clearly, $J[\mathbf{x}(t)]$ is implicitly a function of the current policy of the agent. Therefore, the action outputted by the actor is fed directly to the critic. This is the reason why this algorithm is called "action-dependent." The agent interacts with the environment through outputting the action $\mathbf{a}$, an m-dimensional action vector, to the environment. The environment then responds with $\mathbf{x}$, an n-dimensional state vector, in the next time step according to the dynamics of the system and the action taken by the agent. At the same time, the reward, if any, is also delivered to the agent. Even though any neural network that satisfies the requirement of being a function approximator can be employed for the actor and the critic, one-hidden-layer neural networks are used here. This is also the popular choice in many ADP literatures [11, 31–35].

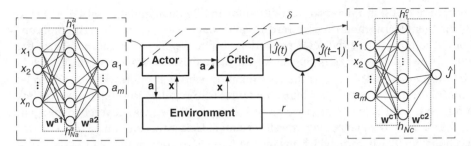

Figure 2.12 Illustration of the actor-critic configuration used in the ADHDP algorithm [27]. Two neural networks, critic network and actor network, are employed in the algorithm to approximate functions that need to be learned. The actor network outputs the action **a** based on the current state **x**. The critic network evaluates the action **a** under the current state and estimates the reward-to-go $\hat{J}$. The critic is updated to maintain a temporal consistency, whereas the actor is updated in order to generate the action that can maximize the reward-to-go. Reproduced with permission of IEEE.

2.2.4.2 On-Line Learning Algorithm

The learning process of the ADHDP algorithm is essentially an optimization process attempting to minimize or maximize certain cost or utility functions through adjusting the synaptic weights in the critic and actor networks, i.e. $\mathbf{w}^{a1}$, $\mathbf{w}^{a2}$, $\mathbf{w}^{c1}$, and $\mathbf{w}^{c2}$. The stochastic gradient descent (SGD) learning, with the help of backpropagation, as presented in Section 2.1.2, is the most popular way of conducting the optimization process. In a forward pass, the actor and critic networks compute the action **a** and the estimated reward-to-go $\hat{J}[\mathbf{x}(t)]$ as follows:

$$h_i^a = \sigma \left(\sum_{j=1}^{n} w_{ij}^{a1} x_j \right) \tag{2.13}$$

$$a_i = \sigma \left(\sum_{j=1}^{N_{ha}} w_{ij}^{a2} h_j^a \right) \tag{2.14}$$

$$\mathbf{p}^c = \begin{bmatrix} \mathbf{a} \\ \mathbf{x} \end{bmatrix} \tag{2.15}$$

$$h_i^c = \sigma \left(\sum_{j=1}^{m+n} w_{ij}^{c1} p_j^c \right) \tag{2.16}$$

$$\hat{J} = \sum_{i=1}^{N_{hc}} w_i^{c2} h_i^c \tag{2.17}$$

In Eqs. (2.13) to (2.17), $\mathbf{h}^a$ and $\mathbf{h}^c$ are N_{ha}-dimensional and N_{hc}-dimensional output vectors from the hidden units in the actor and critic networks, respectively, $\sigma(\cdot)$ is the activation function, and popular choices are the hyperbolic tangent function, the sigmoid function, and the ReLU function, as discussed in Section 2.1.1.

During a backward operation, the critic and actor networks need to update their weights so that the future reward can be maximized. This is done by minimizing two cost functions. The objective of the critic is to learn the reward-to-go $J[\mathbf{x}(t)]$. This can be achieved through minimizing the magnitude of the TD error

$$\delta(t) = \hat{J}[\mathbf{x}(t-1)] - \gamma \hat{J}[\mathbf{x}(t)] - r[\mathbf{x}(t)] \tag{2.18}$$

Clearly, Eq. (2.18) is a necessary condition for the Bellman equation of Eq. (2.11) to hold. The agent learns the correct reward-to-go through maintaining this temporal consistency. The objective of the actor network is to find a policy that can maximize the estimated reward-to-go $\hat{J}[\mathbf{x}(t)]$. Specifically, if there exists a target reward-to-go J_{exp}, one can use $e_a = \hat{J}[\mathbf{x}(t)] - J_{exp}$ as the cost function whose absolute value needs to be minimized. In many control problems, the target is to regulate the states of the plants within a predefined bound. A negative reward, i.e. a punishment, is delivered to the agent when the states of the plants are not well regulated. Under this circumstance, a convenient choice for J_{exp} is zero, which means the hope is to control the plant well within the predefined bound. Therefore, we use $e_a = \hat{J}[\mathbf{x}(t)]$ in this section for the purpose of illustration. However, it is straightforward to implement the algorithm with other forms of cost function for the actor network.

To minimize the cost function $[\delta(t)]^2/2$, the synaptic weights of the critic network can be updated according to

$$\Delta w_i^{c2} = \alpha_c \delta h_i^c \tag{2.19}$$

$$\Delta w_{ij}^{c1} = \alpha_c e_i^{c1} \sigma' \left(\sum_{k=1}^{m+n} w_{ik}^{c1} p_k^c \right) p_j^c \tag{2.20}$$

where $e_i^{c1} = \delta w_i^{c2}$ is the error at the hidden unit h_i^c and α_c is the learning rate for the critic network.

To minimize the cost function $e_a^2/2$, the synaptic weights of the actor network can be updated according to

$$\Delta w_{ij}^{a2} = \alpha_a e_i^{a2} \sigma' \left(\sum_{k=1}^{N_{ha}} w_{ik}^{a2} h_k^a \right) h_j^a \tag{2.21}$$

$$\Delta w_{ij}^{a1} = \alpha_a e_i^{a1} \sigma' \left(\sum_{k=1}^{n} w_{ik}^{a1} x_k \right) x_j \tag{2.22}$$

where

$$e_j^{a1} = \sum_{i=1}^{m} \left[e_i^{a2} \sigma' \left(\sum_{k=1}^{N_{ha}} w_{ik}^{a2} h_k^a \right) w_{ij}^{a2} \right] \tag{2.23}$$

$$e_j^{a2} = \sum_{i=1}^{N_{hc}} \left[e_i^{c1} \sigma' \left(\sum_{k=1}^{m+n} w_{ik}^{c1} p_k^c \right) w_{ij}^{c1} \right] \tag{2.24}$$

$$e_j^{c1} = e_a w_j^{c2} \tag{2.25}$$

In Eqs. (2.23) to (2.25), e_j^{a1}, e_j^{a2}, and e_j^{c1} are the errors at h_j^a, a_j, and h_j^c, respectively.

The complete learning procedure is outlined in Figure 2.13. The first *while* loop corresponds to each trial of the learning. The trial might be terminated because the task has failed, is successful, or the time limit has been reached. The termination is conducted by raising the termination request. Each trial contains many time steps. At each time step, the two networks update the synaptic weights in an alternate fashion. The second and third *while* loops are for critic updating and actor updating, respectively.

Figure 2.13 Pseudocode for the
ADHDP algorithm [27]. Reproduced
with permission of IEEE.

Inputs : w^{a1}, w^{a2}, w^{c1}, w^{c2}: weights for the actor
and critic neural network
I_a, I_c: The maximum number of iterations
allowed for updating actor and critic networks
in one time step
E_a, E_c: Thresholds to control whether an
update can be terminated

1   $t = 0$
2   Actor network forward operation: generating $a(t)$
3   Critic network forward operation: generating $\hat{J}$ [$\mathbf{x}(t)$]
4   Output action a(t) and obtain the updated states
     $\mathbf{x}(t + 1)$, reward $r[\mathbf{x}(t + 1)]$, and termination request
     REQ_{term} from the environment or plant
5   **while** $REQ_{term} \neq 1$ **do**
6 | $t = t + 1$, $i_c = 0$, $i_a = 0$
7 | Actor network forward operation: generating $a(t)$
8 | Critic network forward operation: generating $\hat{J}$ [$\mathbf{x}(t)$]
9 | Compute the temporal difference
         $\delta(t) = \hat{J}$ [$\mathbf{x}(t - 1)$] $- \gamma \hat{J}$ [$\mathbf{x}(t)$] $- r[\mathbf{x}(t)]$
10 | **while** ($i_c < I_c$ && $\delta(t) \geq E_c$) **do**
11 | | Critic network backward operation: updating w^{c2}
            and w^{c1}
12 | | Critic network forward operation: generating
            $\hat{J}$ [$\mathbf{x}(t - 1)$]
13 | | Compute the temporal difference
            $\delta(t) = \hat{J}$ [$\mathbf{x}(t - 1)$] $- \gamma \hat{J}$ [$\mathbf{x}(t)$] $- r[\mathbf{x}(t)]$
14 | | $i_c = i_c + 1$
15 | **while** ($i_a < I_a$ && $e_a \geq E_a$) **do**
16 | | Actor network backward operation: updating w^{a2}
            and w^{a1}
17 | | Actor network forward operation: generating
            $\hat{J}$ [$\mathbf{x}(t)$]
18 | | Compute the cost function $e_a = \hat{J}$ [$\mathbf{x}(t)$] $- J_{exp}$
19 | | $i_a = i_a + 1$
20 | Ouput action $\mathbf{a}(t)$ and obtain the updated states
         $\mathbf{x}(t + 1)$, reward $r[\mathbf{x}(t + 1]$, and termination
         request REQ_{term} from the environment or plant
Output: w^{a1}, w^{a2}, w^{c1}, w^{c2}: updated weights for the
actor and critic neural network

2.2.4.3 Virtual Update Technique

In the ADHDP algorithm outlined in Figure 2.13, multiple update cycles are needed to minimize the corresponding cost function. The update loop for each time step is terminated when either the cost function at that moment is below a preset value, i.e. E_c and E_a, or when the maximum number of cycles allowed is reached. For the ADHDP algorithm in the literature, I_c and I_a are normally in the range of 10–100. In other words, many iterations are carried out for the same input vector, which attempts to minimize the cost function at the current time step. Consequently, one might wonder if it is feasible to preprocess the input vector such that the following process can be conducted more efficiently?

Without loss of generality, the update for the critic network is used as the demonstrating example. The same procedure applies to the actor network as well. Equations (2.20) and (2.16) are rewritten in Eqs. (2.26) and (2.27) with dependence on index i_c explicitly denoted for ease of discussion:

$$\Delta w_{ij}^{c1}(i_c) = \alpha_c e_i^{c1}(i_c)\sigma' \left[\sum_{k=1}^{m+n} w_{ik}^{c1}(i_c)p_k^c \right] p_j^c \tag{2.26}$$

$$h_i^c(i_c + 1) = \sigma \left[\sum_{j=1}^{m+n} w_{ij}^{c1}(i_c + 1)p_j^c \right] \tag{2.27}$$

It is worth noting that the input to the critic network $\mathbf{p}^c$ is not a function of i_c, as it is fixed when the critic network updates its synaptic weights. It is this independence of $\mathbf{p}^c$ on i_c that makes the simplification on the original algorithm possible. By substituting Eq. (2.26) into Eq. (2.27), one can obtain

$$h_i^c(i_c + 1) = \sigma[o_i^c(i_c + 1)] \tag{2.28}$$

where

$$o_i^c(i_c) = \sum_{j=1}^{m+n} w_{ij}^{c1}(i_c)p_j^c \tag{2.29}$$

$$o_i^c(i_c + 1) = o_i^c(i_c) + \epsilon_i(i_c)\Lambda_c \tag{2.30}$$

$$\Lambda_c = \sum_{i=1}^{m+n} (p_i^c)^2 \tag{2.31}$$

$$\varepsilon_i(i_c) = \alpha_c e_i^{c1}(i_c)\sigma'[o_i^c(i_c)] \tag{2.32}$$

The fact that $w_{ij}^{c1}(i_c + 1) = w_{ij}^{c1}(i_c) + \Delta w_{ij}^{c1}(i_c)$ is used in deriving Eqs. (2.28)–(2.32). In the above equations, $\varepsilon_i(i_c)$ is the scaled version of the backpropagated error at the input of the hidden-layer neuron h_i^c, $o_i^c(i_c)$ is the input to the hidden-layer neuron h_i^c, and Λ_c is a constant that needs to be computed at the beginning of each time step. The significance of (2.28)–(2.32) is that instead of computing $\mathbf{h}^c$ for a new iteration from scratch, we can conveniently obtain the result from the information calculated at previous iterations. More conveniently, if we start from the 0th iteration, then we only need to compute $o_i^c(0)$ once with the initial synaptic weight and the input vector as

$$o_i^c(0) = \sum_{j=1}^{m+n} w_{ij}^{c1}(0)p_j^c \tag{2.33}$$

Then, for the ith iteration, Eqs. (2.34) and (2.35) can be used to compute the value of o_i^c:

$$o_i^c(i_c) = o_i^c(0) + E_i(i_c)\Lambda_c \tag{2.34}$$

$$E_i(i_c) = \sum_{k=0}^{i_c} \alpha_c e_i^{c1}(k)\sigma'[o_i^c(k)] \tag{2.35}$$

Therefore, we can evaluate the output of the updated neural network without actually updating its synaptic weights. We call this technique the virtual update technique. The idea of the virtual update is conceptually shown in Figure 2.14. In the conventional training method, two forward iterations are carried out to generate the estimation. The TD error is computed based on the estimation and the error is backpropagated to

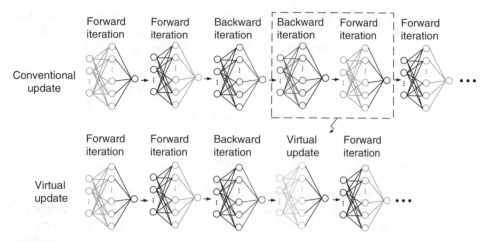

Figure 2.14 Conceptual illustration of the virtual update algorithm. The backward iteration and the forward iteration for the first layer of synapses are lumped into one virtual update operation.

Figure 2.15 Pseudocode for the virtual update algorithm [27]. Only the *while* loop for updating the critic is shown. Reproduced with permission of IEEE.

```
1   while (i_c < I_c && δ(t) ≥ E_c) do
2       if (i_c == I_c − 1) then
3           Normal backward operation, weight update and
                forward operation
4       else
5           Virtual update for the input layer, and forward for
                hidden layer and output layer
6       Compute the temporal difference δ(t)
            i_c = i_c + 1
            if (i_c == I_c || δ(t) < E_c ) then
7               Normal weight update for input layer synapses
```

each synapse through two backward iterations. These forward-backward operations are repeated over and over until the TD error is satisfactory or the maximum number of iterations is reached.

In the proposed training method, the second backward iteration and the first forward iteration are combined into one iteration called the virtual update, which considers that these two iterations are dealing with the same set of inputs. The virtual update algorithm does not rely on any approximation. Therefore, it does not yield any degradation in the precision. The speedup of the algorithm is purely a result of getting rid of ineffectual operations and rearranging the computations more efficiently. The pseudocode for the *while* loop corresponding to the critic-updating phase when the virtual update technique is used is illustrated in Figure 2.15. It is worth mentioning that even though the synaptic weights are not altered during the process of updating the critic network, they have to be updated before exiting the while loop either because the number of iterations reaches the limit or the cost function meets the requirement.

To provide a comparison between the conventional update method and the virtual update algorithm, the complexity of these two algorithms are compared in Table 2.1. In the table, "MAC," "MUL," and "ADD" represent the multiply-accumulate, multiply, and add operations, respectively. Since the original algorithm needs to update the weight and calculate the updated neuron input, the computational complexity for each iteration is on the order of $O(N_i N_h)$. The virtual update algorithm, however, reduces

Table 2.1 Comparison of the computational complexity of a conventional update and a virtual update.

	Conventional update	Virtual update
Forward pass	$N_i N_h L$ **MAC**	$(L-1)N_h$ **MAC** $+ N_i$ **MUL** $+ N_i N_h$ **MAC**
Backward pass	$N_i N_h L$ **MAC**	$(L-1)N_h$ **ADD** $+ N_i N_h$ **MAC**
Total operations (**MAC/MUL/ADD**)	$2N_i N_h L$	$2(L + N_i - 1)N_h + N_i$
Complexity	$O(N_i N_h L)$	$O((L + N_i)N_h)$

Source: data are from [27]. Reproduced with permission of IEEE.

this quadratically scaled complexity to a linearly scaled complexity $O(N_h)$, which significantly increases the throughput of the algorithm. Such a significant reduction in the number of arithmetic operations appears to be counterintuitive at first glance. A simplified diagram is shown in Figure 2.16 to provide some intuition behind the algorithm. In a conventional update, the errors at the hidden layer are multiplied with the input vector to obtain the weight update for the first-layer synapses. The updated synapses are then multiplied with the input vector to obtain the updated hidden layer input. The operation of adding the weight update to the old weights are not shown in the figure for brevity. In the virtual update algorithm, on the other hand, the order of

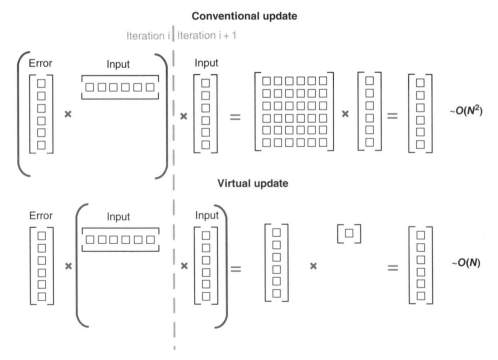

Figure 2.16 Illustration of the intuition behind how the virtual update algorithm can help save arithmetic operations. The virtual update algorithm relies on the loop unrolling and the rearrangement of the matrix multiplication to reduce the computational complexity from $O(N^2)$ to $O(N)$. Additions in the weight update are omitted in the figure for brevity.

conducting matrix multiplication is swapped. Since the input vector stays unchanged for many learning iterations, the inner product of the input vector with itself, which is Λ_c in Eq. (2.31), only needs to be computed once. Again, the operation associated with the old weight is not illustrated in the figure for clarity, as they are not important in this analysis. In other words, the savings in the number of arithmetic operations mainly come from the loop unrolling and reordering of the matrix multiplications.

Even though the virtual update technique is only applicable to synapses between the input layer and the hidden layer, the savings in computational efforts are remarkable, as most weights in neural networks often concentrate in between these two layers. In addition to speeding up the training, the virtual update algorithm can significantly reduce the energy consumption of the hardware as well. The savings in energy consumption mainly come from two sources. The first source is that, since the virtual update algorithm does not update the synaptic weights until the last iteration, it saves the trip to the synapse memory. The second reason that the algorithm is more energy efficient is due to the fact that it involves less arithmetic operations. The actual percentage of savings in the energy consumption is reported in Chapter 3.

2.3 Network Topologies

The neural network is a versatile machine-learning tool that can be employed to solve many real-life problems. Depending on the specific problem, the topology of the neural network can be tailored to provide better leverage of the structure of the data or to provide some unique advantages. In this section, we mainly discuss three major types of network topologies. There are many kinds of neural networks in the literature and there are various ways to categorize them. Therefore, the way we divide the neural networks in this section may not be unique. It is, nevertheless, a practical way to discuss popular neural network structures in the literature.

2.3.1 Fully Connected Neural Networks

A fully connected neural network (FCNN) is one of the most popular and the simplest form of neural networks that is widely used in the literature. As suggested by its name, an FCNN has a full connection between layers. Every neuron in one layer has a connection with every neuron in the adjacent layer. Computations involved in an FCNN are very regular. They are either matrix multiplications or element-wise operations.

An FCNN is particularly useful when there is no obvious correlation between inputs. For any neuron in an FCNN, all other neurons appear to be similar. In other words, there is no spatial structure in this type of neural network. For example, in the case study presented in Section 2.2.4, the input to the neural networks often comprises deliberately selected system states, which are unlikely to be redundant or correlated. Therefore, an FCNN is a proper choice in that task. For some other tasks where the input signals have a certain spatial correlation, like image, other more sophisticated network topologies can be used. This is discussed in the next section. Thanks to its simplicity, FCNNs have achieved many successes in various applications. Even though the simple plain FCNN works well by itself, FCNNs that have special architectures or properties also exist.

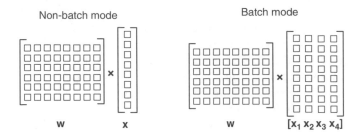

Non-batch mode   Batch mode

w   x   w $[x_1 \ x_2 \ x_3 \ x_4]$

Figure 2.17 Illustration of the matrix-related operations in a non-batch mode and a batch mode. In the batch mode, several input vectors are aggregated to form a matrix. The original matrix–vector multiplication is recast as a matrix–matrix multiplication, which can take advantage of the parallel processing capability of modern CPUs and GPUs.

A radial basis function network is such an example. This network uses a radial basis function as its kernel, as implied by its name [36]. A radial basis function is a function whose value is determined by the distance between the input and the origin. Popular radial basis functions are Gaussian, multiquadric, and so on. Park and Sandberg have shown that the radial basis function network with certain kernel functions can be employed as a universal function approximator [37, 38]. The radial basis function network gained its popularity in the 1990s [39–41].

To utilize the parallel computing power in modern central processing units (CPUs) and graphics processing units (GPUs), both learning and inference of neural networks are often carried out in batch. Figure 2.17 illustrates this idea. In a non-batch mode of the forward operation, each input is evaluated separately. Different input vectors are evaluated serially and only one input vector is processed at a time. In order to utilize the parallel computing to increase the throughput, one can cast multiple input (e.g. multiple images) into a matrix, and the original matrix–vector multiplication becomes a matrix–matrix multiplication. In real implementations of supervised-learning tasks, a batch size of 10–100 is often used, depending on the computing capability of the machine. Another restriction on the batch size may come from the latency requirement. The batch computation can undoubtedly improve the system throughput when there are enough computing resources, yet it increases the latency of evaluation at the same time. Therefore, batch-mode inference is more appropriate for certain non-real-time cloud-based computing applications where the latency is not critical and the inputs from many users can be concatenated together to leverage the parallelism in the servers.

It is worth mentioning that even though we use an FCNN as the example to illustrate the concept of batch, batch processing is not restricted to this type of neural network. Other types of neural networks, e.g. a convolutional neural network (CNN), can also be trained and evaluated in a batch mode to improve the system throughput.

2.3.2 Convolutional Neural Networks

The biggest drawback of an FCNN is that there are too many parameters associated with the neural network as the synapses are densely connected between neurons. For many applications where images and audios are involved, the invariance of the signal in the space and time domain can be readily employed to reduce the number of synapses required.

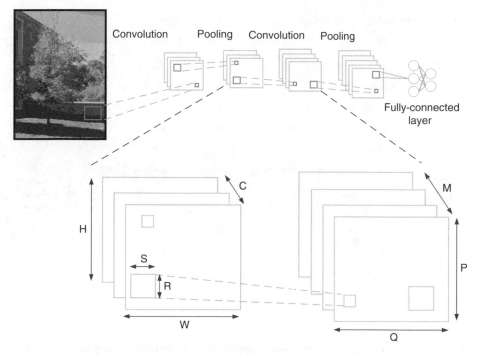

Figure 2.18 Configuration of a typical CNN. For each convolutional operation, a patch in the input map is convolved with several filters. A different filter results in a different output channel. Fully connected layers are often (but not necessarily) used as the final layers of a CNN for classification purposes.

A typical configuration of a CNN is demonstrated in Figure 2.18. In each layer, both the input data and output data are four-dimensional tensors. In a forward pass, output maps at each layer are calculated based on input maps according to the pseudocode shown in Figure 2.19. $O[n][m][p][q]$ is the output element located at the pth row, qth column of the mth output map in the nth batch. $I[n][c][h][w]$ is the input element located at the hth row, wth column of the cth input map in the nth batch. $W[m][c][r][s]$ is the weight element located at the rth row, sth column of the weight filter that correspond to the cth input channel and mth output channel. $B[m]$ is the bias for the output channel m. Detailed meanings for all the parameters are shown in Table 2.2 and Figure 2.18. The stride size U and V are used to indicate how the filters are slid on the input map. A vertical stride size of U means the filter is shifted U units vertically to obtain two vertically adjacent data in the output map. A similar definition applies to the horizontal stride size V as well. It is noted that P and Q are not free parameters, but are determined according to

$$P = \frac{H - R + U}{U} \tag{2.36}$$

$$Q = \frac{W - S + V}{V} \tag{2.37}$$

As suggested by its name, a CNN is a network that involves convolution. CNNs perform particularly well on image and audio inputs. Taking images as examples, there are

```
for(int n = 0; n < N; ++n){                    // nth input in the batch
    for(int m = 0; m < M; ++m){                // mth output channel
        for(int c = 0; c < C; ++c){            // cth input channel
            for(int p = 0; p < P; ++p){        // pth row of the output map
                for(int q = 0; q < Q; ++q){    // qth column of the output map
                    for(int r = 0; r < R; ++r){    // rth row of the filter
                        for(int s = 0; s < S; ++s){    // sth column of the filter
                            O[n][m][p][q] = I[n][c][U*p + r][V*q + s] * W[m][c][r][s] + B[m];
                        }
                    }
                }
            }
        }
    }
}
```

Figure 2.19 Pseudocode for a typical convolution operation involved in a CNN. The input feature map, a four-dimensional tensor, is transformed to the output feature map, which is another four-dimensional tensor, through 7 nested *for* loop.

Table 2.2 Meaning of parameters in Figures 2.18 and 2.19.

Parameter	Meaning
N	Batch size
C	Number of input feature map
M	Number of output feature map
H	Height of the input feature map
W	Width of the input feature map
P	Height of the output feature map
Q	Width of the output feature map
R	Height of the filter
S	Width of the filter
U	Vertical stride size
V	Horizontal stride size

often spatial correlations among pixels. For a pixel located at the left-upper corner of the image, pixels that are around it generally have much stronger correlation with it compared to the pixel located at the right-lower corner. A CNN is able to leverage this spatial correlation by convolving the image with patches of filters. It is worth noting that even though it is not shown in Figure 2.18, non-linear activation functions are often

Figure 2.20 Illustration of average pooling and max pooling. The mean and the maximum value of a patch is outputted for the average pooling and max pooling, respectively.

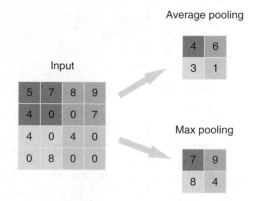

Input

Average pooling

Max pooling

needed in a CNN; otherwise, a multilayer neural network is degenerated into a two-layer linear neural network.

Ideally, each filter in a CNN is attempting to look for one specific feature. There are typically many different features in an input map. Therefore, one normally obtains more output maps than input maps. Convolution operation with a stride of 1, on its own, does not reduce the size of the input map significantly. As a consequence, the amount of data we are dealing with would grow rapidly as the number of layers increases if we did not do anything in between two convolutional layers. Pooling is such an operation that helps reduce the data we need to process. A pooling operation transforms a patch of data into one data. This can be done through either average pooling or max pooling, as shown in Figure 2.20. In the figure, a stride size of 2 is assumed. The average pooling operation outputs the average of the input patch, whereas the max pooling outputs the maximum number in the input patch.

In terms of training, regardless of the complex connections involved in a CNN, training of a CNN is, in essence, the same as the training of an FCNN. One slight difference is that there is weight sharing in a CNN. Therefore, the errors backpropagated from different outputs need to be summed up in order to get the errors for the shared synapses. In terms of evaluation, a CNN is normally compute-bound. For an FCNN, there is generally no weight reuse (except for across different batches) and the weights have to be streamed from the memory to the computing units continuously. A CNN, on the other hand, can reuse the synaptic weights for different input patches and different batches, as shown in Figure 2.19. Therefore, there exist various strategies in implementing CNN accelerators. This is discussed in detail in Chapter 3.

2.3.3 Recurrent Neural Networks

The two aforementioned types of neural networks are both FNNs. An FNN behaves as a memoryless function. Its output solely depends on its current input. In many applications, such as predicting the next letter in a word, outputs of the neural networks depend not only on the current inputs but also the previous inputs. In this case, a simple FNN is not sufficient. One needs to resort to neural networks with memory or feedback mechanisms, which are the recurrent neural networks (RNNs). The main difference between FNNs and RNNs is highlighted in Figure 2.21. For the FNN, only a feedforward connection is allowed, whereas both lateral and sometimes feedback connections are present

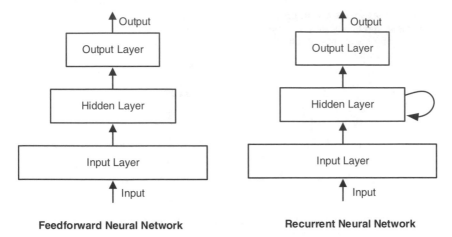

Feedforward Neural Network   **Recurrent Neural Network**

Figure 2.21 Comparison of an FNN and an RNN. While only feedforward connections are allowed in an FNN, both lateral and feedback connections can be used in an RNN. The feedback connections in an RNN provide memory to the network.

in an RNN. Such lateral and feedback connections lead to the memory in RNNs. One of the earliest types of RNN is called a Hopfield network. It is named after John Hopfield, who contributed significantly to make this type of neural network popular [42]. Hopfield networks are RNNs with binary neurons, i.e. neurons that can only provide an output of 0 or 1. The network can serve as a content-addressable memory. The details and applications of the Hopfield network are presented in the Appendix for interested readers.

The memory that exists in an RNN is helpful in dealing with some applications where the input data have temporal or sequential correlations; for example, in a task of predicting the next character in a word. Let us assume the word that the user wants to input is "neuron." For an RNN, one can input the letter "n" to the neural network and then "e" and then "u." The RNN is expected to be able to predict the next character as the typing proceeds. The RNN can perform the prediction because it remembers what characters have been inputted by the user.

Training of RNNs can be achieved through a technique called backpropagation through time [43]. The idea is to unfold the loop in the hidden layer shown in Figure 2.21. For each unfolded loop, errors can be backpropagated to the weight matrix. Since the weight are shared across all unfolded loops, the final error gradients can be summed up and conventional gradient descent learning can be conducted.

One of the problems in training RNNs with the backpropagation through time is the vanishing gradient. As the backpropagation through time essentially unfolds the RNN into a very deep FNN. The gradients tend to vanish or blow up after propagating for several layers as multiplications are involved in the backpropagation. Therefore, many RNNs cannot be trained properly because of this vanishing-gradient problem. With the attempt to tackle this issue, a type of RNN called long short-term memory (LSTM) was proposed by Hochreiter and Schmidhuber in 1997 [44] and has since become very popular. The LSTM successfully solves the vanishing-gradient problem through enforcing a constant error flow. The constant error flow is ensured by the so-called constant error carousels (CECs), which includes an identity feedback to itself in a memory cell [45].

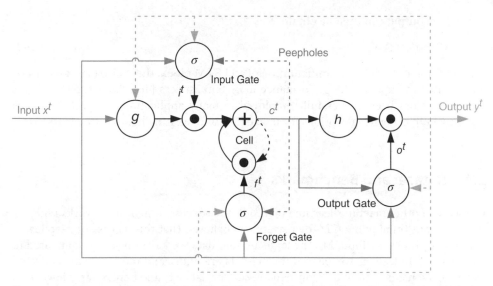

Figure 2.22 Schematic of a typical LSTM block. The input information flows through the LSTM block from left to right. A cell is embedded in an LSTM block to store information. Three gates, namely input gate, output gate, and forget gate, are used to control the information flow. Source: adapted from [46].

The identity feedback keeps the error that propagates through the CEC from vanishing or blowing up. Gates are introduced in an LSTM to help decide when to input new information into the cell memory and when to output the cell memory [44].

One typical configuration of an LSTM block is shown in Figure 2.22. The input vector and output vector of the LSTM cell are denoted as $\mathbf{x^t}$ and $\mathbf{y^t}$, respectively, where t is used to represent the time stamp. The output from the memory cell is represented as $\mathbf{c^t}$, which serves as the core of an LSTM block. The input and output gates decide when to allow information flowing into and flowing out from the cell. Their outputs can be calculated as

$$\mathbf{i^t} = \sigma(\mathbf{W_{ix}x^t} + \mathbf{W_{iy}y^{t-1}} + \mathbf{W_{ic}} \odot \mathbf{c^{t-1}} + \mathbf{b_i}) \tag{2.38}$$

$$\mathbf{o^t} = \sigma(\mathbf{W_{ox}x^t} + \mathbf{W_{oy}y^{t-1}} + \mathbf{W_{oc}} \odot \mathbf{c^t} + \mathbf{b_o}) \tag{2.39}$$

where $\mathbf{W}$ represents a weight matrix or vector, $\odot$ denotes an element-wise multiplication, and $\sigma(\cdot)$ is the activation function. In [47], a forget gate was proposed to be added into an LSTM block in order for the LSTM to reset its internal state. The output of the forget gate can be described similarly to the input and output gate as

$$\mathbf{f^t} = \sigma(\mathbf{W_{fx}x^t} + \mathbf{W_{fy}y^{t-1}} + \mathbf{W_{fc}} \odot \mathbf{c^{t-1}} + \mathbf{b_f}) \tag{2.40}$$

It is noted that the terms $\mathbf{c^{t-1}}$ and $\mathbf{c^t}$ in Eqs. (2.38) to (2.40) were not present in the original LSTM block. It was later proposed in [48] that peephole connections could be added to allow the LSTM to learn precise timing more easily.

With all these three control gates, the cell state can be updated through

$$\mathbf{c^t} = g(\mathbf{W_{cx}x^t} + \mathbf{W_{cy}y^{t-1}} + \mathbf{b_c}) \odot \mathbf{i^t} + \mathbf{c^{t-1}} \odot \mathbf{f^t} \tag{2.41}$$

where the activation function $g(\cdot)$ is typically a hyperbolic tangent function.

The final output can then be obtained as

$$\mathbf{y}^t = h(\mathbf{c}^t) \odot \mathbf{o}^t \tag{2.42}$$

In addition to this classic configuration of an LSTM block, there also exist some variants aiming to improve the performance in certain aspects [46]. Since the LSTM was invented, it has been successfully employed in many applications, such as sequence learning [49], generating captions for image [50], and creating handwritten texts [51].

2.4 Dataset and Benchmarks

Just as with other machine-learning methods, the success of neural networks relies on the abundant training data. Many researchers believed that the success of deep learning can be largely attributed to the large amount data we have nowadays compared to 20 years ago. Indeed, as we entered the big-data era, more and more data are collected by various sensors around us. Consequently, many datasets and benchmarks have been created in the past to facilitate the development and evaluation of various neural networks. With these dataset and benchmark tests, researchers can develop and examine new algorithms conveniently. In this section, several popular datasets and benchmark tests are presented.

As one of the most popular applications of neural networks, image classification has received more and more attention in the past decade. Therefore, many image datasets have been created. The Modified National Institute of Standards and Technology database (MNIST) dataset, is one of the earliest datasets for a benchmarking classifier [52]. It consists of 60 000 training images and 10 000 test images. Each image is a 28×28 pixel grayscale image of a handwritten digit ranging from 0 to 9. A few samples of images from the MNIST dataset are illustrated in Figure 2.23.

Figure 2.23 Illustration of the MNIST dataset. Each image in the dataset contains 28 × 28 pixels. The images are all grayscale handwritten digits from 0 to 9. Source: data are from [52].

The CIFAR-10 and CIFAR-100 dataset were collected by Krizhevsky and Hinton [53]. CIFAR-10 contains 60 000 images. Among these images, 50 000 images can be used as a training set, whereas another 10 000 images can be used as a test set. All the images are 32×32 pixel color images that are categorized into 10 classes, such as airplane, frog, truck, etc. The CIFAR-100 dataset is similar to the CIFAR-10 dataset except that it is divided into 100 classes. These 100 classes can be further grouped into 20 coarser super-classes.

ImageNet is a dataset organized based on the WordNet hierarchy. A "synonym" or a "synset" is a concept in WordNet that can be described by several words. The ImageNet Large Scale Visual Recognition Challenge (ILSVRC) is an annual challenge that attracts many researchers' attention around the world [54]. There are different tasks in the ILSVRC. One task is image classification, where the algorithm produces a list of object classes that exist in the given images. Another task is object detection, where an algorithm produces a list of object classes as well as the bounding box that attempts to locate the positions of the objects. Each year, researchers around the world participated in the ImageNet contest with their well-tuned models.

In addition to the datasets for images, there are also several popular datasets for audio and text. The Texas Instruments/Massachusetts Institute of Technology (TIMIT) is a collection of acoustic-phonetic speech of American English speakers. It contains 630 speakers of both sexes with different dialects. The 20 Newsgroups dataset collects around 20 000 newsgroup documents that have been categorized into 20 different newsgroups. The number of documents for each newsgroup is approximately equal. Other text-based datasets include Yelp dataset, which are abstracted from Yelp reviews, business, and user data [55], WikiText, which contains 100 million tokens from articles on Wikipedia [56], etc. For more information on various datasets for deep learning, interested readers are referred to a webpage that summarizes many different open datasets [57].

Besides the abundant large dataset, various tasks also exist for benchmarking reinforcement-learning algorithms. For example, three popular benchmarks used in the ADP literature are illustrated in Figure 2.24 [58]. The cart-pole balancing task [11, 32–34, 58] is a control task that aims at balancing a pole sitting on a cart. The target is to keep the pole upright while not moving the cart out of a certain range. In other words, the angle between the pole and the vertical direction θ and the distance between the cart and the origin x need to be regulated.

In a beam-balancing problem [58–60], the controller attempts to balance a long beam by applying torque to the beam through a motor installed at the center of the beam. In addition, there is a ball sitting on the beam that complicates this task. The ball can roll along the beam according to the angle between the beam and the horizontal direction, θ. The objective of this benchmark task is to keep the beam as horizontal as possible while maintaining the displacement of the ball within a certain range.

The triple-link inverted pendulum-balancing problem [27, 32, 33] is a more complicated control problem, as shown in Figure 2.24c. It is similar to the cart-pole balancing problem. In this case, however, instead of balancing only one pole, three connected sticks that can rotate with respect to each other need to be balanced. Therefore, four state variables θ_1 to θ_3 and x need to be regulated. Besides these three common benchmarks, other popular benchmarks are pendulum swing up task [33, 61], cart-pole swing up task [61, 62], and the acrobat swing up task [63].

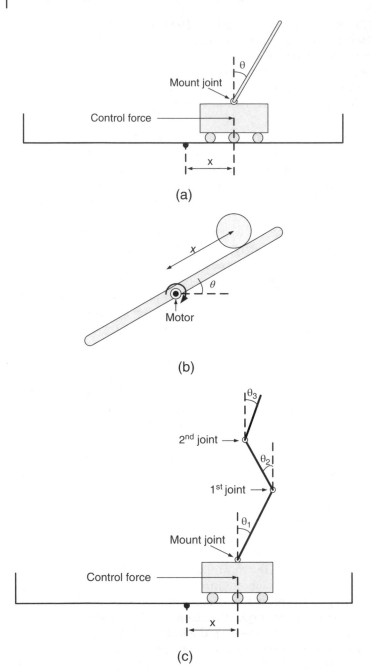

(a)

(b)

(c)

Figure 2.24 Configuration of the (a) cart-pole balancing task, (b) beam balancing task [58], and (c) triple-link inverted pendulum balancing task. In all of these tasks, the objective is to regulate the states of the system within a predefined bound by applying certain control signals. Reproduced with permission of IEEE.

In addition to the above-mentioned classic control benchmarks, many more complicated benchmark tasks, such as Swimmer and Hopper, are also available [64]. These benchmarks are becoming more and more popular as more computational resources are available to solve complicated reinforcement-learning problems.

2.5 Deep Learning

2.5.1 Pre-Deep-Learning Era

Neural networks were built decades ago. In the 1980s and 1990s, many researchers had already studied multilayer perceptrons (MLPs) that were very close to the deep neural networks used nowadays. The popular backpropagation was used at that time for training numerous neural networks. Hornik et al. demonstrated that neural networks with just one hidden layer could be used as universal function approximators [65]. Neural networks have been successfully employed in various fields such as image recognition [52, 66, 67], speech recognition [68, 69], time series prediction [70, 71], and control problems [72, 73].

At that time, however, it was found that the performance of a neural network was not particularly impressive, compared to other popular machine-learning tools, e.g. a support vector machine (SVM). In addition, it was generally believed that one hidden layer was enough for neural networks, as more hidden layers seemed to be unhelpful in improving the classification accuracy. Furthermore, training a deep neural network was a non-trivial task due to many problems such as overfitting and the vanishing gradients. After recognizing these difficulties, researchers in the artificial intelligence (AI) community gradually shifted their attention from neural networks to other machine-learning methods.

2.5.2 The Rise of Deep Learning

After a total silence for nearly a decade, several researchers brought together by the Canadian Institute for Advanced Research (CIFAR) refocused on the training of neural networks. This time, the interests were in training deep neural networks. This has opened the door for deep learning. The success of deep learning cannot be attributed to one or two works. It was several years of effort by numerous researchers that laid down the foundation for deep learning.

As mentioned earlier, the concept of a deep neural network had existed a long time back, yet the rise of deep learning did not occur until late 2000s. Nowadays, it is generally believed that the success of deep learning can be mainly attributed to three factors [74, 75]. The first factor is the unprecedented computing power that researchers now have access to. Driven by the well-known Moore's law, the amount of computations we can perform per unit time are growing rapidly. Many tasks considered impossible to be completed in a reasonable amount of time before 2000 may be a routine task that many personal computers run on a daily basis nowadays. Furthermore, the introduction of GPUs and customized hardware also makes it possible to train large-scale neural networks with a tremendous amount of data.

The second factor is the amount of data available. Thanks to the rapid growth of the Internet, the amount of data we have access to has increased exponentially over the past decades. With more and more data available, researchers can train their deep network more thoroughly. As stated in previous sections, one difficulty of training a deep neural network is the overfitting problem. This can be addressed by providing more training data.

The third factor that leads to the success of deep learning is the improvement in the algorithms. Even though the faster hardware and the larger datasets provided the necessary conditions for the deep learning to take off, researchers still had to figure out smart tricks and techniques that could help train the deep network better. Over the past decade, we have witnessed the explosion of interests in the field of deep learning. Many brilliant ideas and algorithms have been proposed, prototyped, and verified. In the next section, a few representative techniques that help advance deep learning are reviewed.

2.5.3 Deep Learning Techniques

In recent years, numerous deep learning techniques have been developed. We divide these techniques into two categories for ease of discussion. The first group of techniques focuses on improving the performance of the learning. Since our interests are in hardware implementations of neural networks, we also study another group of techniques that can potentially improve the efficiency of the hardware to which the algorithm is mapped.

2.5.3.1 Performance-Improving Techniques

2.5.3.1.1 Unsupervised Pretraining Unsupervised pretraining is one of the earliest techniques employed to conduct effective training on deep neural networks. The motivation behind this technique is that for supervised learning, many labeled data are needed. Labeled data are usually very limited, as the labels are often created manually by experts. The shortage of labeled data is one of the few factors to blame for the mediocre performance of deep neural networks in the early days. Before the technique of unsupervised pretraining was proposed by Hinton and Osindero in 2006 [21], the standard way of training neural networks was to start the synaptic weights randomly. Such an arbitrary starting point posed difficulties in training deep networks, especially when the size of the employed dataset was small. Through a proper unsupervised learning, one can leverage the abundant unlabeled data to pretrain the neural network in order to set up a good initial condition. After unsupervised pretraining, supervised fine-tuning is then carried out to attach each label to the output of the deep network.

Unsupervised pretraining is typically done greedily, layer by layer. In other words, one first conducts unsupervised learning on the first layer alone. After the training is done, the weights for the first layer are fixed and the second layer is stacked on top of the first layer. The unsupervised learning is then conducted on the combination of the first and the second layers. This procedure is repeated until all layers are finished. Then, a conventional backpropagation-based supervised learning is carried out to fine-tune the whole network. Early deep belief networks (DBNs) based on RBMs [21, 26, 76] and stacked denoising autoencoders [25] all used this technique.

Even though the unsupervised pretraining was one of the earliest techniques used in training deep neural networks, it gradually lost its popularity in training many modern deep networks. This is mainly because the efficacy of the unsupervised pretraining is no

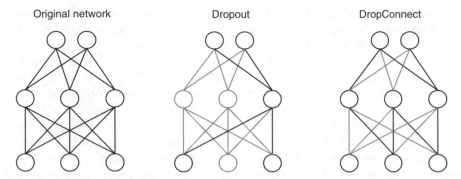

Figure 2.25 Illustration of the dropout and DropConnect techniques. With dropout, activations of some neurons are forced to be zero randomly. With DropConnect, the contributions from some synapses are forced to be zero randomly. The random drop of the activation or synapse is helpful in preventing co-adaptations in the neural network.

longer obvious compared to the direct supervised training, as more and more labeled data are available nowadays. Nevertheless, for applications where the labeled data are in shortage, unsupervised pretraining can still be very effective.

2.5.3.1.2 Dropout Two common problems that are often encountered in training neural networks are underfitting and overfitting. These two cases are illustrated in Figure 2.7. Underfitting occurs when there are many training data while the neural network used to learn the data is not sophisticated enough. In this case, we have more constraints than the degrees of freedom. Therefore, the trained model is not able to represent the data well enough. Another extreme is overfitting. It occurs when the data points available are too few. In this case, we have more degrees of freedom than the constraints. In this case, the trained model might fit the training data well, yet it cannot generalize to represent other data that are not in the training set.

Dropout is a technique used for mitigating the overfitting problem. This method was originally proposed by Hinton et al. [77, 78]. The dropout technique is conceptually illustrated in Figure 2.25. It relies on randomly dropping out activations from some neurons in order to learn a more robust model. The dropout is helpful in breaking co-adaptations, which work for the training data yet not for unseen data. Intuitively, for each input, the neural network needs to work with a random set of neurons to obtain the correct answer. Therefore, each neuron has to work more independently, effectively breaking the co-adaptions that deteriorate the generalization capability of the neural network. During the training, the dropout of the neurons is controlled randomly with a target dropout rate.

Inspired by dropout, DropConnect, as shown in Figure 2.25, is a more general way of regularizing neural networks [79]. Instead of dropping out the activations from certain neurons, DropConnect omits certain connections in the neural network. Clearly, dropout is a special case of DropConnect, where connections from one neuron are all dropped.

2.5.3.1.3 Batch Normalization Batch normalization is a technique recently proposed by researchers at Google [80]. In neural networks, we always want the activations of neurons to be neither too small nor too large. Normalization of data is often performed in the preprocessing phase of learning in order to improve the performance of the learning [81]. Such a normalized input, however, tends to spread as the signals go deeper into the network. This phenomenon is called internal covariate shift in [80].

The idea of batch normalization is that normalization is used at every layer to maintain a good distribution of activations within each layer. The distribution is obtained across the minibatch, which is used anyway in deep learning in order to leverage the parallel nature of modern computing machines. The detailed procedures of batch normalizations are presented in Eqs. (2.43) to (2.46) [80]. The mean and the variance across the batch are calculated in Eqs. (2.43) and (2.44) with standard equations. The normalized activation is calculated in Eq. (2.45) in order to obtain values with a distribution of zero mean and unit variance. A small number ε is added to the equation to address the numerical stability problem. In Eq. (2.46), γ and β are two parameters that need to be learned, as it is hard to know in advance what distributions are preferred for the neural network to achieve good performance. Therefore, these two learnable parameters are helpful in improving the training performance.

$$\mu_B = \frac{1}{m} \sum_{i=1}^{m} x_i \tag{2.43}$$

$$\sigma_B^2 = \frac{1}{m} \sum_{i=1}^{m} (x_i - \mu_B)^2 \tag{2.44}$$

$$\hat{x}_i = \frac{x_i - \mu_B}{\sqrt{\sigma_B^2 + \varepsilon}} \tag{2.45}$$

$$y_i = \gamma \hat{x}_i + \beta \tag{2.46}$$

Batch normalization is not only helpful in training deep networks, but also enables the use of higher learning rates and provides a better regularization. Larger learning rates can be used along with the batch normalization since the batch normalization makes the learning more robust to the parameter scale. The better regularization comes from the fact that batch normalization uses more information across the mini-batch.

2.5.3.1.4 Accelerating Stochastic Gradient Descent Process

Regardless of various newly proposed fancy deep-learning ideas, the core of learning in ANNs is still an SGD scheme. Therefore, how to achieve a faster and better convergence in an SGD process directly determines how good the learning can be. It is because of this importance of the SGD learning that many researchers, in recent years, have devoted to developing techniques and algorithms that can achieve a better convergence.

The simplest way to accelerate the conventional gradient descent as shown in Eq. (2.3) is the momentum method [82–84]. The idea is to accumulate the momentum in the right direction, which can help achieve a faster convergence. Mathematically, the momentum method can be written as

$$v[n] = \beta v[n-1] - \alpha \left. \frac{\partial L(\mathbf{w})}{\partial w} \right|_{w=w[n-1]} \tag{2.47}$$

$$w[n] = w[n-1] + v[n] \tag{2.48}$$

where $v[n]$ is the accumulated momentum at iteration n, β is the momentum coefficient, α is the learning rate, w is a parameter in the network, and $L(\mathbf{w})$ is the loss function that needs to be minimized. The momentum update equation shown in (2.47) can be seen as an infinite impulse response (IIR) filter that filters out noises in the gradient and leaves the low-frequency part that corresponds to the true gradient.

A slightly more advanced momentum method is called the Nesterov accelerated gradient (NAG), which is named after the Russian mathematician Yurii Nesterov [85]. It can be thought of as a look-ahead version of the conventional momentum method. The intuition behind NAG is that since we can foresee that the conventional momentum method will bring the parameter to somewhere near $w[n-1] + \beta v[n-1]$, why do we not directly use the gradient at that point? Formally, NAG can be expressed as

$$v[n] = \beta v[n-1] - \alpha \left. \frac{\partial L(\mathbf{w})}{\partial w} \right|_{w=w[n-1]+\beta v[n-1]} \tag{2.49}$$

$$w[n] = w[n-1] + v[n] \tag{2.50}$$

Clearly, NAG is only slightly different from the conventional momentum method by evaluating the gradient at a look-ahead position.

The conventional momentum method and NAG can perform reasonably well in training many neural networks. One potential problem with these two methods is that we still need to pick the learning rate α and the momentum coefficient β. Choosing β is less tricky because it does not have a strong dependency on the detailed network configurations and the input data. It can be set empirically as 0.5 or 0.9, depending on the stage of learning [86]. Choosing α, unfortunately, can be tricky enough to affect the whole learning process. Therefore, many algorithms have been developed in recent years to avoid the manual setting of the learning rate.

To tackle this problem, Duchi et al. proposed ADAGRAD, which is a method that can help desensitize the convergence of the learning on the choice of the learning rate [87]. The algorithm provides each parameter with an adaptive learning rate that is obtained from the history of the gradients of that component:

$$w[n] = w[n-1] - \frac{\alpha}{\sqrt{G[n-1]} + \varepsilon} \cdot \left. \frac{\partial L(\mathbf{w})}{\partial w} \right|_{w=w[n-1]} \tag{2.51}$$

where

$$G[n] = G[n-1] + \left(\left. \frac{\partial L(\mathbf{w})}{\partial w} \right|_{w=w[n-1]} \right)^2 \tag{2.52}$$

and ε is a small constant to enhance the numerical stability.

Intuitively, the effects of frequently appearing features are diluted by the large $G[n]$, whereas the effects of occasionally appearing features are boosted. In addition, because the actual learning rate is determined by the history of the gradients, the convergence of the learning is less dependent on the selection of α when compared to a vanilla SGD method. The denominator $\sqrt{G[n-1]} + \varepsilon$ in Eq. (2.51) serves as a feedback term. When the gradient is large, one would like to reduce the learning rate to avoid divergence. On the other hand, one would like to increase the learning rate when the gradient is small in order to accelerate the convergence. This can be done by introducing the denominator as shown in Eq. (2.51).

There are many other algorithms similar to ADAGRAD, such as ADADELTA [88], RMSprop [89], and Adam [90]. Since they share similar features, they are not presented here separately. Interested readers are referred to Ruder's review on this topic [82].

2.5.3.2 Energy-Efficiency-Improving Techniques

In addition to improving the performance of deep neural networks, many techniques and algorithms have also been developed to enhance the energy efficiency of the algorithm when being deployed on hardware. From the perspective of algorithms, the efficiency can mainly be improved through reducing the number of operations needed in the neural network model as the energy consumption generally scales with it. Figure 2.26 shows two common ways to reduce the computational efforts in an ANN. A matrix–vector multiplication is used as an illustrating example. The elements in the matrix and the vector are represented by an array of binary bits in hardware implementation, even though mathematically they are scalar. The depth of the cuboid in the figure represents the number of bits needed to represent the data.

The first technique to achieve a more energy-efficient computation is to reduce the number of non-trivial data in a neural network. For example, it has been shown that there are a significant number of redundant weights in a deep neural network [91]. By removing the redundant or unnecessary weights in a network, the original dense matrix becomes a sparse one. Similarly, a low-rank representation of the weight matrix can be

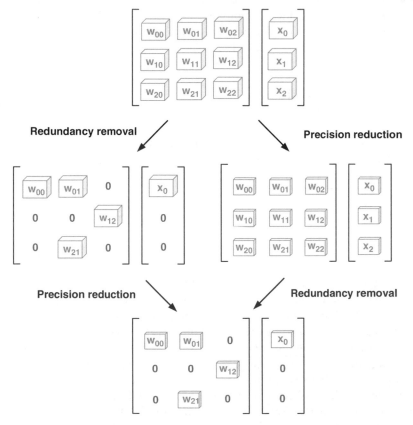

Figure 2.26 Illustration of two popular techniques employed to improve the computational speed as well as the energy efficiency. Eliminating redundancy in a neural network reduces the number of operations needed, whereas precision reduction relies on shortening the bit width needed to represent the data.

formed, which reduces the number of parameters in the network. By doing so, not only the corresponding memory footprint but also the required computational effort can be reduced significantly.

The second technique of boosting the energy efficiency of the hardware is through a precision reduction. Similar to the first technique, reducing precision can also help bring down the memory requirement, which thereby reduces the memory space as well as memory-accessing power. In addition, the power consumption of most components in a neural network accelerator scales with the required precision. The lower the precision, the lower is the power needed to perform the computation.

It is worth noting that these two techniques are, in a sense, orthogonal to each other. One can leverage both in a design to boost the system throughput and energy efficiency. In this section, these two popular ways of reducing computational efforts of an ANN are discussed. Hardware implementation of these methods are presented in Chapter 3.

2.5.3.2.1 Redundancy Removal Synaptic weights of a neural network take the majority of the storage space. All energy-consuming operations such as MAC and memory access operate directly on the weights. Intuitively, the evaluation of a neural network requires less energy and time if the number of parameters in the neural network can be somehow reduced. This can be achieved through weight pruning [92–99], low-rank approximation [100, 101], and so on. All of these methods share the same spirit of removing redundancy in the network. We use weight pruning as an example as it is more commonly used.

The idea of pruning weights can be traced back long before the deep learning became popular [92, 93]. A technique called optimal brain damage (OBD) was proposed by LeCun et al. to reduce the size of the network [93]. The basic idea is to trim away those synaptic weights that do not significantly contribute to the final inference result. The relative importance of each weight can be measured through the so-called "saliency," which is approximated with the help of the second derivative of the objective function. With such an OBD method, a state-of-the-art network can be pruned, and the reduction in the size of the network was shown to be helpful in improving the inference speed and even the recognition rate.

Based on a similar pruning method, network pruning for a deep neural network has been studied extensively by Han et al. [94, 102]. During the pruning phase, both the values and the connections of the synapses are learned. A conventional dense deep network is first trained and small weights are pruned away to compress the network. The pruned network is then trained again to fine-tune the network. Such a procedure can be repeated several times to achieve better learning results. It was shown that through using this pruning method, the reduction in the number of parameters could be as high as 9× and 13× for AlexNet and VGG-16 with no loss of accuracy [102].

Most pruning work in the literature focused on reducing the number of weights as much as possible. Intuitively, there appears to be a positive correlation between the number of parameters of a neural network and the energy spent to evaluate that network. Yang et al., however, demonstrated that this might not always be the case due to the complex memory hierarchy existing in many neural network hardware [99]. In order to maximize the reduction in energy consumption achieved by weight pruning, a method of energy-aware pruning was proposed in [99]. Because the pruning became more and more difficult when the number of deleted synaptic weights grew, Yang et al. started the pruning method from the layer that was most energy-hungry. With such an energy-aware pruning method, a 1.7× reduction in the energy consumption was reported compared to the pruning used in deep compression [99].

One potential problem with a fine-grained weight-pruning method is that the network after pruning may become very irregular. Such an irregularity can incur large overheads in evaluating the network, despite the fact that the size of the network may have shrunk. This is particularly true for general-purpose hardware that is not optimized for conducting a sparse computation. For example, it was shown in [98] that the performance of the pruned networks, where on average 80% of the weights were removed, was actually worse than the baseline network. The loss of a regular weight structure was to be blamed for such a performance degradation. In addition, it was observed in [103] that a coarse-grained pruning can achieve a higher compression ratio because the number of indexes needed to address non-zero data is less with a more regular pruning structure.

Consequently, a significant amount of effort has been devoted to developing pruning with a coarser granularity in recent years [95–98]. For example, structured sparsity learning (SSL) is a technique that attempts to prune the network in a coarser manner [97]. The basic idea of SSL is to utilize group lasso to regularize a set of weights instead individual weight. Such a regularization method can be applied to filter weight, shape of the filter, channel, and even the depth structure in a neural network. In a more recent work, Scalpel was introduced to prune the network adaptively based on the level of parallelism that the target hardware platform had [98]. The basic flow of Scalpel starts with profiling the hardware platform to determine the level of parallelism. Two pruning strategies are then carried out to match the supported level of parallelism in the underlying hardware. With this technique, Scalpel was reported to achieve 3.54×, 2.61×, and 1.25× performance improvement on average when the network was deployed on a microcontroller, a CPU, and a GPU, respectively.

In addition to sparsifying the weight matrix, activations in an ANN can also be sparsified. For example, the ReLU activation function leads to a sparse activation pattern in many deep neural networks [104]. It has been shown in [105] that almost half of the activations in several representative deep networks are zeros. In addition to that, there are network structures that create sparse activation patterns through conditional gating in order to increase the model capacity without incurring a proportionally growing computational cost [106–108]. The basic intuition behind this method is that, for a specific input to the network, only a small portion of the neural network may be needed for the inference, considering that a deep network is often trained to recognize various objects. Therefore, if one can gate off the unneeded portion of the network, the computational cost can be significantly reduced. All these trends in deep learning has led to a sparse activation pattern in the network. Chapter 3 discusses how these sparse activations can be leveraged in an efficient hardware.

2.5.3.2.2 *Precision Reduction*

Another technique that can help shrink the storage space and improve the computing speed is to reduce the precision used to represent data in the network. The most common and effective way in reducing precision is through quantization. Simple linear quantization can be applied to significantly improve the computing speed without introducing too much overhead [109]. More sophisticated quantization methods can also be employed to quantize weights non-linearly. The weights and activations in a deep network are often distributed unevenly, especially when weight decay [86] is used to promote small weights. Under this circumstance, a log quantization might be more appropriate as it provides a wide dynamic range for data when the same number of bits is used. Miyashita et al. looked into the possibility of representing the parameters in a deep network with a logarithmic form in [110]. By doing so, the number of bits needed to represent the weights and activations was significantly reduced. It was reported that only 3-bit precision was needed to encode state-of-the-art deep networks. Such an aggressive quantization strategy only resulted in negligible performance degradation. In addition, representing the data in the logarithmic domain also helped simplify the hardware implementation, as power-hungry multipliers were no longer needed.

In addition to quantizing based on manually picking thresholds, adaptive approaches also exist to learn better quantization strategies from data [94, 111]. A common technique is to use a clustering method, such as K-mean, to divide weights into groups. An originally continuous-value weight is then quantized to the centroid value of the cluster that it belongs to. By doing so, weight sharing can be achieved where all the possible weight values can be stored in a look-up table and only an index is needed to represent the weight. Such a method can greatly reduce the number of parameters needed in a network.

One extreme case for weight compression leads to binary weights, i.e. either +1 or −1. In [112], the concept of BinaryConnect was explored. The idea is to use binary numbers to represent synaptic weights with the hope of benefitting deep learning hardware. Training neural networks with only binary weights can be challenging. One critical step used to ensure successful learning is to only binarize the weights during the forward and the backward propagation phases but not the weight update phase. Such choice is beneficial for small weight changes obtained from SGD to accumulate over time. With binary weights, BinaryConnect was still able to achieve a near state-of-the-art performance. Such a result is promising, as the introduction of binary weights has managed to reduce the number of multiplications by 2/3, not mentioning savings in memory space and bandwidth [112].

A natural extension of BinaryConnect is a binarized neural network where not only the synaptic weights but also the activations are binary [113]. Binarizing both the activations and weights are appealing as the resultant neural network can be very simple, at least in the inference phase. One issue that needs to be addressed with the binarized neural network is how to propagate the gradients through the ill-behaved hard sigmoid activation function used in a binarized neural network. A straight-through estimator was proposed to approximate the gradients [113]. The results obtained with the binarized neural network was only slightly worse compared to the state-of-the-art deep neural network, yet it could achieve almost one order of magnitude improvement in the energy efficiency, due to the reduced bit width of both the activations and the weights.

Over the years, much work has been done to mitigate the performance degradation induced by the extreme compression on the binary weights. Rastegari et al. presented an XNOR-Net in [114]. A scaling factor was introduced and a different binarization method was used. The resultant XNOR-Net was claimed to outperform the BinaryConnect and binarized neural network by large margins. To alleviate the performance penalty induced by an extreme compression of weights, a ternary weight network was proposed in [115] to back off a little from the extreme quantization in synaptic weights. By adding a third state "0" into the original binary weight, the entropy associated with each synapse can be boosted. Nevertheless, a synapse with a 0 weight virtually adds no extra operations as it appears as a disconnected synapse. Therefore, the ternary weight network provides an intermediate solution between a conventional full-precision neural network and an extremely compressed binarized network. To further improve the entropy of the synaptic weight, a trained ternary quantization strategy was introduced in [116]. The basic idea is to use a different scaling factor for excitatory and inhibitory synapses. These two scaling factors can be obtained through learning. Through allowing the positive and negative weights to have different magnitude, Zhu et al. showed that the trained ternary quantization approach could outperform the original ternary weight network by 3%. The main motivation behind a binary/ternary network is to provide a more efficient option for hardware ANNs. Indeed, the introduction of these neural networks has inspired a large number of energy-efficient accelerators. This is discussed in detail in Chapter 3.

Many studies of exploring quantization of data in the network mainly focus on the inference phase. For example, in a binarized network, even though the final weights and activations have only 1-bit precision, full-precision gradients are still required to conduct the learning. To accelerate the learning process as well, the concept of a reduced precision has also been exploited in the learning time [117, 118]. In contrast to the deterministic quantization strategy used for inference-time parameters, stochastic quantization is normally needed in the learning to allow the SGD learning process to converge properly. Intuitively, stochastic quantization serves a similar purpose as the dithering signal that is widely used in the digital signal processing community where dithering is known to help eliminate the data dependency of the quantization noise [119].

2.5.4 Deep Neural Network Examples

With the aforementioned techniques, various successful deep neural networks have been demonstrated over the past decades. In this section, a few of them are briefly discussed. Many of these deep networks are often employed to benchmark neural network accelerators that are discussed in following chapters.

LeNet-5, which was developed by LeCun et al. [52], is one of the earliest deep neural networks. In LeNet-5, the input features are passed through a convolutional layer followed by a pooling layer and then non-linear activation to generate refined features. The deep network is terminated with fully connected layers to form the classifier. Such a classic architecture is the basis for many following popular deep neural networks. Since LeNet-5 was developed to classify the MNIST dataset, it took 28×28 input images. The network uses 5×5 filters and 2×2 pooling layers. The main motivation for LetNet-5 is that it is wasteful to utilize fully connected layers for images, at least for the first few layers. This is because a fully connected layer totally ignores the inherent spatial or temporal structures in images and audios. Through leveraging this structure of the input, only a few parameters are needed, which can improve the training speed.

AlexNet, named after its inventor Alex Krizhevsky [120], is one of the earliest and most popular deep neural networks. It is the winner of 2012 ILSVRC. AlexNet achieved a classification error of 16.4%, which was almost half of the second-best classification error of that year [54]. The astonishing performance of AlexNet soon brought computer-vision researchers' attention to deep neural networks. Starting from then, more and more computer-vision tasks were conducted with the help of deep learning. In AlexNet, five convolutional layers are used to extract useful features before the last few fully-connected layers. The last layer contains 1000 neurons, which corresponds to the 1000 classes in the classification task. The success of AlexNet is mainly attributed to three techniques. The first technique is to utilize GPUs for training the deep network. Two GPUs were employed to accommodate the large network [120]. Thanks to the massively parallel computing capability of GPUs, the AlexNet has benefitted from the large augmented training data, which helped reduce the overfitting. The second technique used in AlexNet was dropout, which is discussed in Section 2.5.3.1.2. It was observed in [120] that the use of dropout significantly mitigated the problem of overfitting. The third technique employed in [120] was the use of a ReLU activation function. It was demonstrated that the ReLU activation function substantially accelerated the learning process by mitigating the vanishing-gradient problem.

GoogLeNet developed by Google researchers is the winner for image classification with provided data in 2014 ILSVRC [121]. The main strategy of GoogLeNet was to arrange the topology of the network smartly such that it could be deeper and wider while still being trained in a reasonable amount of time. The success of GoogLeNet is largely attributed to the introduced inception module, which can be used in the network repeatedly. The network is 22 layers in depth. In spite of being deeper, it was claimed that GoogLeNet winning 2014 ILSVRC used 12 times fewer parameters compared to AlexNet, where 1×1 filters were used in GoogLeNet in order to reduce the number of channels. In addition, the classification error of GoogLeNet was 6.7%, which was much lower than had been achieved by AlexNet in 2012 [54].

VGG is a deep network that won the first and second places in the localization and classification tasks in 2014 ILSVRC [122]. The design philosophy of VGG was to use small and simple 3×3 convolution filters. It was argued that a 5×5 filter can be obtained from cascading two 3×3 filters. Therefore, a similar performance could be achieved while the number of parameters could be reduced. Through using this small convolutional filter, the network could be made very deep, which was helpful in improving the performance. With such a design methodology, VGG achieved a classification error of 7.3% and a 25.3% localization error [54].

ResNet is a 152-layer deep neural network created by researchers at Microsoft Research [123]. The network is 8 times deeper than VGG. ResNet was the winner of 2015 ILSVRC on a classification task with a low classification error of 3.57%. The success of ResNet is mainly attributed to the deep residual learning proposed in [123]. The basic building block of ResNet is illustrated in Figure 2.27. In the network, the input of a few layers of networks are directly forward and added to the output of the network. This direct connection does not require any extra parameter, which is helpful

Figure 2.27 Building block for ResNet. A shortcut is created to bypass a few layers of neurons. Such a bypass connection is helpful in propagating the error. Source: adapted from [123].

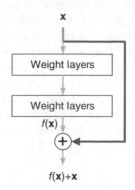

in maintaining the number of parameters that need to be tuned. He et al. examined an interesting performance degradation problem: adding more layers into the network actually made the training error worse. This observation was counterintuitive since a simple deep network could be configured by stacking a shallow network with identity layers. Such a configuration should guarantee the deeper network to perform at least as well as the shallower network. Such a phenomenon suggested that the optimization solver might have trouble in learning identity mapping. Therefore, through providing the shortcut identity connection, the solver might find it easier to come up with a near identity mapping. The identity shortcut in ResNet also shares the same spirit as the identity feedback used in an LSTM, which is helpful in maintaining the backpropagated errors even when the network is very deep.

Considering the large number of applications for edge computing and mobile devices, there is a trend in developing lightweight deep neural networks that can be ported to and run on low-power devices efficiently. For example, SqueezeNet is a deep network that can achieve a similar recognition rate compared to AlexNet, yet with 50× fewer parameters [124]. SqueezeNet mainly consists of a stack of Fire modules, which contains a squeeze layer and an expand layer. There are only 1×1 filters in the squeeze layer, whereas both 1×1 and 3×3 filters are used in the expand layer. The reduction in the number of parameters in SqueezeNet is largely attributed to its liberal use of 1×1 filters, as they demand 9× less parameters compared to 3×3 filters. Another good example of lightweight deep networks is the MobileNet [125]. The main idea behind MobileNet is a depthwise separable convolution. In a conventional weight filter, computation within a channel and across channels are mingled together. In a depthwise separable convolution, the convolution is split into two layers: a depthwise convolution that is separately applied to each channel and a 1×1 filter that combines information across channels. Such a factorization is helpful in bringing down the number of operations needed. In addition, to accommodate various applications with different tolerances to the classification accuracy, MobileNets provides two hyperparameters, namely a width multiplier and a resolution multiplier, to trade off accuracy with a reduced computational cost. It was reported that compared to SqueezeNet, MobileNet could achieve a higher recognition rate on the ImageNet dataset while requiring 22× fewer MAC operations [125].

In addition to the deep networks developed for supervised learning, a deep Q-network was developed by Google researchers to achieve the human-level control in playing Atari games [14]. The architecture of the network is illustrated in Figure 2.28 and is similar to other deep CNNs used as classifiers. In fact, from the perspective of inference, the deep network can be thought as a classifier with its output as a control operation. The main difference here is the learning method. The deep Q-network was trained through deep reinforcement learning. It used techniques and network architectures developed in deep supervised learning to extract features from high-dimensional input data. The refined features were then used to conduct reinforcement learning through Q-learning, which is closely related to the ADHDP algorithm presented in Section 2.2.4. After training such a deep Q-network, the researchers at DeepMind demonstrated that it could outperform all previous algorithms. The network could even achieve a performance that was comparable to a professional human game tester in 49 games.

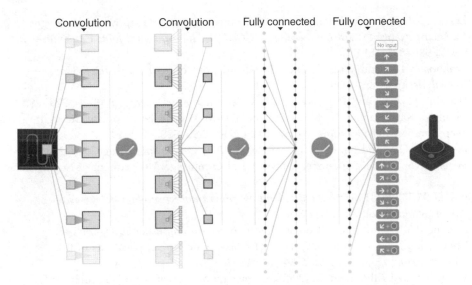

Figure 2.28 Architecture of a deep Q-network [14]. The network itself is similar to most CNNs that are used as classifiers. The high-dimensional inputs, images, are fed into the network, and the control decision is obtained from the last fully connected layer. Reproduced with permission of Springer.

References

1 Gupta, J.N.D. and Sexton, R.S. (1999). Comparing backpropagation with a genetic algorithm for neural network training. *Omega* 27 (6): 679–684.

2 Yu, X.H. (1992). Can backpropagation error surface not have local minima. *IEEE Trans. Neural Networks* 3 (6): 1019–1021.

3 Bianchini, M., Frasconi, P., and Gori, M. (1995). Learning without local minima in radial basis function networks. *IEEE Trans. Neural Networks* 6 (3): 749–756.

4 Sexton, R.S., Dorsey, R.E., and Johnson, J.D. (1999). Optimization of neural networks: a comparative analysis of the genetic algorithm and simulated annealing. *Eur. J. Oper. Res.* 114 (3): 589–601.

5 LeCun, Y., Bengio, Y., and Hinton, G. (2015). Deep learning. *Nature* 521 (7553): 436–444.

6 P. Werbos, "Beyond regression: New tools for prediction and analysis in the behavioral sciences," *Dr. Dissertation,* Appl. Math. Harvard Univ., MA, August, p. 906, 1975.

7 Werbos, P.J. (1982). Applications of advances in nonlinear sensitivity analysis. In: *System Modeling and Optimization,* 762–770. Berlin, Heidelberg: Springer.

8 Rumelhart, D.E., Hinton, G.E., and Williams, R.J. (1986). Learning representations by backpropagation error. *Nature* 323 (6088): 533–536.

9 Sutton, R.S. and Barto, A.G. (1998). *Reinforcement Learning: An Introduction,* vol. 9, no. 5. Cambridge, MA: MIT Press.

10 Bellman, R. (2003). *Dynamic Programming.* Dover Publications.

11 Liu, D., Xiong, X., and Zhang, Y. (2001). Action-dependent adaptive critic designs. In: *Neural Networks, 2001. Proceedings. IJCNN'01. International Joint Conference on*, vol. 2, 990–995. IEEE.

12 Prokhorov, D.V. and Wunsch, D.C. (1997). Adaptive critic designs. *IEEE Trans. Neural Networks* 8 (5): 997–1007.

13 Watkins, C.J.C.H. and Dayan, P. (1992). Q-learning. *Mach. Learn.* 8 (3–4): 279–292.

14 Mnih, V., Kavukcuoglu, K., Silver, D. et al. (2015). Human-level control through deep reinforcement learning. *Nature* 518 (7540): 529–533.

15 Rummery, G.A. and Niranjan, M. (1994). *On-Line Q-Learning Using Connectionist Systems*, vol. 37, no. September. University of Cambridge, Department of Engineering.

16 Powell, W.B. (2011). *Approximate dynamic programming solving the curses of dimensionality*, Hoboken, NJ: Wiley.

17 Kohonen, T. (1998). The self-organizing map. *Neurocomputing* 21 (1–3): 1–6.

18 Kohonen, T., Oja, E., Simula, O. et al. (1996). Engineering applications of the self-organizing map. *Proc. IEEE* 84 (10): 1358–1383.

19 Vesanto, J. and Alhoniemi, E. (2000). Clustering of the self-organizing map. *IEEE Trans. Neural Networks* 11 (3): 586–600.

20 Hinton, G.E. and Salakhutdinov, R.R. (2006). Reducing the dimensionality of data with neural networks. *Science* 313 (5786): 504–507.

21 Hinton, G.E. and Osindero, S. (2006). A fast learning algorithm for deep belief nets. *Neural Comput.* 1554 (7): 1527–1554.

22 Fischer, A. (2015). Training restricted Boltzmann machines. *KI - Künstliche Intelligenz* 29 (4): 441–444.

23 Erhan, D., Bengio, Y., Courville, A. et al. (2010). Why does unsupervised pre-training help deep learning? *J. Mach. Learn. Res.* 11, no. Feb: 625–660.

24 Vincent, P., Larochelle, H., Bengio, Y., and Manzagol, P.-A. (2008). Extracting and composing robust features with denoising autoencoders. In: *Proceedings of the 25th International Conference on Machine Learning*, 1096–1103. ACM.

25 Vincent, P., Larochelle, H., Lajoie, I. et al. (2010). Stacked denoising autoencoders: learning useful representations in a deep network with a local denoising criterion. *J. Mach. Learn. Res.* 11 (Dec): 3371–3408.

26 Bengio, Y., Lamblin, P., Popovici, D., and Larochelle, H. (2007). Greedy layer-wise training of deep networks. In: *Advances in Neural Information Processing Systems* (eds. J.C. Platt and T. Hoffman), 153–160. MIT Press.

27 Zheng, N. and Mazumder, P. (2018). A scalable low-power reconfigurable accelerator for action-dependent heuristic dynamic programming. *IEEE Trans. Circuits Syst. I Regul. Pap.* 65 (6): 1897–1908.

28 Wang, F.Y., Zhang, H., and Liu, D. (2009). Adaptive dynamic programming: an introduction. *IEEE Comput. Intell. Mag.* 4 (2): 39–47.

29 Lewis, F.L., Vrabie, D., and Vamvoudakis, K.G. (2012). Reinforcement learning and feedback control: using natural decision methods to design optimal adaptive controllers. *IEEE Control Syst.* 32 (6): 76–105.

30 Lewis, F.L. and Vrabie, D. (2009). Reinforcement learning and adaptive dynamic programming for feedback control. *IEEE Circuits Syst. Mag.* 9 (3): 32–50.

31 Liu, F., Sun, J., Si, J. et al. (2012). A boundedness result for the direct heuristic dynamic programming. *Neural Networks* 32: 229–235.

32 He, H., Ni, Z., and Fu, J. (2012). A three-network architecture for on-line learning and optimization based on adaptive dynamic programming. *Neurocomputing* 78 (1): 3–13.

33 Si, J. and Wang, Y.T. (2001). On-line learning control by association and reinforcement. *IEEE Trans. Neural Networks* 12 (2): 264–276.

34 Sokolov, Y., Kozma, R., Werbos, L.D., and Werbos, P.J. (2015). Complete stability analysis of a heuristic approximate dynamic programming control design. *Automatica* 59: 9–18.

35 Mu, C., Ni, Z., Sun, C., and He, H. (2017). Air-breathing hypersonic vehicle tracking control based on adaptive dynamic programming. *IEEE Trans. Neural Networks Learn. Syst.* 28 (3): 584–598.

36 Broomhead, D. S. and Lowe, D. "Radial basis functions, multi-variable functional interpolation and adaptive networks," Royal Signals and Radar Establishment, Malvern, United Kingdom, 1988.

37 Park, J. and Sandberg, I.W. (1993). Approximation and radial-basis-function networks. *Neural Comput.* 5 (2): 305–316.

38 Park, J. and Sandberg, I.W. (1991). Universal approximation using radial-basis-function networks. *Neural Comput.* 3 (2): 246–257.

39 Musavi, M.T., Ahmed, W., Chan, K.H. et al. (1992). On the training of radial basis function classifiers. *Neural Networks* 5 (4): 595–603.

40 Chen, S., Billings, S.A., and Grant, P.M. (1992). Recursive hybrid algorithm for non-linear system identification using radial basis function networks. *Int. J. Control* 55 (5): 1051–1070.

41 Chen, S., Mulgrew, B., and Grant, P.M. (1993). A clustering technique for digital communications channel equalization using radial basis function networks. *IEEE Trans. Neural Networks* 4 (4): 570–590.

42 Hopfield, J.J. (1982). Neural networks and physical systems with emergent collective computational abilities. *Proc. Natl. Acad. Sci.* 79 (8): 2554–2558.

43 Werbos, P.J. (1988). Generalization of backpropagation with application to a recurrent gas market model. *Neural Networks* 1 (4): 339–356.

44 Hochreiter, S. and Schmidhuber, J. (1997). Long short-term memory. *Neural Comput.* 9 (8): 1735–1780.

45 Schmidhuber, J. (2015). Deep learning in neural networks: an overview. *Neural Networks* 61: 85–117.

46 Greff, K., Srivastava, R.K., Koutník, J. et al. (2017). LSTM: a search space odyssey. *IEEE Trans. Neural Networks Learn. Syst.* 28 (10): 2222–2232.

47 Gers, F. A., Schmidhuber, J., and Cummins, F. "Learning to forget: Continual prediction with LSTM," IET Conference Proceedings, Institute of Engineering and Technology, pp. 850–855(5), 1999.

48 Gers, F.A., Schraudolph, N.N., and Schmidhuber, J. (2002). Learning precise timing with LSTM recurrent networks. *J. Mach. Learn. Res.* 3 (Aug): 115–143.

49 Sutskever, I., Vinyals, O., and Le, Q.V. (2014). Sequence to sequence learning with neural networks. In: *Advances in Neural Information Processing Systems* (eds. Z. Ghahramani, M. Welling, C. Cortes, et al.), 3104–3112. Curran Associates.

50 Vinyals, O., Toshev, A., Bengio, S., and Erhan, D. (2015). Show and tell: a neural image caption generator. In: *Proceedings of the IEEE Computer Society Conference on Computer Vision and Pattern Recognition*, vol. 07–12–June, 3156–3164.

51 Graves, A. "Generating sequences with recurrent neural networks," *arXiv Prepr. arXiv1308.0850*, 2013.

52 LeCun, Y., Bottou, L., Bengio, Y., and Haffner, P. (1998). Gradient-based learning applied to document recognition. *Proc. IEEE* 86 (11): 2278–2323.

53 Krizhevsky, A. and Hinton, G. "Learning multiple layers of features from tiny images," Citeseer, 2009.

54 Russakovsky, O., Deng, J., Su, H. et al. (2015). Imagenet large scale visual recognition challenge. *Int. J. Comput. Vision* 115 (3): 211–252.

55 "Yelp Dataset." [Online]. Available at: https://www.yelp.com/dataset. [Accessed: 19 April 2018].

56 Merity, S., Xiong, C., Bradbury, J., and Socher, R. "Pointer sentinel mixture models," *arXiv Prepr. arXiv1609.07843*, 2016.

57 "Open datasets for deep learning & Machine learning – Deeplearning4j: Open-source, distributed deep learning for the JVM." [Online]. Available: https://deeplearning4j.org/opendata. [Accessed: 19 Apr 2018].

58 Zheng, N. and Mazumder, P. (2018). A low-power circuit for adaptive dynamic programming. In: *2018 31st International Conference on VLSI Design and 2018 17th International Conference on Embedded Systems (VLSID)*, 192–197. IEEE.

59 Ni, Z., He, H., and Wen, J. (2013). Adaptive learning in tracking control based on the dual critic network design. *IEEE Trans. Neural Networks Learn. Syst.* 24 (6): 913–928.

60 Ni, Z., He, H., Zhong, X., and Prokhorov, D.V. (2015). Model-free dual heuristic dynamic programming. *IEEE Trans. Neural Networks Learn. Syst.* 26 (8): 1834–1839.

61 Doya, K. (1999). Reinforcement learning in continuous time and space. *Neural Comput.* 12 (1): 1–28.

62 Kimura, H. and Kobayashi, S. (1999). Stochastic real-valued reinforcement learning to solve a nonlinear control problem. In: *Systems, Man, and Cybernetics, 1999. IEEE SMC '99 Conference Proceedings. 1999 IEEE International Conference on*, vol. 5, 510–515. IEEE.

63 Murray, R.M. and Hauser, J.E. (1991). *A Case Study in Approximate Linearization: The Acrobot Example*. Electronics Research Laboratory, College of Engineering, University of California.

64 Duan, Y., Chen, X., Houthooft, R. et al. (2016). Benchmarking deep reinforcement learning for continuous control. In: *International Conference on Machine Learning*, 1329–1338.

65 Hornik, K., Stinchcombe, M., and White, H. (1989). Multilayer feedforward networks are universal approximators. *Neural Networks* 2 (5): 359–366.

66 Lawrence, S., Giles, C.L., Tsoi, A.C., and Back, A.D. (1997). Face recognition: a convolutional neural-network approach. *IEEE Trans. Neural Networks* 8 (1): 98–113.

67 Rowley, H.A., Baluja, S., and Kanade, T. (1998). Neural network-based face detection. *IEEE Trans. Pattern Anal. Mach. Intell.* 20 (1): 23–38.

68 Lang, K.J., Waibel, A.H., and Hinton, G.E. (1990). A time-delay neural network architecture for isolated word recognition. *Neural Networks* 3 (1): 23–43.

69 Waibel, A. (1989). Modular construction of time-delay neural networks for speech recognition. *Neural Comput.* 1 (1): 39–46.

70 Azoff, E. (1994). *Neural Network Time Series Forecasting of Financial Markets.* Wiley.

71 Kaastra, I. and Boyd, M. (1996). Designing a neural network for forecasting financial and economic time series. *Neurocomputing* 10 (3): 215–236.

72 Gomi, H. and Kawato, M. (1993). Neural network control for a closed-loop system using feedback-error-learning. *Neural Networks* 6 (7): 933–946.

73 Noriega, J.R. and Wang, H. (1998). A direct adaptive neural network control for unknown nonlinear systems and its application. *IEEE Trans. Neural Networks* 9 (1): 27–34.

74 Sze, V., Chen, Y.H., Yang, T.J., and Emer, J.S. (2017). Efficient processing of deep neural networks: a tutorial and survey. *Proc. IEEE* 105 (12): 2295–2329.

75 Chollet, F. (2017). *Deep Learning with Python.* Manning Publications Co.

76 Lee, H., Grosse, R., Ranganath, R., and Ng, A.Y. (2009). Convolutional deep belief networks for scalable unsupervised learning of hierarchical representations. In: *Proceedings of the 26th Annual International Conference on Machine Learning - ICML '09*, 1–8. ACM.

77 Srivastava, N., Hinton, G.E., Krizhevsky, A. et al. (2014). Dropout: a simple way to prevent neural networks from overfitting. *J. Mach. Learn. Res.* 15 (1): 1929–1958.

78 Hinton, G. E., Srivastava, N., Krizhevsky, A. et al., "Improving neural networks by preventing co-adaptation of feature detectors," *arXiv Prepr. arXiv1207.0580*, 2012.

79 Wan, L., Zeiler, M., Zhang, S. et al. (2013). Regularization of neural networks using dropconnect. In: *Proceedings of the 30th International Conference on Machine Learning (ICML-13)*, 1058–1066.

80 Ioffe, S. and Szegedy, C. (2015). Batch normalization: accelerating deep network training by reducing internal covariate shift. In: *International Conference on Machine Learning*, 448–456.

81 Kotsiantis, S.B., Kanellopoulos, D., and Pintelas, P.E. (2006). Data preprocessing for supervised leaning. *Int. J. Comput. Sci.* 1 (2): 111–117.

82 Ruder, S. "An overview of gradient descent optimization algorithms," *arXiv Prepr. arXiv1609.04747*, 2016.

83 Sutskever, I., Martens, J., Dahl, G., and Hinton, G. (2013). On the importance of initialization and momentum in deep learning. In: *International Conference on Machine Learning*, 1139–1147.

84 Polyak, B.T. (1964). Some methods of speeding up the convergence of iteration methods. *USSR Comput. Math. Math. Phys.* 4 (5): 1–17.

85 Nesterov, Y. (1983). A method for unconstrained convex minimization problem with the rate of convergence O (1/k2). *Doklady an USSR* 269 (3): 543–547.

86 Hinton, G.E. (2012). A practical guide to training restricted Boltzmann machines. In: *Neural Networks: Tricks of the Trade* (eds. G. Montavon, G. Orr and K.-R. Müller), 599–619. Berlin, Heidelberg: Springer.

87 Duchi, J., Hazan, E., and Singer, Y. (2011). Adaptive subgradient methods for online learning and stochastic optimization. *J. Mach. Learn. Res.* 12 (Jul): 2121–2159.

88 Zeiler, M. D. "ADADELTA: An adaptive learning rate method," *arXiv1212.5701 [cs]*, 2012.

89 Tieleman, T., Hinton, G.E., Srivastava, N., and Swersky, K. (2012). Lecture 6.5-rmsprop: divide the gradient by a running average of its recent magnitude. *COURSERA Neural Networks Mach. Learn.* 4 (2): 26–31.

90 Kingma, D.P. and Ba, J. (2014). Adam: a method for stochastic optimization. In: *Proc. International Conference for Learning Representations*, 1–15.

91 Denil, M., Shakibi, B., Dinh, L., and De Freitas, N. (2013). Predicting parameters in deep learning. In: *Advances in Neural Information Processing Systems* (eds. J.C. Burges, L. Bottou, M. Welling, et al.), 2148–2156. Curran Associates, Inc.

92 Hassibi, B. and Stork, D.G. (1993). Second order derivatives for network pruning: optimal brain surgeon. In: *Advances in Neural Information Processing Systems* (eds. S.J. Hanson, J.D. Cohen and C.L. Giles), 164–171. Morgan-Kaufman.

93 LeCun, Y., Denker, J.S., and Solla, S.A. (1990). Optimal brain damage. In: *Advances in Neural Information Processing Systems* (ed. D.S. Touretzky), 598–605. Morgan-Kaufman.

94 Han, S., Mao, H., and Dally, W. J. "Deep compression: Compressing deep neural networks with pruning, trained quantization and Huffman coding," *arXiv Prepr. arXiv1510.00149*, 2015.

95 Li, H., Kadav, A., Durdanovic, I. et al., "Pruning filters for efficient convnets," *arXiv Prepr. arXiv1608.08710*, 2016.

96 Lebedev, V. and Lempitsky, V. (2016). Fast ConvNets using group-wise brain damage. In: *2016 IEEE Conference on Computer Vision and Pattern Recognition (CVPR)*, 2554–2564.

97 Wen, W., Wu, C., Wang, Y. et al. (2016). Learning structured sparsity in deep neural networks. In: *Advances in Neural Information Processing Systems* (eds. D.D. Lee, M. Sugiyama, U.V. Luxburg, et al.), 2074–2082. Curran Associates.

98 Yu, J., Lukefahr, A., Palframan, D. et al. (2017). Scalpel: customizing DNN pruning to the underlying hardware parallelism. In: *Proceedings of the 44th Annual International Symposium on Computer Architecture*, 548–560. Toronto, Canada: ACM.

99 Yang, T.J., Chen, Y.H., and Sze, V. (2017). Designing energy-efficient convolutional neural networks using energy-aware pruning. In: *Proceedings of 30th IEEE Conf. Comput. Vis. Pattern Recognition, CVPR 2017*, vol. 2017, 6071–6079.

100 Denton, E.L., Zaremba, W., Bruna, J. et al. (2014). Exploiting linear structure within convolutional networks for efficient evaluation. In: *Advances in Neural Information Processing Systems* (eds. Z. Ghahramani, M. Welling, C. Cortes, et al.), 1269–1277. Curran Associates.

101 Jaderberg, M., Vedaldi, A., and Zisserman, A. "Speeding up convolutional neural networks with low rank expansions," *arXiv Prepr. arXiv1405.3866*, 2014.

102 Han, S., Pool, J., Tran, J., and Dally, W. (2015). Learning both weights and connections for efficient neural network. In: *Advances in Neural Information Processing Systems* (eds. C. Cortes, N.D. Lawrence, D.D. Lee, et al.), 1135–1143. Curran Associates.

103 Mao, H., Han, S., Pool, J., et al., "Exploring the regularity of sparse structure in convolutional neural networks," *arXiv Prepr. arXiv1705.08922*, 2017.

104 Glorot, X., Bordes, A., and Bengio, Y. (2011). Deep sparse rectifier neural networks. In: *Proceedings of the Fourteenth International Conference on Artificial Intelligence and Statistics*, 315–323.

105 Albericio, J., Judd, P., Hetherington, T. et al. (2016). Cnvlutin: ineffectual-neuron-free deep neural network computing. In: *Proceedings of the 43rd International Symposium on Computer Architecture*, 1–13.

106 Bengio, Y., Léonard, N., and Courville, A., "Estimating or propagating gradients through stochastic neurons for conditional computation," *arXiv Prepr. arXiv1308.3432*, 2013.

107 Bengio, E., Bacon, P.-L., Pineau, J., and Precup, D., "Conditional computation in neural networks for faster models," *arXiv Prepr. arXiv1511.06297*, 2015.

108 Shazeer, N., Mirhoseini, A., Maziarz, K., et al., "Outrageously large neural networks: The sparsely-gated mixture-of-experts layer," *arXiv Prepr. arXiv1701.06538*, 2017.

109 Vanhoucke, V., Senior, A., and Mao, M.Z. (2011). Improving the speed of neural networks on CPUs. In: *Proc. Deep Learning and Unsupervised Feature Learning NIPS Workshop*, vol. 1, 1–8.

110 Miyashita, D., Lee, E. H., and Murmann, B., "Convolutional neural networks using logarithmic data representation," *arXiv Prepr. arXiv1603.01025*, 2016.

111 Gong, Y., Liu, L., Yang, M., and Bourdev, L., "Compressing deep convolutional networks using vector quantization," *arXiv Prepr. arXiv1412.6115*, 2014.

112 Courbariaux, M., Bengio, Y., and David, J.-P. (2015). BinaryConnect: training deep neural networks with binary weights during propagations. In: *Advances in Neural Information Processing Systems* (eds. C. Cortes, N.D. Lawrence, D.D. Lee, et al.), 3123–3131. Curran Associates.

113 Hubara, I., Courbariaux, M., Soudry, D. et al. (2016). Binarized neural networks. In: *Advances in Neural Information Processing Systems* (eds. D.D. Lee, M. Sugiyama, U.V. Luxburg, et al.), 4107–4115. Curran Associates.

114 Rastegari, M., Ordonez, V., Redmon, J., and Farhadi, A. (2016). XNOR-net: ImageNet classification using binary convolutional neural networks. In: *European Conference on Computer Vision*, 525–542.

115 F. Li, B. Zhang, and B. Liu, "Ternary weight networks," *arXiv Prepr. arXiv1605.04711*, 2016.

116 Zhu, C., Han, S., Mao, H., and Dally, W. J., "Trained ternary quantization," *arXiv Prepr. arXiv1612.01064*, 2016.

117 Gupta, S., Agrawal, A., Gopalakrishnan, K., and Narayanan, P. (2015). Deep learning with limited numerical precision. In: *International Conference on Machine Learning*, 1737–1746.

118 Zhou, S., Wu, Y., Ni, Z., et al., "DoReFa-Net: Training low bitwidth convolutional neural networks with low bitwidth gradients," *arXiv Prepr. arXiv1606.06160*, 2016.

119 Schuchman, L. (1964). Dither signals and their effects on quantization noise. *IEEE Trans. Commun.* 12 (4): 162–165.

120 Krizhevsky, A., Sutskever, I., and Hinton, G.E. (2012). Imagenet classification with deep convolutional neural networks. In: *Advances in Neural Information Processing Systems* (eds. F. Pereira, C.J.C. Burges, L. Bottou and K.Q. Weinberger), 1097–1105. Curran Associates.

121 Szegedy, C., Liu, W., Jia, Y. et al. (2015). Going deeper with convolutions. In: *Proceedings of the IEEE Computer Society Conference on Computer Vision and Pattern Recognition*, vol. 07–12–June, 1–9.

122 Simonyan, K., and Zisserman, A., "Very deep convolutional networks for large-scale image recognition," *arXiv Prepr. arXiv1409.1556*, 2014.

123 He, K., Zhang, X., Ren, S., and Sun, J. (2016). Deep residual learning for image recognition. In: *Proceedings of the IEEE Conference on Computer Vision and Pattern Recognition*, 770–778.

124 Iandola, F. N., Han, S., Moskewicz, M. W., et al., "Squeezenet: Alexnet-level accuracy with 50× fewer parameters and <0.5 mb model size," *arXiv Prepr. arXiv1602.07360*, 2016.

125 Howard, A. G., Zhu, M., Chen, B., et al., "Mobilenets: Efficient convolutional neural networks for mobile vision applications," *arXiv Prepr. arXiv1704.04861*, 2017.

3

Artificial Neural Networks in Hardware

Learn as if you were not reaching your goal and as though you were scared of missing it.

–Confucius

3.1 Overview

Neural networks were developed to mimic the salient features of biological brains, such as their ability to recognize patterns and detect motions in the presence of noises. In Chapter 2, many basic concepts related to the learning and inferring with artificial neural networks (ANNs) are discussed. As an algorithm, a neural network needs to be executed on certain hardware platforms before it can be deployed in various applications. In this chapter, neural networks implemented on different hardware platforms are discussed. The advantages and disadvantages of different platforms are considered. There are, in general, three types of hardware that neural network algorithms can be deployed: general-purpose processors, field-programmable gate arrays (FPGAs), and application-specific integrated circuits (ASICs). A comparison of these three hardware platforms is made in Figure 3.1 in terms of energy efficiency and flexibility.

General-purpose processors, such as central processing units (CPUs) and graphics processing units (GPUs), are highly flexible because they can execute instructions compiled from high-level language such as C++ and Python. Most scientists and researchers in the field of artificial intelligence are most familiar with this type of hardware platform due to their wide accessibility. The drawback of a general-purpose processor, however, is its low energy efficiency. The low energy efficiency is apparently a problem for portable devices that are powered by batteries. In addition, more and more datacenters have also started looking for energy-efficient solutions as the cost of electricity power has become a big portion of the total operational cost.

In order to achieve higher energy efficiency, an ASIC solution is often adopted. An ASIC is a customized chip that is designed to serve for a specific type of application, which, in our case, is neuromorphic computing. Since the architecture of an ASIC can be specifically optimized for the target application, an ASIC typically has the highest performance as well as the lowest energy consumption. The drawback of an ASIC is that many logic and arithmetic operations are hard-wired and it typically only implements a fixed function (although some of them do provide certain levels of reconfigurability).

Learning in Energy-Efficient Neuromorphic Computing: Algorithm and Architecture Co-Design,
First Edition. Nan Zheng and Pinaki Mazumder.

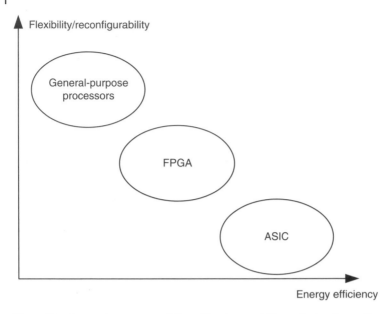

Figure 3.1 Comparison among different hardware platforms for implementing neural networks.

Therefore, when the algorithm or application changes, the original ASIC might not be able to be adapted accordingly. Another problem with the ASIC solution is its long development cycle and cost. It normally takes years and millions of dollars to develop a functional ASIC chip. Such a long development cycle is not affordable by many researchers and companies.

In addition to these two extremes, the FPGA provides an intermediate solution that can provide some levels of programmability and flexibility while achieving better energy efficiency compared to general-purpose processors. Since the main focus of this book is on energy-efficient neuromorphic hardware, most of our discussions are on the ASIC approach as it provides the most energy-efficient solution. Nevertheless, executing ANNs on general-purpose processors and FPGAs are also briefly discussed in order to contrast with the ASIC approach.

3.2   General-Purpose Processors

From cranking up the operating frequency in earlier days to increasing the level of parallelism nowadays, von Neumann architecture-based processors are still the workhorse for continually improved performance. As deep learning starts playing a critical role in more and more applications, there are increasing demands for high-throughput and low-latency computing. Deep learning algorithms can be conventionally realized on general-purpose CPUs. The initial popularity of implementing neural networks on CPUs mainly resulted from the large availability of CPUs and its universal programming capability. However, a CPU might not be the ideal piece of hardware for neural networks. CPUs are famous for providing complex control flows. This might be beneficial for more conventional rule-based computing, but not so much for the data-driven approach like neural networks. In neural network computations, dataflow is the major portion of the computation. Little control is needed in the process of evaluating the neural network.

Upon realizing the inherent parallel nature of the algorithm, more and more researchers started to port the algorithm into general-purpose GPUs to leverage its massive parallel computational power. For example, the winner of 2012 ImageNet large-scale visual recognition challenge, AlexNet, was trained on two GPUs [1]. GPUs were initially designed specifically for image rendering. Millions of pixels in images need to be processed in a short amount of time, which demands a huge throughput, which is practically infeasible for conventional CPUs. Most GPUs are based on the single-instruction multiple-thread (SIMT) architecture, which exploits the data-level parallelism explicitly. Such a computational model is suitable for implementing neural networks, as most operations involved are matrix multiplication, which can benefit greatly from the parallel nature of GPUs [2, 3].

As discussed in Chapter 2, most neural network related tasks can be conducted with low-precision arithmetic operations in order to improve the performance and the energy efficiency of the system. To take advantage of this low-precision requirement, the trend is that many general-purpose processors started to provide low-precision modes for emerging deep-learning applications [4]. In addition, to accelerate the development of deep learning on general-purpose processors, various frameworks have been developed to implement neural networks conveniently and efficiently. Popular frameworks are Torch [5], Caffe [6], and more recently TensorFlow [7] and PyTorch [8]. Through introducing these frameworks, engineers are able to focus more on the high-level implementation of deep-learning algorithms instead of low-level details of the computation.

3.3 Digital Accelerators

3.3.1 A Digital ASIC Approach

Even though the adoption of GPUs has significantly accelerated deep-learning algorithms, the general-purpose nature of GPUs still results in much waste in energy and clock cycles. Many future applications in the field of artificial intelligence, such as self-driving cars and autonomous robots, demand larger and more complicated neural networks to conduct inference with a lower latency and a lower energy consumption. Motivated by this, many academic and industrial efforts have been devoted to developing specialized accelerators solely for deep neural networks. A few representative digital accelerators are presented in this section. In addition, three commonly used strategies in realizing energy-efficient accelerators, namely reducing data movement, scaling precision, and exploiting the sparsity in the network, are also introduced to facilitate the discussions of various neural network accelerators in the literature.

3.3.1.1 Optimization on Data Movement and Memory Access

For most neural network accelerators, memory access often consumes a significant portion of power. The state-of-the-art neural networks are going deeper and larger, resulting in many parameters in the network. Therefore, optimizing the computations

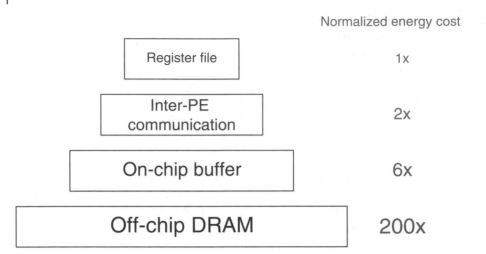

Figure 3.2 An example of the memory hierarchy and the energy cost for data movement in a typical neural network accelerator. Smaller storage units that are located closer to the computing unit are more energy efficient. The normalized energy costs are from [4].

such that the data movements can be minimized has become a critical step in building energy-efficient accelerators. In many neural network accelerators, a memory hierarchy that resembles the cache structure used in modern processors is employed.

Figure 3.2 shows the normalized energy cost for different types of memory access [4]. In the figure, it is assumed that every processing element (PE) has a dedicated register file (RF). Different PEs can also communicate with each other and exchange information. An on-chip global buffer can hold frequently-used information, which is helpful in avoiding access to the large dynamic random-access memory (DRAM) repeatedly. Similar to the memory architecture employed in modern CPUs, the larger the memory, the higher energy cost one has to spend in order to access the data. The off-chip memory access easily costs one or two magnitudes higher energy compared to that of an on-chip memory. In addition, frequently accessed data should be kept as close to the arithmetic logic unit (ALU) as possible in order to reduce the energy consumption for memory access. This shares the same spirit as the CPU design where a multilevel cache structure is often exploited. The difference here is that in a CPU design, the dataflow is often hard to predict. Therefore, statistics and heuristics are often used with an attempt to reduce the cache miss. For a neural network accelerator, however, the dataflow is often fixed and known at the compilation time. Therefore, the designer has more control on how to optimize the data movement and memory access. Over the past few years, many strategies have been proposed to optimize the dataflow of neural network accelerators. A few representative examples are discussed in this section.

3.3.1.1.1 Tiled Matrix Operations Tiling is a well-known technique to leverage the data locality in conducting matrix calculations [9]. It has been widely used in ANN accelerators to increase the level of data reuse. The basic idea of tiling is illustrated in the pseudo code shown in Figure 3.3. The original three loops in calculating the matrix–matrix multiplication are embedded into three tiling loops. These three outermost loops essentially break the large matrix multiplication problem into smaller ones, as shown in Figure 3.4.

```
for(int Bk = 0; Bk < N; Bk += T){                    /*

    for(int Bi = 0; Bi < N; Bi += T){                Tiling

        for(int Bj = 0; Bj < N; Bj += T){            */

            for(int i = Bi; i < Bi + T; ++i){        /*

                for(int k = Bk; k < Bk + T; ++k){    Matrix multiplications within a tile

                    for(int j = Bj; j < Bj + T; ++j){

                        y[i][k] += w[i][j] * x[j][k]  */

                    }

                }

            }

        }

    }

}
```

Figure 3.3 Pseudocode for a matrix multiplication operation with tiling. The outermost three *for* loops are for tiling and the innermost loops are for calculating the matrix multiplications on the small matrices.

By doing so, the data can be reused sufficiently before being evicted from the on-chip data buffer, which has a limited size. The size of the tile can be conveniently adjusted according to the size of the buffer. It is worth noting that the arrangement shown in Figure 3.3 is not unique as there are other ways to arrange the tiling and in-tile calculation. A different arrangement has different implications on the level of data reuse for each matrix. Generally speaking, increasing the data reuse in one matrix decreases that in other matrices.

As an effective method of increasing the level of on-chip data reuse, tiling has been extensively used in building ANN accelerators [10–14]. A good example is the Dian-Nao family, which is a group of machine-learning accelerators developed mainly by researchers at the Institute of Computing Technology, Chinese Academy of Sciences [11–15]. It is one of the earliest works in building specialized hardware for accelerating machine-learning tasks. It has been serving as the state-of-the-art baseline for many subsequent works.

The first member in this family is the DianNao accelerator proposed by Chen et al. in 2014 [14]. It exploits tile-based matrix operations to minimize the memory accesses, which helps lower the power consumption. The main dataflow in the DianNao accelerator is conceptually illustrated in Figure 3.5. The computations are performed by the neural function unit (NFU), which consists of three stages: multiply, add, and

Figure 3.4 Illustration of the tile-based matrix multiplication. Matrices are broken down into small matrices so that the data needed in the computation can fit into an on-chip buffer and can be reused efficiently.

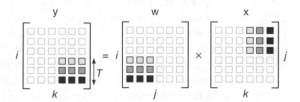

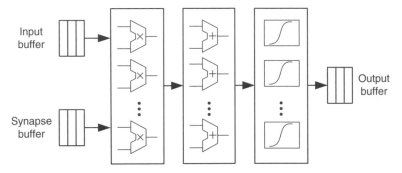

Figure 3.5 Conceptual illustration of the main dataflow employed in a DianNao accelerator [14]. Arithmetic operations needed in neural networks are broken down into three stages. Three buffers are used to exploit the data locality. Source: adapted from [14]. Adapted with permission of ACM.

activate. For different layers in the neural network, the computations conducted by each stage may vary. For example, even though the second stage in the NFU performs additions for fully connected layers, it also contains shifters and max operators for pooling layers. In order to reuse the data efficiently, three data buffers, namely input buffer, synapse buffer, and output buffer, are allocated in the design to exploit the data locality. In addition, a specialized instruction set was developed to make the DianNao accelerator fully programmable in order to accommodate different network configurations and operations. It was shown in [14] that it was possible to achieve 117.87× speedup and 21.08× reduction in energy consumption, when compared to a 128-bit 2 GHz single-instruction multiple-data (SIMD) core.

To overcome the penalty in performance and energy efficiency caused by the frequent main memory access, Chen et al. proposed to distribute storage units across the chip in the upgraded version of the DianNao accelerator: DaDianNao [11]. In the new accelerator, embedded dynamic random-access memory (eDRAM) is leveraged to achieve a high storage density. Through distribution of the memory across the chip, the synaptic weights can be made closer to the computational units compared to the centralized-memory architecture, where a large memory holds all the information. In addition, neuron activations are transferred rather than synaptic weights because the former are many fewer in number. This results in significant savings in the energy and clock cycles. To support a high internal bandwidth, DaDianNao adopts a tile-based design where NFUs are spread out in the chip as tiles, and each of them can perform computing simultaneously. With the new architecture, it was reported in [11] that the DaDianNao accelerator outperformed a single GPU by up to 450.64× while the energy consumption could be reduced up to 150.31×.

3.3.1.1.2 Spatial Architectures Massively parallel computation can be achieved through temporal or spatial architectures [4]. In both of these architectures, a large number of PEs are leveraged to perform computations in parallel. In a temporal architecture, PEs fetch operands directly from a unified memory and there is usually no inter-PE communication. This architecture is often used in CPUs and GPUs. In a spatial architecture, on the other hand, each PE often has its own control logic and memory and can exchange information with other PEs. The overhead of a flexible inter-PE communication could be relatively large when the size of the PEs is small. This is one of the

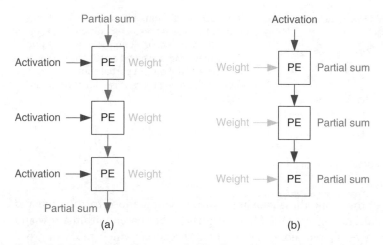

Partial sum   Activation

Activation → PE Weight Weight → PE Partial sum

Activation → PE Weight Weight → PE Partial sum

Activation → PE Weight Weight → PE Partial sum

Partial sum

(a)   (b)

Figure 3.6 Comparison of two popular dataflows that are widely used in neural network accelerators. (a) An example of the weight-stationary dataflow, where the weights are stored locally to the PEs in order to be reused fully. (b) An example of the output-stationary dataflow, where the partial sums are stored locally to PE so that the accumulation of partial sums can occur in place. Source: adapted from [4].

reasons that many conventional architectures that are for general-purpose computing do not employ inter-PE communication at this level. For accelerating ANN, however, the dataflow is known at the compile time and it does not vary very much. Therefore, many accelerators have leveraged a spatial architecture to pass reusable data around in order to save the trip to an on-chip global buffer or an off-chip DRAM [16–20].

Four different strategies of utilizing data locality in computing convolutional neural networks (CNNs), namely weight stationary, output stationary, no local reuse, and row stationary, were analyzed and compared in [4] and [19]. Even though it has been noted in [21] that the categorized dataflows do not cover the entire design space, we borrow them here to discuss common ways of reusing data in the literature as this way of grouping dataflow strategies is easy to understand. The weight-stationary dataflow focuses on the filter weight reuse [22, 23]. An example of the weight-stationary dataflow is illustrated in Figure 3.6a. It fixes the weights in PEs and runs through all computations that need the filters. By reusing the weights, movements of the filter weights can be reduced. Each weight can be sufficiently reused before being evicted from the PE. In order to utilize this dataflow, input activations and partial sums need to be moved between buffers and PEs or between PEs and PEs. Output-stationary dataflow is another architecture where the accumulation of partial sums generated in evaluating the neural networks are made local to PEs [12, 24]. Figure 3.6b shows one example of this type of dataflow. The partial sums generated by the PEs are kept locally to the PEs. By summing the partial sums in place, movements of the partial sums can be minimized.

Another dataflow, called a row-stationary dataflow, was proposed in [18, 19] for Eyeriss, a deep-learning accelerator developed by Chen et al. at Massachusetts Institute of Technology. This dataflow is conceptually illustrated in Figure 3.7. The convolution is broken down into a series of one-dimensional convolution primitives that operate on one row of filter weight and one row of input activations. These one-dimensional primitives can be computed by PEs efficiently. A two-dimensional spatial architecture

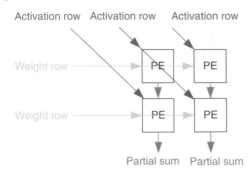

Activation row   Activation row   Activation row

Weight row

Weight row

Partial sum   Partial sum

Figure 3.7 Illustration of the data reuse scheme used in the row-stationary technique [18]. Filter weights can be shared along the row, the input feature maps can be shared across the diagonal direction and the output partial sum is accumulated along the vertical direction. Source: adapted from [18].

is then used on top of these one-dimensional primitive computations to combine and form the final convolutional results. The filter weights can be shared within a row and the input features can be shared along the diagonal direction, which helps improve the data reuse. The generated partial sums can be accumulated conveniently along the vertical direction, which reduces the data movement. In [19], it was demonstrated that compared to the output-stationary, weight-stationary, and no-local-reuse dataflows, the row-stationary dataflow was more energy efficient in convolutional layers when AlexNet [1] was used for benchmarking.

Another example of leveraging a spatial architecture in the design is the Tensor Processing Unit (TPU), an ASIC developed by engineers at Google [20]. The main motivation behind this development was that more and more users started to use neural networks that are executed in datacenters. The projection showed that such a need would double the computation demands of datacenters in 2013. Therefore, Google started looking into building their own customized ASICs that could provide both good performance as well as excellent energy efficiency. In the TPU project, the Google team finished the design, verification, and deployment of the accelerator in datacenters in a mere 15 months.

The heart of the TPU is a matrix multiply unit (MMU), which is responsible for performing efficient matrix multiplication. It contains 256×256 multiply-accumulate (MAC) units performing 8-bit multiply-and-adds. The MMU employs a systolic dataflow, which is conceptually shown in Figure 3.8a. The input activations are fed into the MMU and are multiplied with the preloaded weights. The generated partial sums are then accumulated vertically. To help understand better how the MMU works, Figure 3.8b shows a typical configuration of a systolic array. Weights are stored locally to each PE and the activations are fed to PEs horizontally. To obtain the correct computational results, the input to each row of the systolic array needs to be properly arranged. A group of MAC operations are then conducted along the diagonal direction and it moves like a wave front.

In TPU, instructions are sent from the host over the PCIe bus. The instruction buffer is used to hold instructions temporarily. A complex instruction set computer (CISC) style instruction set is used as the basis for the TPU instruction set, and the average clock cycles per instruction (CPI) of the TPU instructions is typically 10–20. An on-chip weight first-in-first-out (FIFO) helps to buffer data from an off-chip 8 GiB DRAM that holds all the synaptic weights. The intermediate results are stored temporarily in the

Figure 3.8 (a) Illustration of the systolic dataflow employed in TPU. Source: adapted from [20]. (b) Detailed demonstration of how computations are conducted in a systolic array.

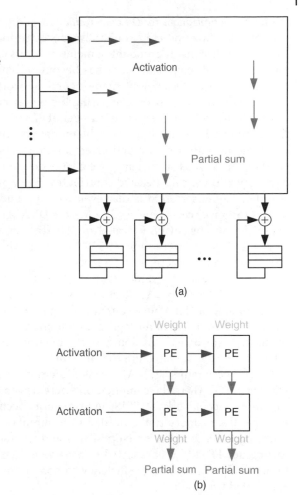

on-chip unified buffer. The TPU was implemented in a 28-nm complementary metal oxide semiconductor (CMOS) technology. The chip was reported to consume 40 W of power while running at a clock frequency of 700 MHz [20].

One of the most valuable contributions in [20] was that since the TPU has been deployed in datacenters, performance metrics on real-life applications were available. It was pointed out that even though most existing neural network architectures were targeted to accelerate CNNs, the portion of workload that CNN applications occupied in the datacenter was actually merely 5%. Therefore, the authors believed that more effort was needed in accelerating fully connected neural networks (FCNNs) and long short-term memories (LSTMs), as they occupy more portions in the usage spectrum. Another trend that was observed in the datacenter was that the demand in response time was actually quite stringent. In the inference phase, latency was strongly preferred over throughput. For example, a 7-ms latency was demanded by the developer. Such a response time requirement naturally puts an upper limit on the batch size that can be used at inference time, as discussed in Section 2.3.1.

3.3.1.1.3   Optimizations for Off-Chip Traffic   In addition to the need of optimizing the on-chip dataflow, off-chip traffic also poses serious challenges for neural network accelerator designs. The amount of memory access is gigantic for neural network applications. The off-chip bandwidth has become a bottleneck for both the performance and the energy efficiency of many neural network accelerators. For example, Gao et al. conducted a study on comparing the power consumption and the peak off-chip DRAM bandwidth needed as the number of PEs increased [25]. The data were obtained from the Eyeriss architecture and the model assumed a VGGNet was running on the accelerator. With a highly optimized on-chip dataflow, data read from an off-chip DRAM can be reused efficiently, yet a large on-chip buffer is required, which occupies a great portion of chip area and contributes significantly to the static power of the chip. Another key observation is that even with an efficient dataflow and a large on-chip static random-access memory (SRAM), the DRAM bandwidth can still exceed 25 GBps when the number of PE is scaled to 400. The interface circuit demanded by such a large off-chip traffic consumes 1.5 W of power, which is much higher than the power consumption of the accelerator chip itself.

To circumvent this difficulty posed by the limited DRAM bandwidth, three-dimensional memory architectures were proposed as an effective approach. A three-dimensional memory consists of multiple DRAM dies stacked together [26]. It can be stacked on top of the logic chip, and the two dies can be connected with the help of through-silicon vias. By doing so, the memory bandwidth can be greatly improved. Neurocube is an architecture that introduces three-dimensional memory into neural network accelerators [27]. A so-called memory-centric neural computing paradigm was proposed where the computations were driven by the data stored in the memory. In a more recent work, TETRIS, an in-memory accumulation technique was proposed to halve the memory traffic needed to accumulate the output feature maps [25]. By reducing the size of the on-chip SRAM and leveraging the in-memory accumulation technique, TETRIS was reported to achieve a 4.1× boost in performance and a 1.5× improvement in the energy efficiency compared to a two-dimensional baseline, which was scaled from Eyeriss.

Another way to avoid the high energy cost associated with the external DRAM access is to reduce the traffic to the DRAM. Many studies have been conducted with the attempt to optimize the on-chip dataflow and maximize the data reuse, such as those methods discussed in the previous section. Many of these techniques focus on optimizing the data reuse within a neural network layer. For each layer, the generated intermediate results, such as output feature maps, still need to be written to an off-chip DRAM and later on read back for the processing in the next layer. This is especially true for the first few layers in a CNN, as the generated intermediate results are often too large to fit into the on-chip buffer. This is illustrated in Figure 3.9a. To reduce this DRAM traffic, a different computing order can be leveraged [28], as shown in Figure 3.9b. By leveraging the locality of a CNN, one can process a pyramid of images that consists of pixels in many layers. By doing so, only the first layer activations need to be brought from DRAM once and all the intermediate results can be generated and consumed on-chip, which saves the trip to the DRAM. With this newly proposed technique, it was reported that 95% of the DRAM traffic could be eliminated compared to the baseline [28].

The DRAM traffic can also be reduced through quantizing and pruning, as discussed in Chapter 2. These methods reduce the external memory bandwidth from the root.

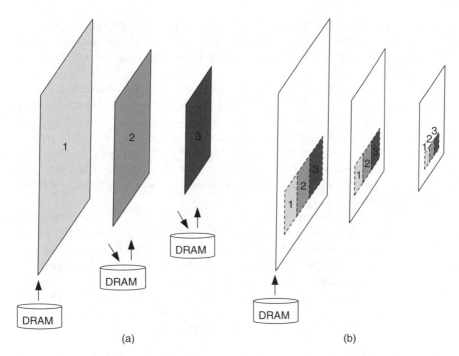

(a)   (b)

Figure 3.9 Comparison of (a) the conventional breadth-first convolution and (b) the depth-first convolution. In the conventional method, the generated intermediate results need to be stored temporarily in a DRAM, whereas in the depth-first convolution method, computations across a few layers are performed together to eliminate the need to buffer the intermediate results.

When being done aggressively enough, the DRAM can even be completely eliminated. Building accelerators to support a reduced precision and pruned network are discussed in the following sections.

3.3.1.2 Scaling Precision

Most general-purpose processors perform floating-point computation, which has a very wide dynamic range that can accommodate the needs of various applications. Such a choice, however, may be suboptimal for most neural network applications, as it is found that a reduced precision has only a small impact on the algorithmic accuracy of the neural network. Therefore, fixed-point computations are employed by most accelerators.

The choice of the number of bits used to represent data in neural networks can serve as a design knob to balance the performance and the power consumption [29–33]. The energy consumption of many components in a neural network accelerator scales with the number of bits used to represent information. Some blocks, such as adders, scale roughly linearly, whereas some blocks, like multipliers, scale superlinearly. Furthermore, when the precision of the network is reduced to a certain level, the energy-consuming external memory, e.g. DRAM, can be eliminated, which results in an even more significant improvement in the energy efficiency.

In general, there are two strategies in reducing the precision in the network. The first one is an open-loop strategy where a well-trained neural network with a full precision (usually floating-point precision) is first employed. The precision is then swept, and the benchmark performances are served as guidelines on how much precision reduction one can tolerate before introducing too much performance penalty. There is no retraining or parameter adjustment for this strategy, which makes this method very simple to implement and use.

The second strategy is a closed-loop approach where a precision reduction is conducted at the training time. In other words, the precision is set as a hyperparameter in the learning. Learning, in this case, serves as a feedback to compensate for the precision reduction. Obviously, the closed-loop approach tends to generate better results, yet it requires a software–hardware co-optimization. The hardware designer might not have the resources or knowledge of retraining the networks. On the other hand, the open-loop approach has the advantage that the boundary between software and hardware is well established.

In terms of hardware implementation, there are also two popular styles in exploiting a reduced precision. The first style is a fixed precision reduction that is set at the design time. The second style is a dynamic precision scaling that can adjust the computing precision at run time. Apparently, the dynamic approach provides more flexibility and may result in more energy saving and performance boosting when the range of the required precision is wide. On the other hand, scaling precision dynamically normally adds additional circuitry to the accelerator and therefore induces overhead in order to support the scalable precision. This might adversely affect the energy efficiency.

3.3.1.2.1 Design-Time Precision Reduction
A reduced precision can often be achieved through either quantizing the data or approximating the results to be computed [34]. For design-time precision reduction, quantization of data can be either uniformly applied to all the data in the system [14, 18] or applied based on the data type [31]. For example, in Minerva, an automated co-design approach developed by researchers at Harvard University to optimize neural network accelerators [31], different quantization strategies are used depending on the type of data. The synaptic weight, neuron activation level, and the intermediate product are quantized into 8-bit, 6-bit, and 9-bit fixed-point numbers, respectively. Such a per-type quantization strategy provided a 1.5× power saving on average for all dataset considered in that work, including MNIST, 20 newsgroups and so on, when compared to a conventional unified 16-bit baseline.

A uniform quantization scheme can help directly reduce the power consumption through reducing the width of arithmetic circuits at the design time. Some special non-uniform quantization schemes can also be harnessed to save power directly. For example, for a logarithmic quantization that is discussed in Section 2.5.3.2.2, multiplications can be performed efficiently as an addition in the log domain [35]. For more general non-uniform quantization, directly applying the quantized data to arithmetic units might not be effective in reducing the power consumption, e.g. for a weight-sharing scheme with only 16 possible weight values. Each of these weight values might still be a 16-bit number, even though the effective number of bits needed to encode the weight is only 4. Therefore, special schemes may be needed to fully leverage this reduced quantization level. For example, a quantization table-based matrix multiplication can be used to conduct efficient multiplications for arbitrarily quantized weights [32]. This technique is conceptually illustrated in Figure 3.10. The first step is to construct the quantization tables. As a result of aggressively quantized weights, one can compute all possible products between the input activations and the

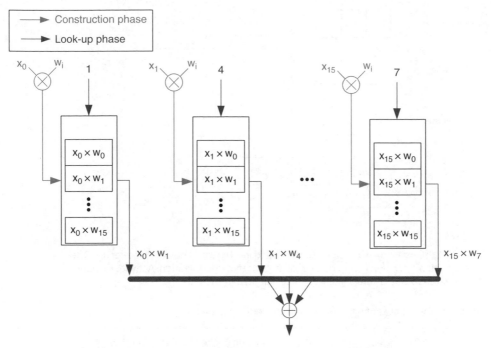

Figure 3.10 Illustration of the quantization table-based matrix multiplication. In the construction phase, the product between the input activations and the weights are pre-computed to construct the quantization table. When the matrix multiplication is conducted, the pre-computed results are read out based on the index of the quantized weight. Source: adapted from [32].

weights. For example, 16 possible weight values w_0-w_{15} are assumed in Figure 3.10. Since an input vector is reused multiple times once it enters the computing engine, the multiplication results can be pre-computed and stored. These pre-computed products are stored in quantization tables in the construction phase. During matrix multiplications, corresponding precomputed products can be read out based on the index of the weight. This technique shares the same spirit as the distributed arithmetic [36] in the sense that storage spaces are traded for less computation at run time.

As the extreme of compressing data, binary-weight network, and binarized network are introduced in Chapter 2. The main advantage of these networks is that they can lead to much more efficient hardware implementation. Such an advantage in energy efficiency was indeed confirmed by hardware implementation of a binary-weight network [37] as well as a binarized network [38, 39]. To demonstrate how a binarized network can simplify the hardware implementation, Figure 3.11 illustrates typical operations that need to be implemented in a binarized network hardware [39]. When both the weights and activations are binary numbers, the multiplication is degenerated to a simple XNOR operation. In addition, the hard sigmoid activation function is essentially taking the sign bit of the final summation. With these simple operations, the computation can be conducted near the memory, reducing the data movement. Furthermore, through simply introducing a mask bit, ternary weight can be achieved naturally without incurring too much overhead. Another big advantage for a binary/ternary network is that by compressing the weights and activations into binary or ternary values, the size of the neural

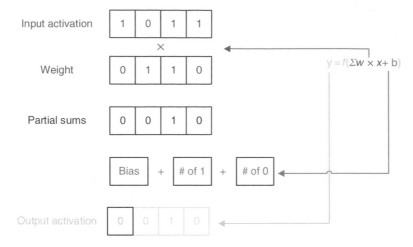

Figure 3.11 Illustration of the operations needed in a binarized network accelerator. The multiplication operation is degenerated to an exclusive-nor operation. Activate operation is degenerated to a hard-thresholding. Source: adapted from [39].

Table 3.1 The number of bits required per convolutional layer.

	Precision needed for each layer (bit)	
Network	When 100% accuracy is preserved	When 1% loss in accuracy is allowed
LeNet	3-3	2-3
AlexNet	9-8-5-5-7	9-7-4-5-7
GoogLeNet	10-8-10-9-8-10-9-8-9-10-7	10-8-9-8-8-9-10-8-9-10-8

Source: data are from [29]. Reproduced with permission of IEEE.

network can be greatly reduced. Such a reduction can eliminate the need for an off-chip DRAM, which in turn saves energy significantly [38, 39].

3.3.1.2.2 *Run-Time Precision Scaling*

In addition to the design-time precision reduction, run-time precision scaling is also an attractive way of leveraging the reduced precision. One of the motivations of conducting run-time precision scaling is that the precision and dynamic range needed in accelerating ANNs not only vary across different network structures but also across different layers in the same network. For example, the required precision for different layers in three common deep network structures are shown in Table 3.1 [29]. It can be observed that the precision requirement in general is not too high, especially for relative simple networks, e.g. LeNet, that target simpler tasks. Moreover, the required precision can be further dropped if slight degradation in the recognition accuracy can be tolerated. Such a per-network and per-layer precision requirement can be readily exploited to improve the performance as well as the energy efficiency of the system if the underlying hardware supports a dynamic precision scaling.

As stated earlier, fixed-point arithmetic is often employed in neural network accelerators to improve energy efficiency. One drawback of a fixed-point number representation is its limited dynamic range. One remedy is to use a dynamic fixed-point number representation [33]. By doing so, the number of bits required to represent numbers in an accelerator can be greatly reduced. Furthermore, a per-layer scaling on the fraction length can be employed to further reduce the bit width. For example, an on-line adaptation scheme was used in [32] to achieve this effect. The number of bits used to represent the fraction part of the numbers in the system were adjusted according to the overflow condition. It was reported that the proposed on-line adaptation method outperformed both the fixed-point number representation as well as the dynamic fixed-point number representation without on-line adaptation. Compared to the 69.9% top-1 accuracy achieved by the 32-bit floating-point baseline, a 66.3% top-1 accuracy could be achieved with a 4-bit word length.

To truly leverage the dynamic precision needed by different networks or different layers in a network, accelerators with a configurable precision are often desired under the assumption that the reduced precision can be traded for either energy efficiency or throughput. To trade precision for power consumption, various techniques from approximate computing can be readily exploited, such as using approximate arithmetic units and voltage over-scaling [40, 41]. One intuitive way to reduce the precision in computing is to directly discard the lower bits [30, 42]. To take a multiplier as an example, by preventing the lower bits of the inputs from toggling, the switching activities can be reduced, which can be helpful in reducing power consumption. Furthermore, in many arithmetic circuits such as adders and multipliers, the critical path starts from the lowest bit of the input. Consequently, if we can ignore the lower bits, the critical path of the circuit can be shortened, which provides an opportunity to lower the supply voltage. Such a reduction in the supply voltage has a quadratic effect on bringing down the power consumption. This method is called dynamic voltage accuracy scaling (DVAS) in [30].

To trade precision for throughput, one popular way is to divide a high-precision computation into multiple low-precision ones. With this method, the basic building blocks in the system are often arithmetic units with a low bit width. When a low-precision computation is needed, these building blocks can directly serve the purpose. On the other hand, when a high-precision computation is on demand, each of these low-bit-width arithmetic units is employed to compute a portion of the result. The outputs of several units are then aggregated either temporally [29, 35, 43, 44] or spatially [45, 46] to form the final result. These two aggregation methods are conceptually compared in Figure 3.12.

In a temporal aggregation, bits in the operands are fed into the arithmetic units serially. For example, to exploit the difference in numerical precision needed by each layer in a deep neural network, a bit-serial strategy was employed in Stripes [29]. The binary number is processed bit by bit to accommodate a variable-length number representation. The partial results are aggregated temporally. The bit-serial computation inevitably prolongs the computational time if the level of parallelism remains the same. Fortunately, the inherent parallelism that exists in the evaluation of neural networks enables the Stripes to utilize parallelism in other dimensions to compensate for the compromised throughput. In [43], instead of a direct bit-serial computation, precomputing similar to that used in distributed arithmetic [36] is employed to amortize the computational cost over many reuses of same-input vectors. In addition to accumulating the partial result in the time domain, the accumulation loop can be unfolded, and the aggregation can be performed in the spatial domain [45, 46], as shown in Figure 3.12.

Multplication #1   Multplication #2

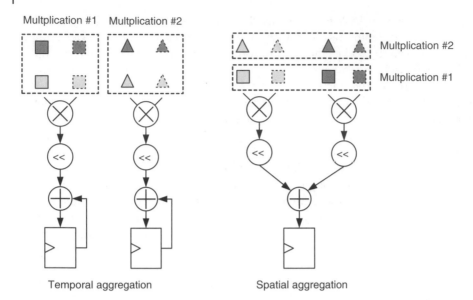

Temporal aggregation   Spatial aggregation

Figure 3.12 Comparison of two aggregation modes in a bit-precision scalable architecture. With a temporal aggregation, bits are fed into the multiplier serially and the result takes several cycles to be computed. With a spatial aggregation, bits can be fed into the multipliers in parallel. Source: adapted from [46]. Adapted with permission of ACM.

3.3.1.3 Leveraging Sparsity

Many natural signals are sparse in nature. Some forms of sparsity might be directly visible in the high-dimensional raw form, whereas others may only be present in low-dimensional feature spaces. This type of sparsity depends on the actual input and is not known until run time, so in some literature it is also called dynamic sparsity. Another type of sparsity comes from the network structure, which may result from, for example, network pruning. Several famous examples of this kind have already been presented in Section 2.5.3.2.1. This type of sparsity does not change with the input signal and is known at compile time. Therefore, it is often called static sparsity. For both of these two kinds of sparsity, special handling in the hardware is needed in order to fully take advantage of them.

3.3.1.3.1 Sparse Connections Sparse connections can have various forms. For example, a convolutional layer is one kind of sparse connection where only input neurons in the receptive field are connected to the output neurons. This type of structural sparsity in connections can be readily leveraged without too much adaptation in hardware. Similarly, structured pruning methods, as discussed in Chapter 2, can often also result in regular structures in the synaptic weights, which leads to an efficient hardware implementation.

One direct benefit of sparse connections is the reduced parameters that need to be stored. A reduction in the number of synaptic weights may help remove the power-hungry external DRAMs. For example, it was shown by Han et al. [47] that going from DRAM to SRAM helped achieve a 120× reduction in the energy by leveraging the sparsity in the network. Unlike a structural sparsity, sparsity in a random fashion generally requires some encoding and decoding scheme to store the sparse weight matrix efficiently. Popular choices are a compressed sparse row (CSR) and a compressed sparse column (CSC) format [4, 47, 48].

One difficulty of taking advantage of the sparsity of synaptic weights in computing is that sparse connections obtained from a non-structured pruning generally do not have a regular pattern. Therefore, it generally demands a specially designed hardware to exploit this scattered sparsity. For example, in Cambricon-X, an accelerator specially designed for accelerating sparse neural networks [49], weight compression is used where zero weights are skipped by encoding. An indexing module is used to select neurons that are needed for effective computation in the neural network. Compared to the baseline DianNao accelerator, Cambricon-X was able to achieve more than 7× improvement in performance and more than 6× improvement in the energy efficiency [49].

3.3.1.3.2 *Sparse Activations*   In contrast to sparse connections that are often being seen in a pruned network, sparse activations can be encountered in virtually all modern deep network that use a rectified linear unit (ReLU) activation function. It was shown in [50] that around 40–50% of neuron activations are zero across various famous deep network structures. Zeros in the activation vectors carry little information, yet they may still take a lot of storage space and communication bandwidth. Therefore, one immediate way to leverage the sparsity in the activations is to compress the activation information so that the power spent to store and move it can be reduced [18, 30]. For example, in Eyeriss, a run-length compression scheme was employed to compress the sparse activation vectors in order to save the traffic power.

In addition to reducing the memory footprint and communication bandwidth, another opportunity of leveraging the sparse activations is in the computation. A zero-value activation in a typical ANN does not propagate to following layers. Therefore, zero activations can be safely removed without altering the final results. Noticing this, zero activations have been leveraged in many ANN accelerators in order to improve the energy efficiency or throughput. There are two approaches to leveraging this sparsity: to bypass or to skip the zero activations.

A relatively intuitive approach to harness the sparse activations is to data-gating the datapath [18, 30, 31]. For example, data gating is employed in Eyeriss to disable the filter weight fetching and prevent the MAC unit from switching once a zero-input activation is detected in order to save energy. With such a data gating logic, 45% of power saving was reported [18]. In addition to zero activations, small activations can also be pruned away to enhance the efficiency. In Minerva, small activations are exploited as sparsity in order to skip unnecessary computations [31]. Small activations have little impact on the final output compared to other large activation values. Therefore, the small activations can be discarded without noticeably deteriorating the performance of the system. In the datapath of Minerva, a comparator is used to determine if an activation is larger than a predefined threshold. If the activation is too small, no synaptic weight is fetched from the SRAM memory, saving memory-access power. In addition, the following registers can be clock-gated when small activations are detected, which reduces the dynamic power of the system. Such a pruning and data-gating strategy yielded an average power saving of 2× over all datasets employed in that work [31].

Another more complex approach that can potentially yield a higher return is to skip zero activations. Skipping zero activations tend to be harder than skipping zero weights. This is because sparsity in synaptic weights is static information that can be known at compile time. Therefore, the burden to achieve an efficient mapping and computing can be shifted to the compiler side as long as the target accelerator is flexible enough. On the other hand, sparsity in the activations can only be observed at run time, which demands that the hardware itself figures out how to optimally arrange the computation and skip zeros. To effectively skip zeros in the activations, many hardware architectures have been proposed in recent years. Some of them focus on skipping zero activations [50], whereas others can skip both zero activations as well as zero synaptic weight [47, 51, 52].

For example, the efficient inference engine (EIE) was developed by Han et al. to accelerate the evaluation of compressed deep neural networks through exploiting both the dynamic and static sparsity [47]. EIE uses a variation of the CSC format to store the compressed weight matrix. By doing so, zeros in the weight matrix can be effectively skipped. In addition to the sparsity that exists in the network structure, data-level sparsity can also be exploited through computing only non-zero activation levels. The non-zero activations can be detected by the leading non-zero detection module and are passed to the activation queues in PEs.

To accommodate more layer structures, e.g. the convolutional layer, the sparse convolutional neural network (SCNN) architecture was proposed by Parashar et al. [52]. The SCNN is based on an input-stationary dataflow where both the neuron activations and the synaptic weights are encoded in a compressed form in order to eliminate unnecessary computations. The dataflow delivers non-zero input activations and non-zero synaptic weights to the multiplier arrays to form effective computations. By doing so, the complexity is shifted to the summation of the resultant partial sums. This can be done based on the coordinates of the sparse partial sums.

One common difficulty faced by many accelerators that attempt to skip both zero weights and zero activations is the load-balancing problem induced by the uneven distribution of zeros in either activations or synaptic weights [47, 51, 52]. The zero-imbalance in the synaptic weight can be solved through assigning filter weights with a similar number of zeros to a group of PEs since the weights are known at the compile time [51]. For imbalance in the input activations, a queue structure can often be used to mitigate this problem [47].

In [53], Albericio et al. took the concept of skipping zeros further. In that work, not only zero activations but also zeros in the non-zero activations were skipped with the help of a bit-serial processing. Two types of zeros, static zero and dynamic zero, can be identified. These two kinds of zeros are illustrated in Figure 3.13. In the figure, it is assumed that the system has an 8-bit precision and the task at hand only requires a precision of 5 bits. The statically ineffectual zeros refer to the zeros that can be spotted at the compile time. For example, in Figure 3.13, the first two zeros and the last zero are

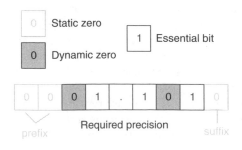

Figure 3.13 illustration of the static and dynamic zeros defined in [53]. Static and dynamic zeros refer to those zeros that can and cannot be determined *a priori*, respectively. It is assumed that the system uses an 8-bit fixed-point number representation and the required precision is only 5 bits. Source: adapted from [53]. Adapted with permission of ACM.

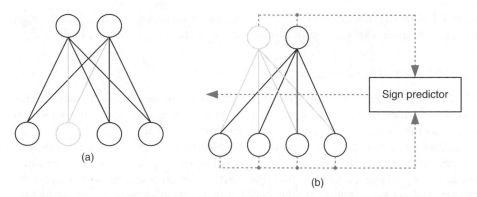

(a)

(b)

Figure 3.14 Comparison of two typical ways to leverage (a) sparsity in the input activations and (b) sparsity in the output activations. Input zero-value activations can be directly harnessed by skipping the computations that involve those activations. Skipping output activations, on the other hand, normally requires speculation.

static zeros because we know beforehand that these bits will be zeros for the task we are interested in. The dynamically ineffectual zeros, on the other hand, refer to the zeros that can only be detected at run time. Being able to skip zero bits can be a very effective technique in improving the energy efficiency as well as the throughput of an accelerator, as it was shown in [53] that around 92% of the bits of the activations were zero in modern CNNs. The idea behind Pragmatic is conceptually simple: the original parallel multiplication can be broken down into serial–parallel shift-and-add operations, and consequently zero bits can be skipped. Several design techniques have been proposed to make the bit-serial processing practical. With all the proposed strategies, 4.31× speedup and 1.70× increase in the energy efficiency compared to the 16-bit DaDianNao baseline were reported with 1.68× area overhead.

Even though a lot of work in the literature focuses on exploiting the sparsity in the input activations, sparsity in the output activations can also be leveraged. Indeed, output activations in the current layer is the input activations for the next layer. Nevertheless, the ways in which these two types of sparsity can be used are quite different, as illustrated in Figure 3.14. For an input zero activation, the multiplicative edges connected to the corresponding presynaptic neurons can be removed without affecting the final results. In hardware implementation, this can be achieved through either data gating the MAC operation or directly skipping that activation, as mentioned previously. Zero activations in a deep ANN are normally a consequence of applying a ReLU activation function where negative partial sums are clamped to zero. Therefore, if we know in advance that the partial sum for a postsynaptic neuron is negative, we can safely remove the connected synapses and directly output zero without altering the final result. The difficulty here, however, is the causality. How can one find out the sign of the partial sum without actually calculating it? It turns out that there are ways to know this information before all the computations are performed. For example, it was shown in [54] that by arranging the weights in a particular order, the data pipe may be terminated early if a negative partial sum is observed, assuming that all the input activations are positive. Such a relatively conservative way does not lead to any loss in the classification accuracy of the neural network. A more aggressive way is to perform speculation on the sign of the partial sum. The speculation can be achieved based on certain rules [54] or through a trial computation with a low-rank approximation of the weight matrix [55–57].

3.3.2 FPGA-Based Accelerators

FPGA is a piece of hardware that can be reconfigured to conduct different types of computations and operations. It usually serves as the prototype used to verify the algorithm and architecture before a costly ASIC is built. As shown in Figure 3.1, an FPGA provides an attractive alternative to general-purpose processors and ASICs. It typically provides more performance per watt compared to general-purpose processors while enjoying more flexibility compared to fixed-function ASICs. Therefore, it has been adopted by many researchers for prototyping their initial designs and even some major companies have used it as the platform for conducting deep learning.

The implementation of neural networks on FPGAs started a long way back even before deep learning became popular [58–60]. In many cases, FPGAs were used as a rapid prototype to demonstrate the efficacy of the algorithm. Recently, with the revival of deep learning, there is a tremendous need for energy-efficient high-throughput hardware for conducting deep learning tasks. The strategies of building low-power high-performance FPGA-based ANN accelerators are similar to those introduced in the previous section. For example, optimizing the data access pattern [61], fused-layer convolution [62, 63], quantizing the data in the system [64, 65], as well leveraging the sparsity in the network [66, 67] have also been demonstrated to be very effective in the context of FPGA-based accelerators.

Nevertheless, many design constraints of an FPGA are different from those of an ASIC. Even though an FPGA has a higher programming flexibility, its resources, such as the number of available programming units and DRAM bandwidth, are generally hard-limited. Therefore, how to achieve the optimal performance and energy efficiency under resource constraints has been a topic studied by many researchers [61, 65, 68]. For example, in [61], Zhang et al. introduced a design methodology for implementing CNNs on an FPGA. For each specific FPGA platform, a roofline model can be built and the corresponding design space can be explored. The basic idea of a roofline model is that the performance of a computing engine is either constrained by the available computing resources (compute bound) or the available memory bandwidth (memory bound). An optimal design should choose a level of parallelism that can exploit all the computing resources in the system but with a memory bandwidth as low as possible. With such a design methodology, an FPGA-based accelerator was demonstrated and a 4.8× speedup compared to a16-thread software solution was reported.

As mentioned earlier, one of the biggest advantages of an FPGA compared to an ASIC is its flexibility. However, flexibility in programming does not equal ease in programming. In fact, compared to general-purpose processors, programming on an FPGA can be much more challenging, especially for designers who do not have much knowledge about the underlying hardware. Therefore, to facilitate the deployment of ANNs on to an FPGA fabric, many studies have been presented recently to develop a methodology of mapping a neural network on FPGAs [69–71]. Programming an FPGA with high-level programming tools is, in general, a challenging problem [72]. Nevertheless, most state-of-the-art deep neural networks have very regular network structures and involve only a few types of regular arithmetic operations. Therefore, one can take advantage of a modular design approach to map a network described in a high-level programming language to the underlying FPGA platforms [69, 71].

In addition to academic efforts, the FPGA-based neural network accelerators have also been adopted in the commercial world. Microsoft Research has reported [73] that they were in the process of accelerating deep CNNs by leveraging the infrastructure developed in Catapult, a project that Microsoft announced in 2014 that aimed to accelerate various datacenter-level tasks with FPGAs [74]. It was stated that the FPGA approach was able to provide a good throughput that was comparable to what could be achieved with GPUs yet with only a fraction of the power consumption. Such an improvement in the energy efficiency is promising for datacenters as the power consumption has grown to be one of the major issues that needs to be addressed in recent years.

More recently, Microsoft has announced its Brainwave project, which targets building hardware platforms to accelerate deep-learning tasks [75, 76]. The philosophy there is to use a batch size of 1 or, in other words, no batching, to avoid the latency penalty induced by a large batch number. The throughput, on the other hand, can be optimized through exploiting the parallelism within a service request. With such a strategy, it was reported that a large-scale gated recurrent unit model with five times the computational cost compared to ResNet-50 could be served within 1 ms with just one FGPA. Such a low latency translated to an effective throughput of 39.5 TFLOPS [75].

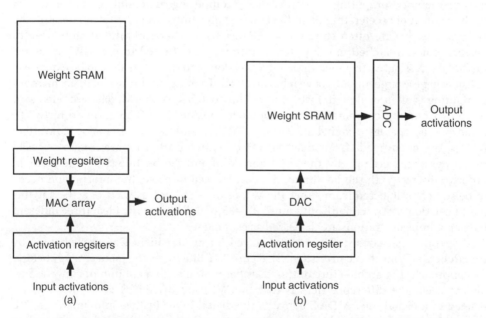

Figure 3.15 Comparison of (a) the conventional digital computing and (b) the in-SRAM analog computing. In the conventional approach, synaptic weights stored in the SRAM array need to be read out and transferred to registers before the actual computations can take place. The large number of memory accesses can be expensive in terms of the energy cost. When the approach of in-memory computing is employed, on the other hand, the MAC operations are performed directly within the memory, thereby saving the memory access power.

3.4   Analog/Mixed-Signal Accelerators

Since the birth of ANNs, researchers have started looking into possibilities of building analog neural network hardware. In the 1990s, many researchers believed that analog-based neural network hardware might be the right way to go [77–80], as the brain, the inspiration for ANNs, conducts analog computations. However, as technology advanced around 2000, the new technology node favored a digital circuit much more than an analog circuit. In addition, the trend of implementing hardware-based neural networks were not that popular any more, as the usefulness of neural networks were rather limited at that time.

Consequently, work on analog-based neural networks appeared to be few and far between in the literature during that period of time. Recently, interest in hardware-based neural networks has revived due to the huge success achieved by deep learning. Analog computing is again refocused, as the matrix–vector multiplication needed in neural networks can be conveniently built in analog circuits with the help of physical laws. In this section, some recent work in analog neural networks is presented.

3.4.1   Neural Networks in Conventional Integrated Technology

3.4.1.1   In/Near-Memory Computing

One of the approaches that many analog implementations exploit is in/near-memory computing. The conventional digital-based computing and one popular type of in/near-memory computing, in-SRAM analog computing, are compared in Figure 3.15 in the context of accelerating neural network applications. In a digital accelerator, the synaptic weights are often stored in an SRAM array. Therefore, the weights need to be read out and latched into registers before they can be fed into a MAC array for further processing. Such a memory content readout and data transfer can be expensive, considering the large amount of weights involved. To mitigate this problem, in-memory computing is often exploited. Instead of reading data from SRAM, data are often sent to SRAM and the computations are conducted inside the SRAM array directly. To achieve this, analog-to-digital converters (ADCs) and digital-to-analog converters (DACs) are often needed. For most SRAM cells, two ports that are usually exposed to the peripheral circuits are the word-line (WL) and the bit-line (BL). Therefore, in order to interact with the bit stored in an SRAM cell in place, the information needs to be somehow injected from either the WL or the BL. Various ways of achieving this have been demonstrated in the literature [81–85]. To illustrate the idea, three different flavors of implementations are discussed as examples.

A classifier that consisted of many boosted linear classifiers was implemented by researchers at Princeton University [82]. Figure 3.16 illustrates the main strategy behind this approach. The architecture of this classifier is similar to a column of conventional SRAM cells. The difference here is that the WLs are driven by an analog voltage, instead of a digital one. A DAC converts the digital input feature into an analog WL voltage through a current DAC followed by a cell replica circuit. The cell current is then approximately proportional to the magnitude of the input feature. The binary weight stored in the SRAM cell then determines from which BL the charge is sunk, which serves as a role of multiplication. The products between the input features and the binary weights are summed up on the BLs. A rail-to-rail comparator is then employed to hard-threshold the result to form the classification result. Such an in-SRAM processing

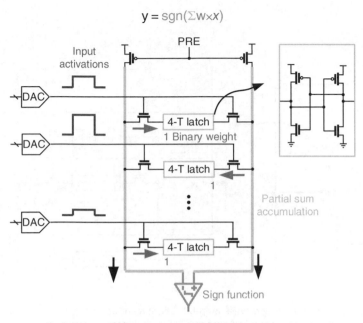

$$y = \mathrm{sgn}(\Sigma w \times x)$$

Figure 3.16 Architecture of the in-SRAM classifier. The word-lines of the SRAM cells are driven by DACs whose inputs are determined by the input features. The binary weights stored in the cells are then multiplied with the input features to form cell currents. The cell currents are summed up on the bit-lines to obtain partial sums. Source: adapted from [82].

resulted in an energy metric of 630 pJ per classification, which was significantly lower than a discrete system where the memory and the arithmetic units were separated.

A different approach was taken by Kang et al. [84, 85]. Since the target there was for conventional ANN where both the input activations and the weights are multibit numbers, a scheme for combining information in multiple SRAM cells was developed. The idea is briefly explained with the help of Figure 3.17. The hardware architecture is called the deep in-memory architecture (DIMA), where multiple rows of SRAM cells in the array are logically grouped together to form a word-row. Each word-row holds the bits used to represent one weight. The multibit weight information can be read out by asserting the corresponding WLs. Binary-weighted pulse-width-modulated (PWM) pulses are generated by digital-to-time converters (DTCs) and are used to drive the WLs of a word-row. The bits stored in the bit cells are then read out and summed on the BLs. Since the cell current is proportional to the content stored in the bit cells, the voltage drop on the BLs is also approximately proportional to the weight. This voltage drop is then sensed and multiplied with the input activations in a mixed-signal multiplier to get the partial sums and then output activations.

The third approach was taken in conv-RAM, a computational technique developed by the researchers at Massachusetts Institute of Technology [83]. The target of that work was to accelerate binary-weight networks presented in Chapter 2. Figure 3.18 illustrates this technique. In contrast to the approach shown in Figure 3.16, where the input feature is fed into the SRAM cells by modulating the WL voltage, the conv-RAM injects the input through the BLs of the SRAM array. The injection is done through charging the BL with a current source. The duration during which the current source is on is controlled by a DTC whose input is the input activation. Therefore, after the precharging, the

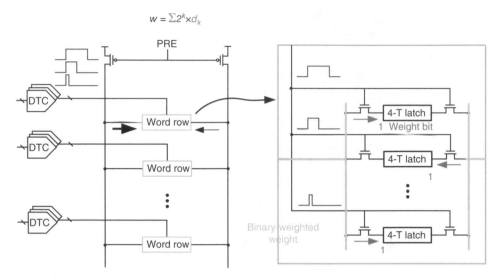

Figure 3.17 Techniques employed in DIMA to achieve a multi-bit weight readout. Each weight is stored in a word row, which consists of several physical rows of SRAM cells. Binary-weighted PWM pulses are injected into WLs by DTCs. The current flowing on the BLs and the voltage drops on the BLs are then approximately proportional to the weight stored in the SRAM cells. Source: adapted from [85].

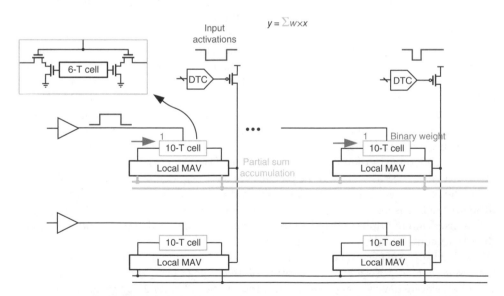

Figure 3.18 Techniques used in conv-RAM. Input activations are injected through pre-charging BLs to certain voltage levels. The binary weights stored in the SRAM cells are read out and multiplied by the BL voltage. Charge sharing is then used to sum all the partial sums. Source: adapted from [83].

voltage on the BL is proportional to the input feature. It was argued in [83] that such an approach could result in a more variation-tolerant design compared to that presented in [82]. In addition, since a 10T cell is used, the dynamic range of the signal on the BLs can be boosted without worrying about a write disturb, which may happen when a 6T SRAM cell is used. During the read phase, one of the two BLs is discharged to ground, depending on the information stored in the bit cell. This is equivalent to multiplying the 1-bit weight information with the input activations. The partial sums are then summed up across different columns with the help of multiply-and-average (MAV) circuits. Finally, charge-sharing-based ADCs are used to convert the analog results into digital form.

In addition to digitally stored weights, weights stored in analog form have also been leveraged in the past in analog accelerators [86–88]. An analog accelerator for an extreme learning machine (ELM), called a machine-learning co-processor (MLCP), was reported by Chen et al. [87]. ELM is one type of neural network that was proposed by Huang et al. in 2004 [89]. Its unique feature compared to other neural networks is that its input weights do not have to be tuned [90]. Only synaptic weights associated with the output layer need to be learned. Such a configuration provides some benefits for hardware implementations. For example, one appealing advantage of utilizing an ELM in an analog design is that the fixed random weights in the input layer can be readily implemented by a current mirror array with fabrication mismatch [87].

Figure 3.19 shows the basic ideas of the main data path in MLCP. A current mirror array is used to implement MAC operations. The random weights needed in the hidden layer are obtained naturally with the mismatches between transistor pairs due to local process variations. Chen et al. biased the current mirrors in the subthreshold region, which not only achieved a low power consumption but also large variations in the current amplification ratio. The input to the network is converted into current and is amplified according to the gains of the current mirror array. The obtained partial sums are in the form of current and are summed together vertically. The currents are then fed to current-controlled oscillators (CCOs), which convert input currents into digital pulse trains. The pulses are collected by the counters, which help perform the analog-to-digital conversion. The activation function can be achieved by the non-linear transfer function of the CCO, as shown in Figure 3.19.

Another example of using analog weight for in/near-memory computing was presented by Lu et al., where a floating-gate-based analog deep-learning engine (ADE) was demonstrated [88]. The ADE consists of a floating-gate memory used to provide non-volatile storage for the parameters needed in learning and inference, reconfigurable analog computation units, distance processing units, and training controlling circuitry. The learning algorithm employed was K-mean clustering. Current-mode circuits as well as subthreshold biasing of the transistors are used extensively to implement the neural network efficiently. The floating-gate memory provides current outputs, which can be directly utilized by the current-mode circuit. The accelerator was able to process 8300 input vectors per second while only consuming $11.4\,\mu W$ from a 3 V supply voltage.

3.4.1.2 Near-Sensor Computing

For in-memory computing, the power spent to read memory can be reduced. Similarly, for systems where the inference is performed on the sensed data in situ, analog computing can also be leveraged to directly process the collected data from sensors in order to save the energy consumed in ADCs. Figure 3.20 illustrates this idea. In a conventional approach, the analog signal is first sampled and quantized and a digital neural network

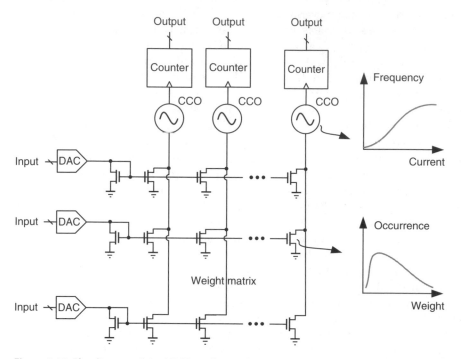

Figure 3.19 The diagram of the MLCP implementing an ELM [87]. The input layer is implemented as current mirrors. The mismatches between transistors are readily leveraged to implement the random weights in an ELM. The hidden-layer neurons are realized in the form of current-controlled oscillators that drive output counters. Source: adapted from [87].

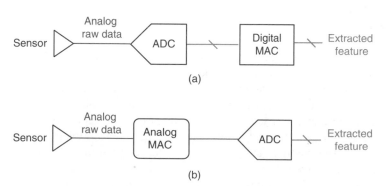

Figure 3.20 Comparison of two approaches employed to perform inference directly on data collected by sensors. (a) In a conventional approach, the analog signal is first sampled and quantized by the ADC, and the backend digital accelerator is used to perform the MAC operations. (b) In an alternative approach, analog/mixed-signal computing is carried out directly on the sensed data in the analog domain. By doing so, the workload to the power-hungry ADC can usually be lowered significantly.

Figure 3.21 Implementing mixed-signal MAC operations in (a) the charge domain (adapted from [91]) and (b) the time domain (adapted from [92]). For computing in the charge domain, input activations are multiplied with the weights encoded by the capacitance values and are converted into charges. For computing in the time domain, the input activation is converted into pulse trains, and the phase of the pulse train is accumulated in a time window whose duration is controlled by the synaptic weight.

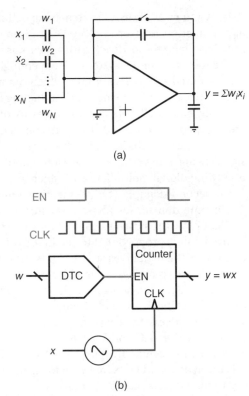

(a)

(b)

accelerator can be employed to carry out the inference. An alternative is to directly operate on the analog signal gathered by the sensor, and the extracted features in the analog domain are then converted into digital form for further processing. The main motivation of this approach is to reduce the power consumption of the ADCs. Indeed, in many neural networks, high-dimensional raw data are gradually refined to form low-dimensional features. Therefore, even after one layer of feature extraction, the amount of data that needs to be processed can be greatly reduced. Consequently, by leveraging the energy-efficient analog processing in front of the ADC, the workload of the power-hungry ADC can be remarkably reduced. Near-sensor computing usually involves analog raw input data and synaptic weights that are stored in a digital memory. Therefore, mixed-signal MAC operation is often needed. There are many ways of conducting MAC in a mixed-signal fashion, and two popular choices are shown in Figure 3.21.

The first approach is to realize the MAC operation in the charge domain [91, 93, 94]. The analog voltage input to the mixed-signal MAC unit is first converted into charges that are stored on the capacitor. The input capacitors are programmed according to the weights that need to be multiplied. The multiplication is naturally conducted during this voltage-to-charge conversion through the classic equation $Q = CV$, where Q is the charge stored on each input capacitor, C is the capacitance of the input capacitor, which is programmed to be the value of the weight to be multiplied, and V is the input voltage. In Figure 3.21a, the computed result is still in the analog domain. It is also possible to merge the mixed-signal MAC unit with the following ADC [93, 94]. For example, in

[94], Wang et al. added a switched-capacitor feedback divider block into a successive approximation register (SAR) ADC to implement the multiplication. The idea is that the division on the feedback path appears as multiplication to the external world as long as the loop gain of the system is high enough. The division circuit can be conveniently implemented as a passive circuit, which can normally achieve better linearity compared to an active circuit. One merit of conducting computation in the charge domain is that it is usually energy efficient as the charges are often reused as much as possible before they are dumped to the ground. This is in contrast to the digital circuit, where the capacitors in the circuit are charged and discharged repeatedly. In addition, the relatively small signal swing at the charge conservation node can also help save energy compared to the full-swing digital multipliers and adder trees [95].

The second example of implementing a mixed-signal MAC unit is to encode the signal in time domain (or phase domain) [92, 96]. In Figure 3.21b, a voltage-controlled oscillator (VCO) is used to convert an analog input into pulse trains whose frequency is approximately proportional to the input voltage within a certain range. A DTC is used to convert the digitally stored weights into time-domain signals. By doing so, the two operands are converted into a pulse train and a pulse-width modulated pulse, as shown in Figure 3.21b. The multiplication result can be obtained as the accumulated phase of the VCO during the period of time when the output of the DTC is high. To help understand this process, the frequency of the VCO can be described as $f = K_{VCO}x$, where x is the input to the VCO and K_{VCO} is a constant relating the input voltage of the VCO and its oscillating frequency. The pulse width of the DTC can be modeled as $T = K_{DTC}w$, where w is the synaptic weights in digital form and K_{DTC} is the gain of the DTC. Therefore, the accumulated phase can be written as

$$\phi = 2\pi \int_0^T f dt = 2\pi \int_0^{K_{DTC}w} K_{VCO}x dt = 2\pi K_{DTC}K_{VCO}wx \qquad (3.1)$$

Clearly, the accumulated phase of the VCO is proportional to the product of the synaptic weight and the input activation. To read out the phase accumulation, we still need a time-to-digital converter (TDC) to convert the information in the time/phase domain back to the digital domain. This can be conveniently achieved through a counter, as shown in Figure 3.21b. With this counter, the accumulation needed in a MAC operation can occur naturally within the counter. In addition, an ADC implemented with a VCO and a counter naturally implements the first-order $\Sigma - \Delta$ modulation [97]. In other words, quantization residue in quantizing the phase of the oscillator is accumulated in the VCO, which is helpful in improving the resolution of the ADC. One of the appealing reasons to adopt a time-domain mixed-signal circuit is that it generally benefits more from technology scaling compared to a conventional voltage-domain mixed-signal circuit [98].

3.4.2 Neural Network Based on Emerging Non-volatile Memory

As modern technologies favor digital circuits more than analog circuits, many computationally intensive applications have tended to be implemented in digital form in recent years. However, by blending some more advanced nanotechnologies, analog accelerators can sometimes show certain advantages. One good example is using a memristor in analog/mixed-signal accelerators. Since memristors were first demonstrated in

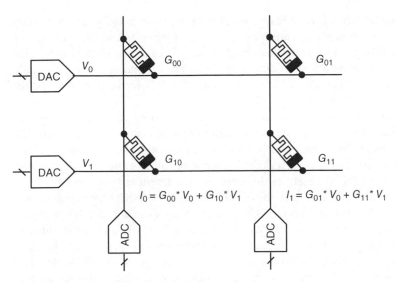

Figure 3.22 Illustration of using memristive devices for conducting matrix–vector multiplication in the analog domain. Voltage is applied on the word-lines (horizontal lines). The memristor devices located at the cross-point convert the voltage into current with a weight proportional to their conductance values. The current on the same bit-lines (vertical lines) are summed to form the computed results.

2008 [99], they have drawn the attention of many researchers. As the missing fourth circuit component, the existence of memristors was hypothesized four decades ago by Chua [100]. A memristor can be treated as a resistor with memory. The resistance of a memristive device is controlled by the amount of charge that passes through the device [101]. To date, many types of memristive devices have been fabricated and demonstrated, such as a ferroelectric memristor [102], a memristor based on a single nanowire [103], a TiO_2-based memristor [104], memristive devices based on oxide heterostructures [104], and so on. In addition to memristors, there are also many other types of emerging non-volatile memory (NVM) technologies. Over the years, much effort has been spent to realize neuromorphic hardware based on these emerging NVM technologies [105, 106].

3.4.2.1 Crossbar as a Massively Parallel Engine

A high-density NVM-based ANN can be implemented in the form of a crossbar, which is very popular in the literature. One example of a memristor crossbar is shown in Figure 3.22. The computation is conducted with the help of Ohm's law and Kirchhoff's laws. With the conductances of the memristors in the array being programmed to the values of the elements in a weight matrix, a matrix–vector multiplication can be naturally carried out. When the input vector is presented on the word-lines in the form of voltage, the magnitude of the current obtained at each column corresponds to each element in the output vector. Such a computational model is massively parallelized, as an entire matrix–vector multiplication can be finished in one shot. Since input and output of the crossbar are often in digital form, ADCs and DACs are usually needed to perform the conversions.

Compared to the SRAM cells that are often used to hold the synaptic weights for a digital ANN accelerator, emerging NVM-based cells are often smaller and have the potential to store multibit values. Therefore, synaptic weights can be conveniently stored directly on an NVM-based crossbar, which eliminates the need of other non-volatile storage or DRAM. Consequently, power-consuming off-chip traffic can be avoided, which helps reduce the power consumption. In addition, in-memory computing can be naturally conducted in a crossbar structure, which can reduce the data movement and further improve the energy efficiency of the system.

However, matrix multiplication based on an NVM crossbar does pose some challenges for designers. The first challenge is how to program a crossbar accurately. It is not trivial to program the conductance value of each NVM cell in a crossbar to the desired value, considering various non-idealities such as device variation, interconnect resistance, and so on. One solution to this problem is through on-chip learning that can adjust the weights in NVM cells directly. Such a method might be too heavyweight for applications where only inference is needed. Alternatively, a closed-loop approach that keeps monitoring the conductance of each target NVM cell is often needed [105, 107–110]. Hu et al. proposed a conversion algorithm for mapping arbitrary matrices on to a crossbar fabric [109]. The mapping method relies on a high-efficiency solver that can help simulate the voltage and current in the array. The final mapping can be obtained through several steps of tuning and optimization. With the proposed mapping algorithm, a neural network was mapped on to a memristor-based crossbar, and no performance degradation was observed compared to the software baseline. Furthermore, thanks to the high-efficiency crossbar structure, three to four orders of magnitude improvement in the speed-efficiency product compared to that of a digital ASIC approach was reported.

In addition to the initialization of the crossbar array, as the inference goes on, the small voltages applied to the crossbar may make the conductance of memristor cells drift away from their original values. Such a gradual deviation from the desired synaptic weights might potentially deteriorate the inference accuracy. Therefore, special handling is sometimes needed to counteract this resistance drift. For example, in Harmonica, a framework of heterogenous computing systems with a memristor crossbar [111], on-chip calibration is used to periodically refresh the synaptic weights to tackle this weight-drifting problem.

The second challenge associated with a crossbar-based computing comes from the relatively low accuracy of analog computing. For example, a linear constant conductance is assumed in Figure 3.22 for conducting the matrix–vector multiplication. In reality, however, the conductance of each NVM cell is likely to be dependent on the input voltage. Such a non-linearity inevitably introduces error in the final computed results. In addition, it is normally difficult to achieve a high computing precision in analog computing. Even though neural networks are known to tolerate inaccuracy in the computation, many neural network accelerators in the literature still attempt to keep a 16-bit fixed-point precision. To achieve this resolution, a straightforward approach requires each memristor in the array to be able to store 2^{16} different conductance levels and each ADC to have a resolution higher than 16 bits. Such a high precision in an analog computing system is hard and expensive to achieve. Therefore, in order to counteract the non-linearity in the memristor as well as the low accuracy in an analog approach, usually only a few bits are stored in a memristor device. For example, In ISAAC, a neural network accelerator with a memristor-based crossbar [112], each memristor only stores 2-bit information, and multiple memristors are employed to store one synaptic weight. In addition, the input vector is fed into the crossbar array bit by bit, and the final results are obtained through shifting and adding. Such a method essentially replaces one high-precision computation with multiple low-precision ones. The low-precision operations are distributed spatially and temporally to leverage the high density and massive parallelization of a crossbar array. Similarly, In PRIME, another memristor crossbar-based neural network accelerator [113], each memristor is used to store 4-bit information, and two cells are combined to represent one 8-bit synaptic weight.

Interestingly, the two aforementioned problems with the NVM crossbar can be greatly mitigated if binary-weight or binarized networks are implemented. When the weights of the neural network are limited to binary values, the two stable states of an NVM cell can naturally serve as the binary weights. Programming of the crossbar array in this case is to simply set and reset the bit cells. This binary programming is much easier to achieve compared to the multibit case, where careful closed-loop programming is needed. When the input and output of the crossbar array are binary vectors, the non-linear characteristic of an NVM device is no longer problematic. Furthermore, the A/D and D/A conversions needed in a conventional crossbar array often take a significant portion of power. In [114], it was observed that the overhead associated with ADCs and DACs were around half of the system power, even when low-precision 3-bit ADCs and DACs were used. The use of a binarized neural network can eliminate the need for multibit ADCs and DACs, which improves the energy efficiency remarkably. Because of these advantages, many NVM crossbar-based binarized networks have been demonstrated recently through either simulations [115, 116] or experiments on fabricated prototypes [114, 117, 118].

A binarized network is a good example of algorithm-architecture co-design, where the design complexity is greatly shifted toward the offline training side such that the inference on the hardware can be made to be both fast and low-power. An interesting work that shares a similar spirit was presented by Li et al. [119]. The main motivation of that work is to merge the power-hungry A/D and D/A interfaces into the crossbar so that the system power consumption can be reduced. This is done by directly training the neural network on binary patterns. Since with a binary coding scheme most significant bits (MSBs) have a higher impact on the inference results compared to least significant bits (LSBs), the loss function is modified such that the errors associated with MSBs have larger weights. With such a technique, the input and output to the crossbar can be made binary, which eliminates the need for ADCs and DACs.

3.4.2.2 Learning in a Crossbar

As mentioned in the previous section, one challenge faced by NVM crossbar-based neural engines is how to program the desired synaptic weights accurately. Such a difficulty can be overcome through on-chip learning. Indeed, learning is a closed-loop feedback process that can calibrate out many non-idealities in the circuits. The conventional backpropagation-based gradient descent learning also involves a large number of matrix multiplications. Therefore, an NVM crossbar that is optimized for matrix multiplication, in theory, can be used to accelerate the learning process as well. However, learning itself is a hard task to achieve. Conceptually, training is more complicated than inference as more complex data dependencies exist in this phase. In order to address this issue, in [120], a pipelined architecture, called PipeLayer, was proposed to accelerate both the training and inference phases. To accelerate the training phase, innovations on both the inter-layer and intra-layer pipeline schemes were made to

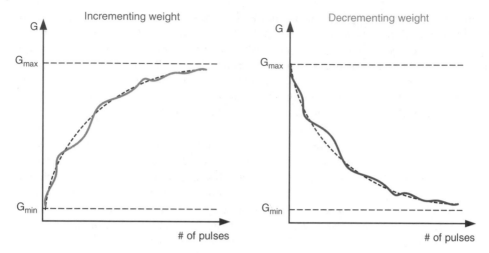

Figure 3.23 Demonstration of a few typical non-idealities in programming NVM cells. The minimum and maximum achievable conductances are bounded and are likely to vary across devices. The increment and decrement of the conductance of the device are also asymmetrical.

exploit the parallelism in the algorithm and the inherent dataflow more efficiently. With the newly proposed architecture, more than 42× improvement in speed compared to a baseline GPU was observed for tasks where both training and inference were involved.

Another difficulty in online training of an NVM-based crossbar is the device non-ideality. Figure 3.23 illustrates a typical profile of the conductance change when a train of programming pulses are applied to an NVM device [105, 121, 122]. Three common non-idealities shown in the figure are: (i) a limited achievable conductance range, (ii) randomness in updating the weight, and (iii) asymmetry in incrementing and decrementing the conductance. These three non-idealities have different levels of impact on the learning performance.

The noise associated with the weight update is probably the most harmless one as most ANNs are trained with a stochastic gradient descent method, which has a good tolerance to unbiased noise. Indeed, the weight update obtained at each iteration is accumulated. Such a process behaves in a similar way to a low-pass filtration, which can help average out noise as long as the noise is not biased.

The asymmetry in incrementing and decrementing the conductance, on the other hand, tends to have a large negative impact on the learning result. It is shown in Section 5.3.5 that such an asymmetry introduces a bias term in the weight update, which prevents the synaptic weights from converging to the optimal values. To make things worse, certain NVM cells, such as those based on phase change memory (PCM) and bipolar filamentary resistive random-access memory (RRAM), have a highly asymmetrical conductance-changing characteristic where the weight update is gradual in one direction and abrupt in another direction.

To make the learning feasible on a real NVM crossbar-based neural engine, many studies have been conducted recently to cope with the non-idealities in the NVM devices [123–128]. For example, two crossbars are often needed to store synaptic weights for a neural network, as one NVM cell naturally can only store positive weight. One crossbar stores the weight matrix $\mathbf{G}^+$, whereas the other crossbar stores the matrix $\mathbf{G}^-$.

The actual weight matrix of the neural network is the difference between these two matrices $G = G^+ - G^-$. Under this circumstance, for devices that can only be gradually tuned toward one conductance-changing direction, a common technique is to only update toward the direction where the conductance can be smoothly changed [123, 125, 127, 128]. The weight update toward the other direction can be done by changing the complementary weight matrix. Of course, periodic refreshing is needed to abruptly change both G^+ and G^- in order for them to stay within the tunable range. Such a refresh, however, was shown to be infrequent [123].

Besides the above-mentioned common non-idealities, another important non-ideality that is often overlooked is the endurance of an NVM cell. It was shown in [129] that training ResNet-50 required half a million iterations. Consequently, for an RRAM crossbar with an endurance limit of five million, only 10 times of training could be conducted before the RRAM array wore out. To prolong the lifetime of an NVM crossbar, a structured gradient sparsification can be employed to reduce the number of write operations to the crossbar. In other words, only rows or cells with the most significant gradients need to be updated. In addition, an aging-aware row swapping scheme is also effective in extending the lifetime of a crossbar array by balancing the write operations that are needed for each row. This scheme swaps heavily written rows in a crossbar array with rows that are updated less frequently. It was reported that these techniques were helpful in prolonging the lifetime of the NVM crossbar by approximately two orders of magnitude while achieving a similar performance [129].

In addition to the architecture-level and circuit-level exploration, another research direction that is also pursued by many researchers is the device optimization [130–133]. If one can tackle the non-idealities of the devices directly at the source, many circuit- and architecture-level workaround and compensation are no longer needed, which has the potential to improve the system energy efficiency significantly. In general, an NVM device with a bidirectional, symmetrical, and linear conductance-changing characteristic is highly desired, as such a device can help achieve the same classification accuracy as can be obtained with the conventional software approach when implemented as a crossbar [122, 127].

3.4.3 Optical Accelerator

Attempts to exploit light as the computational media have never stopped. Even before many electronic versions of ANNs have been made, optical learning systems were demonstrated [134, 135]. Matrix–vector multiplication has previously been demonstrated to be possible with integrated optoelectronics [136]. With the breakthrough in deep learning, many researchers in recent years have also started attempting to harness the power of optical devices.

In [137], bio-inspired angle-sensitive pixel (ASP) sensors were employed as both the sensor and the computing devices for the first convolutional layer. The ASP sensor was employed to perform optical convolution directly on the input. The main motivation is to move some computation efforts in front of ADCs in order to reduce the system power consumption, which is similar to the spirit of near-sensor processing discussed in Section 3.4.1.2. Since the ASP sensor extracts edges from the raw image, the effective data that need to be quantized by the ADC and fed into the rest of the system can be reduced significantly.

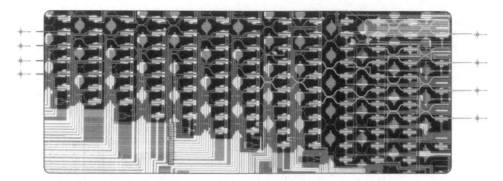

Figure 3.24 Illustration of the optical neural interface unit [138]. The portion of the chip that is responsible for matrix multiplication is highlighted with red color. The portion that is responsible for attenuating the signal is highlighted with blue color. Reproduced with permission of Springer.

More recently, an effort was made to extend the optical computing into more layers in a deep network and to make the parameters of the network programmable [138]. A neural network was implemented in the photonic circuit, as shown in Figure 3.24, where 56 programmable Mach-Zehnder interferometers (MZIs) are implemented on the chip. The weight matrix in a neural network is implemented with the help of singular value decomposition (SVD). There are two appealing features for this type of optical neural network. The first is that it can provide both a high throughput and a low latency. At a typical 100 GHz photodetection rate, an N-node optical neural network can provide a throughput of 10^{11} N-dimensional matrix–vector multiplications per second. The second feature is that the optical neural network can theoretically be low-power, as the matrix–vector multiplication, which is accomplished as the light traveling in the photonic circuit, is totally passive [138].

3.5 Case Study: An Energy-Efficient Accelerator for Adaptive Dynamic Programming

In this case study, a customized accelerator developed in [139] is demonstrated for the adaptive dynamic programming (ADP) algorithm presented in Section 2.2.4. Even though the ADP algorithm has gained its popularity in recent years for solving various real-life control and decision-making problems, the highly iterative algorithm running on general-purpose processors fails to provide energy-efficient solutions to many applications where power consumption is an important design consideration. Recently, more and more microrobots [140–143] and internet-of-things (IoT) devices [144–146] have been demonstrated. For these microrobots and IoT devices, which chiefly rely on energy scavenging from the environment or energy stored in a tiny battery, general-purpose processors might not be a feasible option. Consequently, there is a strong need to develop customized accelerators for the highly iterative and interactive ADP algorithm to facilitate its deployment in these low-power platforms.

As reviewed in Sections 3.3 and 3.4, many customized accelerators have been demonstrated recently, mainly for supervised-learning tasks. Despite the fact that both deep supervised-learning accelerators and ADP accelerators involve neural networks, certain unique challenges exist for the ADP accelerators. For instance, most existing accelerators only implement the function of inference, whereas the learning of the synaptic weights is assumed to occur offline. Such an assumption indeed works well for classification tasks where the energy- and time-consuming training can be accomplished by power-hungry GPUs with high-precision computational capability. The trained synaptic weights are then downloaded to the accelerator for classification purposes. Such an operating model, unfortunately, is unlikely to work for ADP accelerators. Most ADP algorithms aim at controlling plants or making optimal decisions in a dynamic environment. Consequently, an ADP accelerator needs to learn the optimal policy for the plant that it controls or for the environment that it interacts with in an online fashion. Therefore, learning in an ADP accelerator is likely to be a lifelong and real-time task. With such a demand in mind, a hardware architecture as well as design methodologies for ADHDP accelerators were introduced in [139], which help to conduct both inference and learning effectively.

The proposed architecture in [139] is flexible, scalable, and energy-efficient. The flexibility is supported by a set of customized instructions that can be employed to program the accelerator for different tasks and applications. To provide scalability, a tile-based computing strategy is adopted. Low-power operations are achieved through reducing the data movements by utilizing and partitioning data buffers. In addition, the virtual update technique, which is outlined in Section 2.2.4.3 to accelerate the learning process, is also supported in the architecture.

3.5.1   Hardware Architecture

The diagram of the overall architecture of the ADP accelerator is illustrated in Figure 3.25. The architecture is divided into three main parts, as shown in the figure: on-chip memory, datapath, and controller. The on-chip memory stores all the synaptic weights, neuron activations, and intermediate results. The datapath is the core of the ADP accelerator and handles all the arithmetic operations needed in an ADP algorithm. The controller oversees all the operations in the accelerator and is in charge of the instruction flow. In the following sections, we discuss these three partitions separately in more detail.

3.5.1.1   On-Chip Memory

Memory in our system physically consists of an SRAM array as well as registers. Logically, the on-chip memory can be further divided into synapse memory, neuron memory, input buffers, and scalar registers. The dense SRAM array is used to store synaptic weights of the neural networks because the number of synapses grows quadratically with the size of the neural network. The neuron activations are stored in an array of registers. To leverage the locality of the data, the frequently used data can be stored in the input buffers for reuse. In order to increase the throughput of the accelerator, the inherent parallelism in the neural network is exploited through adopting an SIMD architecture. In other words, for most instructions executed on the accelerator, the processed data is in vector form. Consequently, most storage units are arranged in vector form for ease of access. Nevertheless, the ADP algorithm does generate some intermediate scalar results in certain steps of the algorithm. Therefore, a bank of scalar register is also needed.

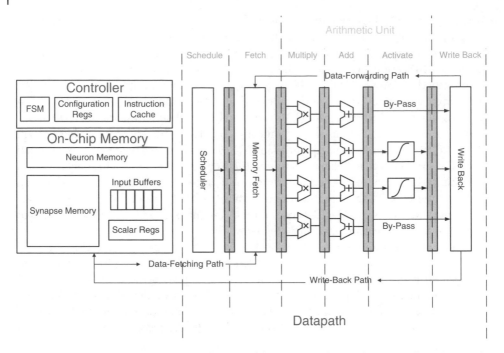

Figure 3.25 Hardware architecture for the ADP accelerator [139]. The accelerator can be partitioned into three parts: (i) on-chip memory that stores all the synaptic weights, neuron activations, and intermediate results, (ii) datapath that performs various arithmetic operations needed in the inference and learning phases, and (iii) controller that programs and oversees all the operations performed by the accelerator. Data-level parallelism is exploited through utilizing multiple datapath lanes. A reconfigurable five/six-stage pipeline is used for each lane. Reproduced with permission of IEEE.

As discussed in Section 3.3.1.1, one important strategy in building energy-efficient accelerators is to optimize the dataflow and memory access. Figure 3.26 illustrates the dataflow and memory access pattern employed in the proposed accelerators. The computations in the forward operations shown in Eqs. (2.13) to (2.17) in Chapter 2 are mostly matrix multiplication operations. Similar to many other machine-learning accelerators [11, 15], the tile-based matrix-multiplication strategy discussed in Section 3.3.1.1.1 is adopted by partitioning the matrix into several smaller blocks. A tile size of 4 is chosen because such a size is enough to accommodate all the targeted benchmark tasks. Nevertheless, larger tile sizes can be used by increasing the number of lanes in the datapath of the design in order to accommodate larger problems.

For the forward operation, multiplications are conducted sequentially for each row in the matrix. Since the column vector that consists of neuron activations needs to be reused several times, they are first loaded into the activation buffers. Doing so can avoid accessing the relatively large neuron memory repeatedly, thereby reducing energy consumption. The activation buffer is a circular buffer that rotates one circle for multiplying one matrix and one vector.

For the backward operation, two tasks, namely error backpropagation and weight update, are scheduled alternately. The error backpropagation operation is also a matrix–vector multiplication. The propagated error is stored in a linear buffer for data reuse. Similar to the forward operation, a tile-based multiplication is used. The

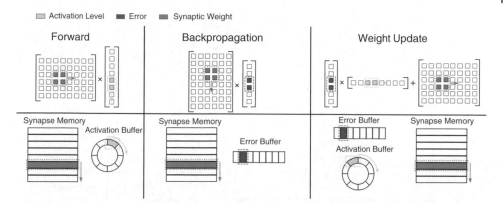

Figure 3.26 Illustration of the dataflow and memory access pattern in the ADP accelerator [139]. Data buffers are employed to exploit the locality of the data. Reproduced with permission of IEEE.

difference is that the multiplication in this case is done sequentially for each column in the matrix. The motivation for this arrangement is that the memory access pattern is always sequential for both forward and backward operation, as illustrated in Figure 3.26.

In the design, synaptic weights needed in one tile operation are stored in one memory word, so that they can be read or written together. Therefore, a sequential memory access pattern is helpful when off-chip memory is used. For the weight update operation, elements in the row vector are stored in the circular buffer, whereas elements associated with the column vector are stored in the error buffer. To reuse the same tile of synaptic weights, the error backpropagation and weight update are scheduled alternately. By doing so, the propagated error store in the error buffer can be reused efficiently as well.

3.5.1.2 Datapath

The datapath in the accelerator performs the arithmetic operations needed in a neural network. In addition, to increase the throughput of the system, the datapath is partitioned into six-stage reconfigurable pipelines: schedule, fetch, multiply, add, activate, and write back.

3.5.1.2.1 *Datapath Operations* The first stage in the datapath is to plan and schedule the operation. Instructions are fetched from the instruction memory in the controller. The decoded instructions are used as guidelines for following datapath stages. In addition to generating and latching the source addresses for the following fetch stage, the main responsibility of the scheduler is to detect any potential data hazard. Once a potential data hazard is detected, the scheduler looks for a possibility of conducting data forwarding directly, as shown in Figure 3.25. In the case that even data forwarding is not effective, a STALL operation is inserted in the pipe as a null operation in order to wait for the dependent data to be ready. Once an operation is scheduled, it flows through the datapath pipeline. In the fetch stage, source data are fetched and latched according to the addresses latched in the schedule phase.

The arithmetic unit, which consists of multiply, add, and activate stages, is the core for conducting inference and learning in the neural network. To reuse the same hardware for different operations such as forward, backpropagation, and weight update, the adders in the add stage can be configured as parallel adders, adder trees, or a mix of both.

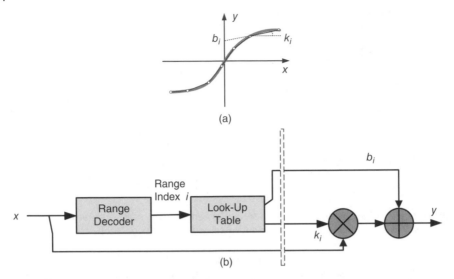

(a)

(b)

Figure 3.27 (a) Piecewise linear approximation of the hyperbolic tangent activation function. (b) Circuit diagram implementing the approximate activation function [154]. Reproduced with permission of IEEE.

For the activate stage, the hyperbolic tangent function is selected, as it is the most popular choice for the ADP algorithm in the literature [147–152]. In the proposed design, the activation function is implemented with a piecewise linear interpolation, similar to those employed in [14] and [153]. The operational principle of the activate stage is illustrated in Figure 3.27 [154]. The activation function is first approximated by a piecewise linear function. A set of coefficients k_i and b_i are obtained from the approximation and are stored in a look-up table. During the activation operation, the corresponding coefficients are read out and are multiplied and added to form the activation level. To reduce the critical-path length, the readout of the coefficients is actually implemented in the preceding add stage to balance the delay. Depending on the operation conducted, the activation stage may be bypassed, as the activation operation is only needed in the forward phase. In this case, the six-stage pipeline is reduced to a five-stage pipeline. The last write-back stage is in charge of writing the computed results back to the storage units.

3.5.1.2.2 Datapath Quantization As discussed in Section 3.3.1.2, one important consideration in designing customized accelerators is the choice of bit width used to represent data in the system. It is very popular to use a fixed-point number representation in machine-learning accelerators [11, 14, 18, 30–32, 47] because of its ease of implementation and good computational efficiency.

Three popular benchmark tasks presented in Section 2.4, the cart-pole balancing problem, the beam-balancing problem, and the triple-link inverted pendulum problem, are employed to provide some guidelines in choosing the number of bits needed to represent data in the system. For each task, data used in the algorithm, including synaptic weights, neuron states, and other temporary results, are quantized to numbers with a fractional bit width of Q_f.

For each benchmark task, 50 runs of simulations are conducted, where a run of simulation contains several trials. The agent learns to perform the task in each trial, and

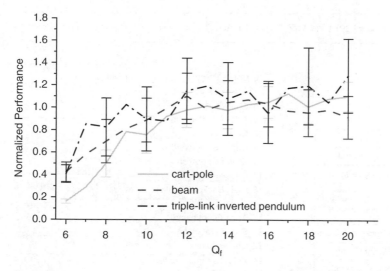

Figure 3.28 Comparison of the learning performance achieved with different levels of data quantization for three classic ADP benchmarks [139]. The obtained performances are normalized with respect to those obtained from double-precision floating-point computations. Reproduced with permission of IEEE.

a trial is a complete process from start to end. A trial may be terminated due to failure or when the maximum allowed time steps is reached. In the simulations, the performance of the learning is measured by the accumulated time that the agent is able to regulate the states of the plant within a desired bound. The obtained results, which are shown in Figure 3.28, are normalized with respect to the baseline performance that is achieved by computations with a double-precision floating-point number representation. Error bars in the figure are used to indicate a 95% confidence interval. Q_f represents the number of bits used to represent the fractional part of the data. From the figure, one can conclude that as the number of bits used to represent the information in the system increases, the performance achieved by the reduced-precision computation starts to match the baseline. In the figure, it is shown that a Q_f of 12 can result in a reasonably good performance. To leave some margins for the design, a 6-bit integer part (including a 1-bit sign information) and an 18-bit fractional part are used to represent data in our accelerators.

3.5.1.3 Controller

To provide more flexibility, a customized instruction set is developed for the ADP accelerator. Through programming the accelerator with the instructions provided, various tasks can be performed with different network configurations. The main role of the controller is to determine the instruction flow. The format of instructions is shown in Figure 3.29a.

The operation code field specifies the type of instruction. There are six basic types of instructions in our baseline accelerator, as shown in Figure 3.29b. To reduce the unnecessary control overhead, each instruction may contain the workload that spans multiple clock cycles. Conveniently, the operations associated with one matrix operation can be grouped together and performed with one instruction. The code "FF" represents the forward operation, which is the most common instruction. The code "SCA" is for scalar operation such as calculating the temporal difference. The code "BP_WU" is used for

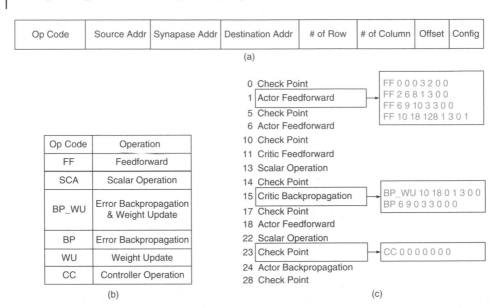

Op Code	Source Addr	Synapase Addr	Destination Addr	# of Row	# of Column	Offset	Config

(a)

Op Code	Operation
FF	Feedforward
SCA	Scalar Operation
BP_WU	Error Backpropagation & Weight Update
BP	Error Backpropagation
WU	Weight Update
CC	Controller Operation

(b)

0 Check Point
1 Actor Feedforward
5 Check Point
6 Actor Feedforward
10 Check Point
11 Critic Feedforward
13 Scalar Operation
14 Check Point
15 Critic Backpropagation
17 Check Point
18 Actor Feedforward
22 Scalar Operation
23 Check Point
24 Actor Backpropagation
28 Check Point

FF 0 0 0 3 2 0 0
FF 2 6 8 1 3 0 0
FF 6 9 10 3 3 0 0
FF 10 18 128 1 3 0 1

BP_WU 10 18 0 1 3 0 0
BP 6 9 0 3 3 0 0 0

CC 0 0 0 0 0 0 0

(c)

Figure 3.29 Illustration of the instructions used in the accelerators [139]. (a) Format of the instruction. (b) List of the main operation codes and their corresponding operations. (c) A sample program for implementing the ADHDP algorithm. Reproduced with permission of IEEE.

layers where both error backpropagation and weight update are needed, such as hidden layers. The code "BP" is used for error backpropagation, whereas "WU" is used for conducting a weight update. They are used in layers where only one operation is needed. For example, the "WU" code can be used for the input layer where error backpropagations are not needed. "BP" can be used when errors are propagating from the critic network to the actor network. In this case, the synaptic weights in the critic network do not need to be updated. The code "CC" calls for a controller operation. It can be used, for example, to implement the conditional jumps in the algorithm. The fields "Source Addr," "Synapse Addr," and "Destination Addr" are used to specify the addresses for the source data, the address for the synaptic weights, and the addresses to write back, respectively.

The synapse memory has its own address space, whereas all other registers share a unified address space. Such an arrangement is based on the consideration that off-chip memory might be used in the next generation of the accelerator. The "# of Row" and "# of Column" fields indicate the size of the matrix on which the accelerator currently operates. The field "Offset" specifies any offset in computing the matrix multiplication. For example, backpropagation for the actor network only needs to be done for $\mathbf{a}(t)$. Therefore, the elements associated with $\mathbf{x}(t)$ may be skipped. This can be easily done through specifying the "Offset" field in the instruction. Finally, the "Config" field is used for configuration purposes: for example, to specify whether the activation stage in the datapath should be bypassed.

The scheduler can schedule operations based on the information provided in the instruction. In the proposed architecture, one instruction specifies all operations conducted on one matrix. To illustrate how the customized instructions can be employed as well as the flexibility and reconfigurability of the accelerator, an example of the instructions corresponding to the pseudocode shown in Figure 2.13 in Chapter 2 is illustrated in Figure 3.29c. Only a portion of the instructions is shown for the purpose of brevity. All instructions in the figure correspond to a series of operations conducted by the datapath except for the "Check Point" operation, where the controller conducts a conditional jump with the help of a simple finite-state machine (FSM). Clearly, with the customized instruction set, the accelerator can be conveniently programmed to conduct various tasks with different network configurations.

In order to exploit the proposed virtual update technique, one extra instruction "VU" is added to the baseline instruction set. The newly added instruction implements the virtual update algorithm in two sets of operations. The first set of operations compute and store Λ_c when the current input vector is presented for the first time. The second set of operations are the multiply-and-add operations. The operation of accumulating $E_i(i_c)$ is merged to the normal "BP" or "BP_WU" operations without introducing any overheads in computing time. This is because the adders are idle anyway during the error backpropagation operations when the conventional update is employed.

The virtual update algorithm requires some memory spaces to hold the newly added intermediate results, $o_i^c(0)$ and $E_i(i_c)$. Based on the memory access pattern of the architecture, these two vectors can be stored conveniently in the synapse memory, noticing that the weight memory is not utilized during the virtual update operation. The memory overhead due to storing these two temporary terms is negligible, especially when the size of the neural network is large. Indeed, the size of these two terms scale linearly with the size of the network, whereas the memory requirement for the synaptic weights scales quadratically.

3.5.2 Design Examples

To provide a comparison between the conventional update and the virtual update, two accelerators are demonstrated in this section. The baseline accelerator conducts learning based on the conventional update. It employs the basic instruction set illustrated in Figure 3.29. The upgraded accelerator employs the virtual update technique. The augmented instruction set with the newly added "VU" instruction is employed in this accelerator. The designs are implemented in a 65-nm technology. Since two versions of accelerators have a similar layout, only the one with the virtual update is shown in Figure 3.30. As shown in the figure, the on-chip memory takes most of the area. The arithmetic unit is the second largest block. The controller and the scheduler take the rest of the area. The specifications of the accelerator equipped with the virtual update algorithm are shown in Table 3.2. The accelerators are designed to operate at 175 MHz. With such a clock frequency and the chosen level of parallelism, all the target benchmark tasks can be conducted in real time.

To examine the architecture and design techniques for the ADP accelerators, two simulators were implemented to model the accelerators. One simulator models the behavior of the accelerator chip by providing the same input–output mapping. This simulator is used to examine how well the ADP accelerator can perform various benchmark tasks under certain adaptations made in order to simplify the design. One example of the adaptations that were made is the reduced precision in the data. Another simulator is used to model the number of cycles needed to perform certain tasks. This simulator is used to simulate the throughput of the final accelerator. Other metrics such as area, speed, and power consumption are estimated based on the postlayout simulation.

550 μm

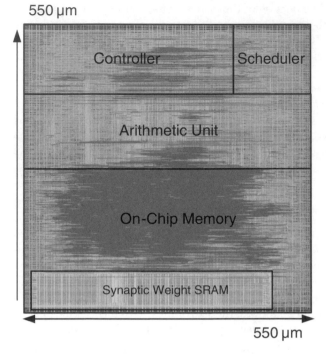

550 μm

Figure 3.30 Chip layout and floorplan of the accelerator chip with the virtual update algorithm [139]. Reproduced with permission of IEEE.

Table 3.2 Summary of Specifications of the ADHDP accelerator.

Technology	65 nm
Area	550 μm × 550 μm
Number of lanes	4
Arithmetic precision	24-bit fixed-point
Supply voltage	1.2 V
Clock frequency	175 MHz
Power consumption	25 mW

Source: data are from [139]. Reproduced with permission of IEEE.

The first thing that needs to be ensured is that the use of the accelerator does not degrade the performance of conducting various control tasks when it is compared with the software approach. The three popular benchmark tasks employed in Section 3.5.1.2.2 are used to test the performance of the accelerator in terms of performing control tasks. The same metric, accumulated time steps, is used for comparison here.

The obtained results are shown in Figure 3.31. In this figure, the accumulated time steps achieved by the accelerators are normalized with respect to that achieved by the software approach for ease of comparison. The results in the figure are generated by the behavior-level model of the chip. Mathematical models used for simulating these

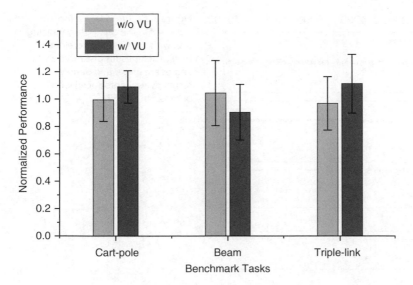

Figure 3.31 Comparison of learning performances achieved by the accelerators and the software approach for three commonly used benchmark tasks [139]. The results obtained from the accelerators are normalized to those obtained from software. Error bars correspond to a 95% confidence interval. Reproduced with permission of IEEE.

benchmark tasks can be found in [147, 148, 150, 151, 155, 156]. Clearly, the efficacy of the ADHDP algorithm is not degraded by the use of the designed accelerators.

One thing worth noting is that the performance of the baseline accelerator and the updated one are not identical. Ideally, since the virtual update algorithm does not use any approximation, the results produced by these two accelerators should be the same. However, due to the use of a fixed-point number representation, a slight difference exists between these two accelerators as a result of different quantization orders. Despite this slight difference, the two accelerators should be able to produce similar results, as shown in Figure 3.31.

To provide a closer look at how well the proposed accelerator performs on complex control problems, the transient waveforms of the states of the plants in the benchmark tasks are illustrated in Figures 3.32 to 3.34. For the first two tasks, four state variables are used as the input to the actor network: namely the offset x, the angle θ, and the corresponding velocities x' and θ'. The action in the cart-pole task is a binary quantity, which determines the polarity of the force. This binary action is obtained by hard-thresholding the output from the actor network. The action in the beam-balancing task, on the other hand, can vary continuously. The target of the cart-pole balancing task is to maintain the states x and θ within the range of $[-2.4\,\text{m}, 2.4\,\text{m}]$ and $[-12°, 12°]$, respectively. Similarly, the target of the beam-balancing task is to regulate the states such that x and θ can stay in the range of $[-0.48\,\text{m}, 0.48\,\text{m}]$ and $[-13.75°, 13.75°]$.

For a more complex triple-link task, x, $\theta_1 - \theta_3$ and their corresponding derivative x' and $\theta'_1 - \theta'_3$ are the eight state variables that are needed by the actor network, u is the applied control voltage, J is the estimated reward-to-go, $\sum |w|$ is the sum of absolute values of all the weights, and r is the reward signal, which is -1 if the states of the plant exceed the target range. The target of this task is to control x to be within the range of

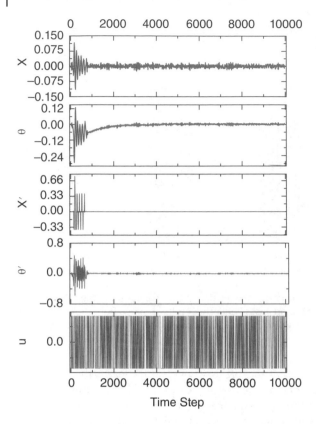

Figure 3.32 Typical waveforms obtained in the cart-pole balancing task with the baseline accelerator [154]. The units for distances and angles are meter and degree, respectively. Reproduced with permission of IEEE.

[−1 m, 1 m] and to maintain $\theta_1 − \theta_3$ in the range [−20°, 20°]. Another constraint is that the applied voltage should be bounded by ±30 V. This can be easily achieved by using a hyperbolic tangent function as the activation function for the output neuron in the actor network.

From Figures 3.32 to 3.34, it can be observed that the accelerator successfully learns an effective control policy and maintains the states of the system well within the target range. These results demonstrate that the ADP circuit can successfully learn how to regulate the states of the plant as desired.

To show how the virtual update algorithm can have an impact on the throughput and the energy efficiency of the accelerator, Figures 3.35 to 3.37 compare the throughput, the power consumption, and the energy efficiency of the two accelerators, respectively. To provide some insight into how the sizes of the neural network affect the level of improvement, three different neural networks are employed. In the figures, the results labeled "4-6-1, 5-6-1" and "4-10-1, 5-12-1" are obtained from the cart-pole task, where the two sets of numbers refer to the sizes of the critic and actor networks, respectively. For example, "4-6-1, 5-6-1" refer to an actor network with four input neurons, six hidden-layer neurons, and one output neuron and a critic network with five input neurons, six hidden-layer neurons and one output neuron. The results with the label "8-20-1, 9-20-1" are obtained from the triple-link inverted pendulum task.

The average number of clock cycles needed for each critic and actor update iteration is compared in Figure 3.35 for the two accelerators. Clearly, the forward and backward

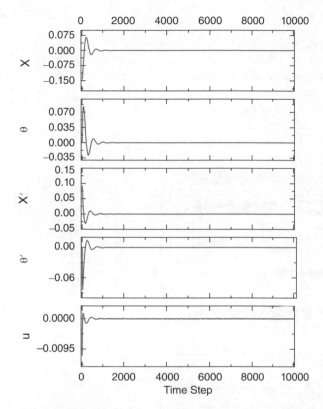

Figure 3.33 Typical waveforms obtained in the beam-balancing task with the baseline accelerator [154]. The units for distances and angles are meter and degree, respectively. Reproduced with permission of IEEE.

operations take most of the clock cycles, and the portions that these two operations occupy increase as the sizes of the neural network grow. The reason is that as the neural network becomes larger, the number of matrix–vector operations that can flow through the datapath without being interrupted by the control or branch operation increases. The reduced control overhead is one of the main reasons that the customized accelerator is able to speed up certain applications compared to the general-purpose processors where many cycles are wasted in the unnecessary and complex control flow. Thanks to the virtual update technique, the improvement in the throughput is observed for all three cases.

In addition, it is shown that the improvement is more obvious for larger networks. The main reason for the growing improvement is that the virtual update algorithm effectively replaces quadratically scaled operations with linearly scaled operations. Therefore, the savings in the number of clock cycles increases with the size of the problem. A 1.47× improvement in the throughput is achieved for the largest network size.

The power consumptions of the two accelerators are compared in Figure 3.36. The baseline accelerator has a slightly higher power consumption, as shown in the figure. The reduction in the power consumption is mainly because of the reduced memory operations, as illustrated by the shrunk green portion in Figure 3.36. It is also observed

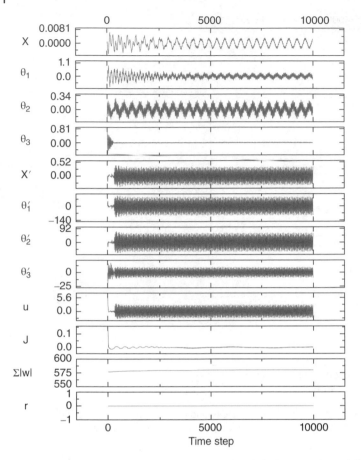

Figure 3.34 Typical waveforms obtained in the triple-link inverted pendulum task with the baseline accelerator [139]. The units for distances and angles are meter and degree, respectively. Reproduced with permission of IEEE.

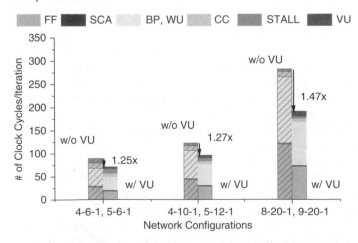

Figure 3.35 Comparison of the numbers of clock cycles needed for every critic/actor update iteration [139]. The first two groups of data are obtained from the cart-pole balancing task, whereas the third group of data are from the triple-link inverted pendulum task. Forward and backward operations consume most clock cycles. The overheads of scalar operations and control operations are quickly diluted as the sizes of the neural networks increase. Reproduced with permission of IEEE.

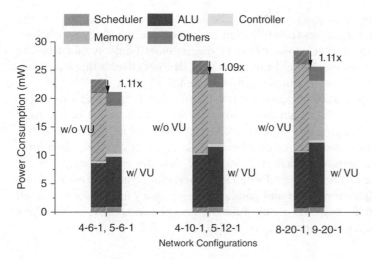

Figure 3.36 Comparison of the power consumption breakdown for the two accelerators [139]. The first two groups of data are obtained from the cart-pole balancing task, whereas the third group of data are from the triple-link inverted pendulum task. The arithmetic unit and the memory consume most of the power. The virtual update technique improves the power consumption slightly. Reproduced with permission of IEEE.

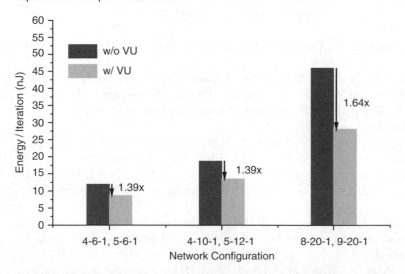

Figure 3.37 Comparison of the energy consumption for every critic/actor update iteration [139]. The first two groups of data are obtained from the cart-pole balancing task, whereas the third group of data are from the triple-link inverted pendulum task. The virtual update technique effectively improves the energy efficiency. The improvement is more significant for larger neural networks. Reproduced with permission of IEEE.

that the power consumption of the arithmetic unit is slightly increased. This trend can be attributed to two reasons. The first reason could be the additional multiplexers added in the arithmetic unit. In order to implement the operations associated with the virtual update algorithm, dataflow in the datapath needs to be altered slightly through the introduction of multiplexers to accommodate the change. This additional multiplexing logic induces a small overhead. The second reason could be that the virtual update algorithm increases the utilization of the arithmetic unit. Nevertheless, the slight increase in the power consumption of the arithmetic unit is offset by the decrease in the power consumption of the memory unit. The net effect is that the power consumed by the upgraded accelerator is slightly lowered.

By combining the advantages obtained from both the throughput and the power consumption, the virtual update algorithm boosts the energy efficiency of the ADP algorithm effectively, as illustrated in Figure 3.37. An improvement as high as 1.64× has been achieved for the most complicated benchmark task we employ in the simulations, the triple-link inverted-pendulum task. It is expected that the enhancement in the energy efficiency can be further increased when larger and more complicated problems are involved. Such a high throughput and good energy efficiency enable sophisticated ADP algorithms to be deployed in many different energy-constrained applications where optimal decision-making or control is needed.

References

1 Krizhevsky, A., Sutskever, I., and Hinton, G.E. (2012). Imagenet classification with deep convolutional neural networks. In: *Advances in Neural Information Processing Systems* (eds. F. Pereira, C. Burges, L. Bottom and K.Q. Weinberger), 1097–1105. Curran Associates.

2 Owens, J.D., Houston, M., Luebke, D. et al. (2008). GPU computing. *Proc. IEEE* 96 (5): 879–899.

3 Fatahalian, K., Sugerman, J., and Hanrahan, P. (2004). Understanding the efficiency of GPU algorithms for matrix-matrix multiplication. In: *Proceedings of the ACM SIGGRAPH/EUROGRAPHICS Conference on Graphics Hardware – HWWS '04*, 133–137. ACM.

4 Sze, V., Chen, Y.H., Yang, T.J., and Emer, J.S. (2017). Efficient processing of deep Neural networks: a tutorial and survey. *Proc. IEEE* 105 (12): 2295–2329.

5 Collobert, R., Kavukcuoglu, K., and Farabet, C. (2011, no. EPFL-CONF-192376). Torch7: a matlab-like environment for machine learning. In: *BigLearn, NIPS Workshop*, 1–6.

6 Jia, Y., Shelhamer, E., Donahue, J. et al. (2014). Caffe: convolutional architecture for fast feature embedding. In: *Proceedings of the 22nd ACM International Conference on Multimedia*, 675–678.

7 Abadi, M., Barham, P., Chen, J. et al. (2016). TensorFlow: a system for large-scale machine learning. In: *12th USENIX Symposium on Operating Systems Design and Implementation (OSDI 16)*, 265–283.

8 Paszke, A., Gross, S., Chintala, S., et al., "Automatic differentiation in pytorch," 2017.

9 Patterson, D.A. and Hennessy, J.L. (2013). *Computer Organization and Design MIPS Edition: The Hardware/Software Interface*. Newnes.

10 Bang, S., Wang, J., Li, Z. et al. (2017). 14.7 a 288µW programmable deep-learning processor with 270KB on-chip weight storage using non-uniform memory hierarchy for mobile intelligence. In: *2017 IEEE International Solid-State Circuits Conference (ISSCC)*, 250–251.

11 Chen, Y., Luo, T., Liu, S. et al. (2014). DaDianNao: a machine-learning supercomputer. In: *2014 47th Annual IEEE/ACM International Symposium on Microarchitecture*, 609–622.

12 Du, Z., Fasthuber, R., Chen, T. et al. (2015). ShiDianNao: shifting vision processing closer to the sensor. In: *Proceedings of the 42nd Annual International Symposium on Computer Architecture*, 92–104.

13 Liu, D., Chen, T., Liu, S. et al. (2015). PuDianNao: a polyvalent machine learning accelerator. In: *Proceedings of the Twentieth International Conference on Architectural Support for Programming Languages and Operating Systems*, 369–381.

14 Chen, T., Du, Z., Sun, N. et al. (2014). Diannao: a small-footprint high-throughput accelerator for ubiquitous machine-learning. In: *Proceedings of the 19th International Conference on Architectural Support for Programming Languages and Operating Systems*, 269–284.

15 Chen, Y., Chen, T., Xu, Z. et al. (2016). DianNao family: energy-efficient hardware accelerators for machine learning. *Commun. ACM* 59 (11): 105–112.

16 Wang, S., Zhou, D., Han, X., and Yoshimura, T. (2017). Chain-NN: an energy-efficient 1D chain architecture for accelerating deep convolutional neural networks. In: *2017 Design, Automation & Test in Europe Conference & Exhibition*, 1032–1037.

17 Jo, J., Cha, S., Rho, D., and Park, I. (2018). DSIP: a scalable inference Accelerator for convolutional Neural networks. *IEEE J. Solid-State Circuits* 53 (2): 605–618.

18 Chen, Y.H., Krishna, T., Emer, J.S., and Sze, V. (Jan. 2017). Eyeriss: an energy-efficient reconfigurable accelerator for deep convolutional neural networks. *IEEE J. Solid-State Circuits* 52 (1): 127–138.

19 Chen, Y.H., Emer, J., and Sze, V. (2016). Eyeriss: a spatial architecture for energy-efficient dataflow for convolutional neural networks. In: *Proceedings – 2016 43rd International Symposium on Computer Architecture, ISCA*, 367–379.

20 Jouppi, N.P., Young, C., Patil, N. et al. (2017). In-datacenter performance analysis of a tensor processing unit. In: *Proceedings of the 44th Annual International Symposium on Computer Architecture*, 1–12.

21 Yang, X., Gao, M., Po, J., et al., "DNN Dataflow Choice Is Overrated," *arXiv Prepr. arXiv1809.04070*, 2018.

22 Chakradhar, S., Sankaradas, M., Jakkula, V., and Cadambi, S. (2010). A dynamically configurable coprocessor for convolutional neural networks. In: *Proceedings of the 37th Annual International Symposium on Computer Architecture*, 247–257.

23 Park, S., Bong, K., Shin, D. et al. (2015). 4.6 A1.93TOPS/W scalable deep learning/inference processor with tetra-parallel MIMD architecture for big-data applications. In: *Digest of Technical Papers – IEEE International Solid-State Circuits Conference*, vol. 58, 80–81.

24 Peemen, M., Setio, A.A.A., Mesman, B., and Corporaal, H. (2013). Memory-centric accelerator design for convolutional neural networks. In: *2013 IEEE 31st International Conference on Computer Design, ICCD 2013*, 13–19.

25 Gao, M., Pu, J., Yang, X. et al. (2017). TETRIS: scalable and efficient neural network acceleration with 3D memory. In: *Proceedings of the Twenty-Second International Conference on Architectural Support for Programming Languages and Operating Systems*, 751–764.

26 Loh, G.H. (2008). 3D-stacked memory architectures for multi-core processors. In: *Proceedings of the 35th Annual International Symposium on Computer Architecture*, 453–464.

27 Kim, D., Kung, J., Chai, S. et al. (2016). Neurocube: a programmable digital neuromorphic architecture with high-density 3D memory. In: *Proceedings of the 43rd International Symposium on Computer Architecture*, 380–392.

28 Alwani, M., Chen, H., Ferdman, M., and Milder, P. (2016). Fused-layer CNN accelerators. In: *The 49th Annual IEEE/ACM International Symposium on Microarchitecture*, 22:1–22:12.

29 Judd, P., Albericio, J., Hetherington, T. et al. (2016, vol. 16, no. 1). Stripes: bit-serial deep neural network computing. In: *2016 49th Annual IEEE/ACM International Symposium on Microarchitecture (MICRO)*, 1–12.

30 Moons, B. and Verhelst, M. (2017). An energy-efficient precision-scalable ConvNet processor in 40-nm CMOS. *IEEE J. Solid-State Circuits* 52 (4): 903–914.

31 Reagen, B., Whatmough, P., Adolf, R. et al. (2016). Minerva: enabling low-power, highly-accurate deep neural network accelerators. In: *Proceedings of the 43rd International Symposium on Computer Architecture*, 267–278.

32 Shin, D., Lee, J., Lee, J., and Yoo, H.-J. (2017). 14.2 DNPU: an 8.1 TOPS/W reconfigurable CNN-RNN processor for general-purpose deep neural networks. In: *2017 IEEE International Solid-State Circuits Conference (ISSCC)*, 240–241.

33 Gysel, P., "Ristretto: Hardware-oriented approximation of convolutional neural networks," *arXiv Prepr. arXiv1604.03168*, 2016.

34 Kung, J., Kim, D., and Mukhopadhyay, S. (2015). A power-aware digital feedforward neural network platform with backpropagation driven approximate synapses. In: *2015 IEEE/ACM International Symposium on Low Power Electronics and Design (ISLPED)*, 85–90.

35 Ueyoshi, K., Ando, K., Hirose, K. et al. (2018). QUEST: A 7.49TOPS multi-purpose log-quantized DNN inference engine stacked on 96MB 3D SRAM using inductive-coupling technology in 40 nm CMOS. In: *2018 IEEE International Solid-State Circuits Conference (ISSCC)*, 216–218.

36 White, S.A. (1989). Applications of distributed arithmetic to digital signal processing: a tutorial review. *IEEE ASSP Mag.* 6 (3): 4–19.

37 Andri, R., Cavigelli, L., Rossi, D., and Benini, L. (2018). Yodann: an architecture for ultralow power binary-weight CNN acceleration. *IEEE Trans. Comput. Des. Integr. Circuits Syst.* 37 (1): 48–60.

38 Moons, B., Bankman, D., Yang, L. et al. (2018). BinarEye: an always-on energy-accuracy-scalable binary CNN processor with all memory on chip in 28nm CMOS. In: *2018 IEEE Custom Integrated Circuits Conference, CICC*, 1–4.

39 Ando, K., Ueyoshi, K., Orimo, K. et al. (2018). BRein memory: a single-chip binary/ternary reconfigurable in-memory deep Neural network accelerator achieving 1.4 TOPS at 0.6 W. *IEEE J. Solid-State Circuits* 53 (4): 983–994.

40 Venkataramani, S., Chakradhar, S.T., Roy, K., and Raghunathan, A. (2015). Approximate computing and the quest for computing efficiency. In: *Proceedings of the 52Nd Annual Design Automation Conference*, 120:1–120:6.

41 Han, J. and Orshansky, M. (2013). Approximate computing: an emerging paradigm for energy-efficient design. In: *2013 18th IEEE European Test Symposium (ETS)*, 1–6.

42 Venkataramani, S., Ranjan, A., Roy, K., and Raghunathan, A. (2014). AxNN: energy-efficient neuromorphic systems using approximate computing. In: *Proceedings of the 2014 International Symposium on Low Power Electronics and Design*, 27–32.

43 Lee, J., Kim, C., Kang, S. et al. (2018). UNPU: a 50.6TOPS/W unified deep neural network accelerator with 1b-to-16b fully-variable weight bit-precision. In: *2018 IEEE International Solid – State Circuits Conference – (ISSCC)*, 218–220.

44 Sharify, S., Lascorz, A.D., Siu, K. et al. (2018). Loom: exploiting weight and activation precisions to accelerate convolutional Neural networks. In: *Proceedings of the 55th Annual Design Automation Conference*, 20:1–20:6.

45 Moons, B., Uytterhoeven, R., Dehaene, W., and Verhelst, M. (2017). 14.5 Envision: a 0.26-to-10TOPS/W subword-parallel dynamic-voltage-accuracy-frequency-scalable convolutional neural network processor in 28nm FDSOI. In: *2017 IEEE International Solid-State Circuits Conference (ISSCC)*, 246–247.

46 Sharma, H., Park, J., Suda, N. et al. (2018). Bit fusion: bit-level dynamically composable architecture for accelerating deep neural network. In: *2018 ACM/IEEE 45th Annual International Symposium on Computer Architecture (ISCA)*, 764–775.

47 Han, S., Liu, X., Mao, H. et al. (2016). EIE: efficient inference engine on compressed deep neural network. In: *Proceedings of the 43rd International Symposium on Computer Architecture*, 243–254.

48 Vuduc, R.W. and Demmel, J.W. (2003). *Automatic Performance Tuning of Sparse Matrix Kernels*, vol. 1. University of California, Berkeley.

49 Zhang, S., Du, Z., Zhang, L. et al. (2016). Cambricon-X: an accelerator for sparse neural networks. In: *The 49th Annual IEEE/ACM International Symposium on Microarchitecture*, 20:1–20:12.

50 Albericio, J., Judd, P., Hetherington, T. et al. (2016). Cnvlutin: ineffectual-neuron-free deep neural network computing. In: *Proceedings of the 43rd International Symposium on Computer Architecture*, 1–13.

51 Kim, D., Ahn, J., and Yoo, S. (2017). A novel zero weight/activation-aware hardware architecture of convolutional neural network. In: *2017 Design, Automation & Test in Europe Conference & Exhibition*, 1462–1467.

52 Parashar, A., Rhu, M., Mukkara, A. et al. (2017). SCNN: an Accelerator for compressed-sparse convolutional neural networks. In: *Proceedings of the 44th Annual International Symposium on Computer Architecture*, 27–40.

53 Albericio, J., Judd, P., Delmás, A. et al. (2016). Bit-pragmatic deep neural network computing. In: *Proceedings of the 50th Annual IEEE/ACM International Symposium on Microarchitecture*, 382–394.

54 Akhlaghi, V., Yazdanbakhsh, A., Samadi, K. et al. (2018). SnaPEA: predictive early activation for reducing computation in deep convolutional neural networks. In: *2018 ACM/IEEE 45th Annual International Symposium on Computer Architecture (ISCA)*, 662–673.

55 Zhu, J., Jiang, J., Chen, X., and Tsui, C. (2018). SparseNN: an energy-efficient neural network accelerator exploiting input and output sparsity. In: *2018 Design, Automation & Test in Europe Conference & Exhibition*, 241–244.

56 Zhu, J., Qian, Z., and Tsui, C.-Y. (2016). LRADNN: high-throughput and energy-efficient deep neural network accelerator using low rank approximation.

In: *2016 21st Asia and South Pacific Design Automation Conference (ASP-DAC)*, 581–586.

57 Davis, A., and Arel, I., "Low-rank approximations for conditional feedforward computation in deep neural networks," *arXiv Prepr. arXiv1312.4461*, 2013.

58 Maeda, Y. and Tada, T. (2003). FPGA implementation of a pulse density neural network with learning ability using simultaneous perturbation. *IEEE Trans. Neural Networks* 14 (3): 688–695.

59 Cox, C.E. and Blanz, W.E. (1992). GANGLION – a fast field-programmable gate array implementation of a connectionist classifier. *IEEE J. Solid-State Circuits* 27 (3): 288–299.

60 Zhu, J. and Sutton, P. (2003). FPGA implementations of neural networks – a survey of a decade of progress. In: *International Conference on Field Programmable Logic and Applications*, 1062–1066.

61 Zhang, C., Li, P., Sun, G. et al. (2015). Optimizing FPGA-based accelerator design for deep convolutional neural networks. In: *Proceedings of the 2015 ACM/SIGDA International Symposium on Field-Programmable Gate Arrays – FPGA '15*, 161–170.

62 Li, G., Li, F., Zhao, T., and Cheng, J. (2018). Block convolution: towards memory-efficient inference of large-scale CNNs on FPGA. In: *2018 Design, Automation & Test in Europe Conference & Exhibition*, 1163–1166.

63 Xiao, Q., Liang, Y., Lu, L. et al. (2017). Exploring heterogeneous algorithms for accelerating deep convolutional neural networks on FPGAs. In: *Proceedings of the 54th Annual Design Automation Conference 2017*, 62:1–62:6.

64 Li, Y., Liu, Z., Xu, K. et al. (2017). A 7.663-TOPS 8.2-W energy-efficient FPGA accelerator for binary convolutional neural networks. In: *Proceedings of the 2017 ACM/SIGDA International Symposium on Field-Programmable Gate Arrays*, 290–291.

65 Qiu, J., Wang, J., Yao, S. et al. (2016). Going deeper with embedded FPGA platform for convolutional neural network. In: *Proceedings of the 2016 ACM/SIGDA International Symposium on Field-Programmable Gate Arrays – FPGA '16*, 26–35.

66 Aimar, A., Mostafa, H., Calabrese, E. et al. (2018). NullHop: a flexible convolutional neural network accelerator based on sparse representations of feature maps. *IEEE Trans. Neural Netw. Learn. Syst.*: 1–13.

67 Han, S., Kang, J., Mao, H. et al. (2017). ESE: efficient speech recognition engine with sparse LSTM on FPGA. In: *Proceedings of the 2017 ACM/SIGDA International Symposium on Field-Programmable Gate Arrays*, 75–84.

68 Motamedi, M., Gysel, P., Akella, V., and Ghiasi, S. (2016). Design space exploration of FPGA-based deep convolutional Neural networks. In: *2016 21st Asia and South Pacific Design Automation Conference (ASP-DAC)*, 575–580.

69 Sharma, H., Park, J., Mahajan, D. et al. (2016). From high-level deep neural models to FPGAs. In: *The 49th Annual IEEE/ACM International Symposium on Microarchitecture*, 17:1–17:12.

70 Venieris, S.I. and Bouganis, C.-S. (2017). fpgaConvNet: automated mapping of convolutional neural networks on FPGAs. In: *Proceedings of the 2017 ACM/SIGDA International Symposium on Field-Programmable Gate Arrays*, 291–292.

71 Wang, Y., Xu, J., Han, Y. et al. (2016). DeepBurning: automatic generation of FPGA-based learning accelerators for the neural network family. In: *Proceedings of the 53rd Annual Design Automation Conference*, 110:1–110:6.

72 Nane, R., Sima, V.M., Pilato, C. et al. (2016). A survey and evaluation of FPGA high-level synthesis tools. *IEEE Trans. Comput. Des. Integr. Circuits Syst.* 35 (10): 1591–1604.

73 Ovtcharov, K., Ruwase, O., Kim, J., et al., "Accelerating deep convolutional neural networks using specialized hardware," Microsoft Whitepaper, vol. 2, no. 11, pp. 3–6, 2015.

74 Putnam, A., Caulfield, A.M., Chumg, E.S. et al. (2014). A reconfigurable fabric for accelerating large-scale datacenter services. In: *2014 ACM/IEEE 41st International Symposium on, Computer Architecture (ISCA)*, 13–24.

75 Chung, E., Flowers, J., Ovtcharov, K. et al. (2018). Serving DNNs in real time at datacenter scale with project brainwave. *IEEE Micro* 38 (2): 8–20.

76 Fowers, J., Ovtcharov, K., Papamichael, M. et al. (2018). A configurable cloud-scale DNN processor for real-time AI. In: *Proceedings of the 45th Annual International Symposium on Computer Architecture*, 1–14.

77 Almeida, A.P. and Franca, J.E. (1996). Digitally programmable analog building blocks for the implementation of artificial neural networks. *IEEE Trans. Neural Netw.* 7 (2): 506–514.

78 Satyanarayana, S., Tsividis, Y.P., and Graf, H.P. (1990). A reconfigurable analog VLSI neural network chip. In: *Advances in Neural Information Processing Systems*, 758–768.

79 Verleysen, M., Thissen, P., Voz, J.-L., and Madrenas, J. (1994). An analog processor architecture for a neural network classifier. *IEEE Micro* 14 (3): 16–28.

80 Holler, M., Tam, S., Castro, H., and Benson, R. (1989). An electrically trainable artificial neural network (ETANN) with 10240 'floating gate' synapses. In: *International 1989 Joint Conference on Neural Networks*, vol. 2, 191–196.

81 Khwa, W.S., Chen, J.J., Li, J.F. et al. (2018). A 65 nm 4 kb algorithm-dependent computing-in-memory SRAM unit-macro with 2.3 ns and 55.8 TOPS/W fully parallel product-sum operation for binary DNN edge processors. In: *2018 IEEE International Solid – State Circuits Conference – (ISSCC)*, vol. 61, 496–498.

82 Zhang, J., Wang, Z., and Verma, N. (2017). In-memory computation of a machine-learning classifier in a standard 6T SRAM Array. *IEEE J. Solid-State Circuits* 52 (4): 915–924.

83 Biswas, A. and Chandrakasan, A.P. (2018). Conv-RAM: an energy-efficient SRAM with embedded convolution computation for low-power CNN-based machine learning applications. In: *2018 IEEE International Solid – State Circuits Conference – (ISSCC)*, 488–490.

84 Kang, M., Gonugondla, S.K., Keel, M.-S., and Shanbhag, N.R. (2015). An energy-efficient memory-based high-throughput VLSI architecture for convolutional networks. In: *2015 IEEE International Conference on Acoustics, Speech and Signal Processing (ICASSP)*, 1037–1041.

85 Kang, M., Lim, S., Gonugondla, S., and Shanbhag, N.R. (2018). An in-memory VLSI architecture for convolutional neural networks. *IEEE J. Emerg. Sel. Top. Circuits Syst.* 8 (3): 494–505.

86 Merrikh-Bayat, F., Guo, X., Klachko, M. et al. (2018). High-performance mixed-signal neurocomputing with nanoscale floating-gate memory cell arrays. *IEEE Trans. Neural Netw. Learn. Syst.* 29 (10): 4782–4790.

87 Chen, Y., Yao, E., and Basu, A. (2015). A 128 channel extreme learning machine based neural decoder for brain machine interfaces. *IEEE Trans. Biomed. Circuits Syst.* 10 (3): 679–692.

88 Lu, J., Young, S., Arel, I., and Holleman, J. (2015). A 1 TOPS/W analog deep machine-learning engine with floating-gate storage in 0.13 μm CMOS. *IEEE J. Solid-State Circuits* 50 (1): 270–281.

89 Huang, G.B., Zhu, Q.Y., and Siew, C.K. (2004). Extreme learning machine: a new learning scheme of feedforward neural networks. In: *2004 IEEE International Joint Conference on Neural Networks*, vol. 2, 985–990. vols.2-985–990 2.

90 Huang, G.B., Zhu, Q.Y., and Siew, C.K. (2006). Extreme learning machine: theory and applications. *Neurocomputing* 70 (1): 489–501.

91 Likamwa, R., Hou, Y., Gao, Y. et al. (2016). RedEye: analog ConvNet image sensor architecture for continuous mobile vision. In: *Proceedings of the 43rd International Symposium on Computer Architecture*, 255–266.

92 Anvesha, A. and Raychowdhury, A. (2017). A 65 nm 376 nA 0.4 V linear classifier using time-based matrix-multiplying ADC with non-linearity aware training. In: *2017 IEEE Asian Solid-State Circuits Conference (A-SSCC)*, 309–312.

93 Lee, E.H. and Wong, S.S. (2016). 24.2 A 2.5 GHz 7.7 TOPS/W switched-capacitor matrix multiplier with co-designed local memory in 40 nm. In: *2016 IEEE International Solid-State Circuits Conference (ISSCC)*, 418–419.

94 Wang, Z., Zhang, J., and Verma, N. (2015). Realizing low-energy classification systems by implementing matrix multiplication directly within an ADC. *IEEE Trans. Biomed. Circuits Syst.* 9 (6): 825–837.

95 Moons, B., Bankman, D., Yang, L. et al. (2018). An always-on 3.8μJ/86% CIFAR-10 mixed-signal binary CNN processor with all memory on chip in 28nm CMOS. In: *2018 IEEE International Solid – State Circuits Conference – (ISSCC)*, 222–224.

96 Aurangozeb, Hossain, A.D., Ni, C. et al. (2017). Time-domain arithmetic logic unit with built-in interconnect. *IEEE Trans. Very Large Scale Integr. Syst.* 25 (10): 2828–2841.

97 Iwata, A., Sakimura, N., Nagata, M., and Morie, T. (1998). An architecture of delta-sigma A-to-D converters using a voltage controlled oscillator as a multi-bit quantizer. In: *ISCAS '98. Proceedings of the 1998 IEEE International Symposium on Circuits and Systems (Cat. No.98CH36187)*, vol. 1, 389–392.

98 Miyashita, D., Yamaki, R., Hashiyoshi, K. et al. (2014). An LDPC decoder with time-domain analog and digital mixed-signal processing. *IEEE J. Solid-State Circuits* 49 (1): 73–83.

99 Strukov, D.B., Snider, G.S., Stewart, D.R., and Williams, R.S. (2008). The missing memristor found. *Nature* 453 (7191): 80–83.

100 Chua, L. (1971). Memristor-the missing circuit element. *IEEE Trans. Circuit Theory* 18 (5): 507–519.

101 Kavehei, O., Iqbal, A., Kim, Y.-S. et al. (2010). The fourth element: characteristics, modelling and electromagnetic theory of the memristor. *Proc. R. Soc. London A Math. Phys. Eng. Sci.* 466 (2120): 2175–2202.

102 Chanthbouala, A., Garcia, V., Cheriff, R.O. et al. (2012). A ferroelectric memristor. *Nat. Mater.* 11 (10): 860–864.

103 Johnson, S.L., Sundararajan, A., Hunley, D.P., and Strachan, D.R. (2010). Memristive switching of single-component metallic nanowires. *Nanotechnology* 21 (12): 125204.

104 Yang, J.J., Pickett, M.D., Li, X. et al. (2008). Memristive switching mechanism for metal/oxide/metal nanodevices. *Nat. Nano.* 3 (7): 429–433.

105 Yu, S. (2018). Neuro-inspired computing with emerging nonvolatile memorys. *Proc. IEEE* 106 (2): 260–285.

106 Burr, G.W., Shelby, R.M., Sebastian, A. et al. (2017). Neuromorphic computing using non-volatile memory. *Adv. Phys. X* 2 (1): 89–124.

107 Alibart, F., Gao, L., Hoskins, B.D., and Strukov, D.B. (2012). High precision tuning of state for memristive devices by adaptable variation-tolerant algorithm. *Nanotechnol.* 23 (7): 75201.

108 Gao, L., Chen, P., and Yu, S. (2015). Programming protocol optimization for analog weight tuning in resistive memories. *IEEE Electron Device Lett.* 36 (11): 1157–1159.

109 Hu, M., Strachan, J.P., Li, Z. et al. (2016). Dot-product engine for neuromorphic computing: programming 1T1M crossbar to accelerate matrix-vector multiplication. In: *Proceedings of the 53rd Annual Design Automation Conference*, 19.

110 Xia, L., Gu, P., Li, B. et al. (2016). Technological exploration of RRAM crossbar array for matrix-vector multiplication. *J. Comput. Sci. Technol.* 31 (1): 3–19.

111 Liu, X., Mao, M., Liu, B. et al. (2016). Harmonica: a framework of heterogeneous computing systems with memristor-based neuromorphic computing accelerators. *IEEE Trans. Circuits Syst. I Regul. Pap.* 63 (5): 617–628.

112 Neural, C. and Accelerator, N. (2016). ISAAC: a convolutional neural network accelerator with in-situ analog arithmetic in crossbars. In: *Proceedings of the 43rd International Symposium on Computer Architecture*, 14–26.

113 Chi, P., Li, S., Xu, C. et al. (2016). PRIME: a novel processing-in-memory architecture for neural network computation in ReRAM-based main memory. In: *Proceedings of the 43rd International Symposium on Computer Architecture*, 27–39.

114 Su, F., Chen, W.H., Xia, L. et al. (2017). A 462GOPs/J RRAM-based nonvolatile intelligent processor for energy harvesting IoE system featuring nonvolatile logics and processing-in-memory. In: *2017 Symposium on VLSI Circuits*, C260–C261.

115 Tang, T., Xia, L., Li, B. et al. (2017). Binary convolutional neural network on RRAM. In: *Design Automation Conference (ASP-DAC), 2017 22nd Asia and South Pacific*, 782–787.

116 Sun, X., Peng, X., Chen, P. et al. (2018). Fully parallel RRAM synaptic array for implementing binary neural network with (+1, −1) weights and (+1, 0) neurons. In: *2018 23rd Asia and South Pacific Design Automation Conference (ASP-DAC)*, 574–579.

117 Yu, S., Li, Z., Chen, P.Y. et al. (2016). Binary neural network with 16 Mb RRAM macro chip for classification and online training. In: *2016 IEEE International Electron Devices Meeting (IEDM)*, 16.2.1–16.2.4.

118 Chen, W.H., Li, K.X., Lin, W.Y. et al. (2018). A 65nm 1Mb nonvolatile computing-in-memory ReRAM macro with sub-16ns multiply-and-accumulate for binary DNN AI edge processors. In: *2018 IEEE International Solid – State Circuits Conference – (ISSCC)*, vol. 61, 494–496.

119 Li, B., Xia, L., Gu, P. et al. (2015). MErging the interface: power, area and accuracy co-optimization for RRAM crossbar-based mixed-signal computing system. In: *2015 52nd ACM/EDAC/IEEE Design Automation Conference (DAC)*, 1–6.

120 Song, L., Qian, X., Li, H., and Chen, Y. (2017). PipeLayer: a pipelined ReRAM-based accelerator for deep learning. In: *2017 IEEE International Symposium on High Performance Computer Architecture (HPCA)*, 541–552.

121 Querlioz, D., Bichler, O., Dollfus, P., and Gamrat, C. (2013). Immunity to device variations in a spiking neural network with memristive nanodevices. *IEEE Trans. Nanotechnol.* 12 (3): 288–295.

122 Burr, G.W., Shelby, R.M., Sidler, S. et al. (2015). Experimental demonstration and tolerancing of a large-scale neural network (165 000 synapses) using phase-change memory as the synaptic weight element. *IEEE Trans. Electron Devices* 62 (11): 3498–3507.

123 Eryilmaz, S.B., Kuzum, D., Jeyasingh, R.G. et al. (2013). Experimental demonstration of array-level learning with phase change synaptic devices. In: *Electron Devices Meeting (IEDM), 2013 IEEE International*, 25.

124 Liu, B., Hu, M., Li, H. et al. (2013). Digital-assisted noise-eliminating training for memristor crossbar-based analog neuromorphic computing engine. In: *Proceedings of the 50th Annual Design Automation Conference*, 7:1–7:6.

125 Narayanan, P., Fumarola, A., Sanches, L.L. et al. (2017). Toward on-chip acceleration of the backpropagation algorithm using nonvolatile memory. *IBM J. Res. Dev.* 61 (4/5): 11:1–11:11.

126 Cai, Y., Tang, T., Xia, L. et al. (2018). Training low bitwidth convolutional neural network on RRAM. In: *Proceedings of the 23rd Asia and South Pacific Design Automation Conference*, 117–122.

127 Sidler, S., Boybat, I., Shelby, R.M. et al. (2016). Large-scale neural networks implemented with non-volatile memory as the synaptic weight element: impact of conductance response. In: *2016 46th European Solid-State Device Research Conference (ESSDERC)*, 440–443.

128 Cheng, M., Xia, L., Zhu, Z. et al. (2017). TIME: a training-in-memory architecture for memristor-based deep neural networks. In: *Proceedings of the 54th Annual Design Automation Conference 2017*, 26:1–26:6.

129 Cai, Y., Lin, Y., Xia, L. et al. (2018). Long live TIME: improving lifetime for training-in-memory engines by structured gradient sparsification. In: *Proceedings of the 55th Annual Design Automation Conference*, 107:1–107:6.

130 van de Burgt, Y., Lubberman, E., Fuller, E.J. et al. (2017). A non-volatile organic electrochemical device as a low-voltage artificial synapse for neuromorphic computing. *Nat. Mater.* 16 (4): 414.

131 Adam, G.C., Hoskins, B.D., Prezioso, M. et al. (2017). 3-D memristor crossbars for analog and neuromorphic computing applications. *IEEE Trans. Electron Devices* 64 (1): 312–318.

132 Fuller, E.J., Gabaly, F.E., Léonard, F. et al. (2016). Li-ion synaptic transistor for low power analog computing. *Adv. Mater.* 29, no. SAND-2017-0895J.

133 Prezioso, M., Merrikh-Bayat, F., Hoskins, B.D. et al. (2015). Training and operation of an integrated neuromorphic network based on metal-oxide memristors. *Nature* 521 (7550): 61–64.

134 Fisher, A.D., Lippincott, W.L., and Lee, J.N. (1987). Optical implementations of associative networks with versatile adaptive learning capabilities. *Appl. Opt.* 26 (23): 5039–5054.

135 Farhat, N.H. (1987). Optoelectronic analogs of self-programming neural nets: architectures and methods for implementing fast stochastic learning by simulated annealing. *Appl. Opt.* 26 (23): 5093–5103.

136 Yang, L., Zhang, L., and Ji, R. (2013). On-chip optical matrix-vector multiplier. *Proc.SPIE* 8855, pp. 8855–8855–5.

137 Chen, H., Jayasuriya, S., Yang, J. et al. (2016). ASP vision: optically computing the first layer of convolutional Neural networks using angle sensitive pixels. In: *Proceedings of the IEEE Conference on Computer Vision and Pattern Recognition*, 903–912.

138 Shen, Y., Harris, N.C., Skirlo, S. et al. (2016). Deep learning with coherent nanophotonic circuits. *Nat. Photonics* 11 (7): 441–446.

139 Zheng, N. and Mazumder, P. (2018). A scalable low-power reconfigurable accelerator for action-dependent heuristic dynamic programming. *IEEE Trans. Circuits Syst. I Regul. Pap.* 65 (6): 1897–1908.

140 Hu, D., Zhang, X., Xu, Z. et al. (2014). Digital implementation of a spiking neural network (SNN) capable of spike-timing-dependent plasticity (STDP) learning. In: *14th IEEE International Conference on Nanotechnology, IEEE-NANO 2014*, 873–876.

141 Wood, R.J. (2008). The first takeoff of a biologically inspired at-scale robotic insect. *IEEE Trans. Robot.* 24 (2): 341–347.

142 Mazumder, P., Hu, D., Ebong, I. et al. (2016). Digital implementation of a virtual insect trained by spike-timing dependent plasticity. *Integr. VLSI J.* 54: 109–117.

143 Pérez-Arancibia, N.O., Ma, K.Y., Galloway, K.C. et al. (2011). First controlled vertical flight of a biologically inspired microrobot. *Bioinspiration Biomimetics* 6 (3): 036009.

144 Lee, I., Kim, G., Bang, S. et al. (2015). System-on-mud: ultra-low power oceanic sensing platform powered by small-scale benthic microbial fuel cells. *IEEE Trans. Circuits Syst. I Regul. Pap.* 62 (4): 1126–1135.

145 Lee, Y., Bang, S., Lee, I. et al. (2013). A modular $1\,mm^3$ die-stacked sensing platform with low power I2C inter-die communication and multi-modal energy harvesting. *IEEE J. Solid-State Circuits* 48 (1): 229–243.

146 Chen, Y.P., Jeon, D., Lee, Y. et al. (2015). An injectable 64 nW ECG mixed-signal SoC in 65 nm for arrhythmia monitoring. *IEEE J. Solid-State Circuits* 50 (1): 375–390.

147 Si, J. and Wang, Y.T. (2001). On-line learning control by association and reinforcement. *IEEE Trans. Neural Netw.* 12 (2): 264–276.

148 Liu, D., Xiong, X., and Zhang, Y. (2001). Action-dependent adaptive critic designs. In: *International Joint Conference on Neural Networks, 2001. Proceedings. IJCNN'01*, vol. 2, 990–995.

149 Liu, F., Sun, J., Si, J. et al. (2012). A boundedness result for the direct heuristic dynamic programming. *Neural Netw.* 32: 229–235.

150 He, H., Ni, Z., and Fu, J. (2012). A three-network architecture for on-line learning and optimization based on adaptive dynamic programming. *Neurocomputing* 78 (1): 3–13.

151 Sokolov, Y., Kozma, R., Werbos, L.D., and Werbos, P.J. (2015). Complete stability analysis of a heuristic approximate dynamic programming control design. *Automatica* 59: 9–18.

152 Mu, C., Ni, Z., Sun, C., and He, H. (2017). Air-breathing hypersonic vehicle tracking control based on adaptive dynamic programming. *IEEE Trans. Neural Netw. Learn. Syst.* 28 (3): 584–598.

153 Larkin, D., Kinane, A., Muresan, V., and O'Connor, N.E. (2006). An efficient hardware architecture for a neural network activation function generator. *Adv. Neural Netw.* 3973: 1319–1327.

154 Zheng, N. and Mazumder, P. (2018). A low-power circuit for adaptive dynamic programming. In: *2018 31st International Conference on VLSI Design and 2018 17th International Conference on Embedded Systems (VLSID)*, 192–197.

155 Ni, Z., He, H., and Wen, J. (2013). Adaptive learning in tracking control based on the dual critic network design. *IEEE Trans. Neural Netw. Learn. Syst.* 24 (6): 913–928.

156 Ni, Z., He, H., Zhong, X., and Prokhorov, D.V. (2015). Model-free dual heuristic dynamic programming. *IEEE Trans. Neural Netw. Learn. Syst.* 26 (8): 1834–1839.

4

Operational Principles and Learning in Spiking Neural Networks

I never teach my pupils, I only provide the conditions in which they can learn.
— Albert Einstein

4.1 Spiking Neural Networks

Spiking neural networks (SNNs) were inspired by biological neural networks [1, 2]. The study of SNNs can be traced back to long before artificial neural networks (ANNs) were widely known. The original research on SNNs aimed to model biological neural networks [3–6]. In recent years, there has been more and more interest in using SNNs for the purposes of computing, led by pioneers in these fields, such as Gerstner [7–9], Maass [1, 10, 11], etc. Even though the early development of ANNs was inspired by SNNs, SNNs differ significantly from ANNs, which were discussed in Chapters 2 and 3 in the following ways:

 (i) The ways in which information is encoded in SNNs and ANNs are different. A non-spiking neuron uses real-value activations to convey information, whereas a spiking neuron modulates information on spikes.
 (ii) A non-spiking neuron in an ANN does not have any memory, yet spiking neurons typically have memory.
(iii) The output generated by many ANNs, especially feedforward ANNs, is not a function of time, yet most SNNs are time-varying in nature.

In this chapter, basic operational principles and learning in SNNs are discussed. We start by presenting a few popular neuron models that are widely used in SNNs for various purposes. Compared to artificial neurons, spiking neurons often have much more complicated dynamics. Therefore, many researchers believe that this can potentially make SNNs more powerful compared to their ANN counterpart. It is the learning capability that empowers a neural network with the capability to solve many real problems. In this chapter various methods are presented for training SNNs. The learning in a shallow network is first introduced. Inspired by recent progress in deep learning, it is desired to train and exploit deep SNNs as well. Even though classical learning algorithms for SNNs work well for a shallow network, many of them do not naturally apply to a deep network. Therefore, at the end of this chapter we study how to train a deep SNN effectively.

Learning in Energy-Efficient Neuromorphic Computing: Algorithm and Architecture Co-Design,
First Edition. Nan Zheng and Pinaki Mazumder.
© 2020 John Wiley & Sons Ltd. Published 2020 by John Wiley & Sons Ltd.

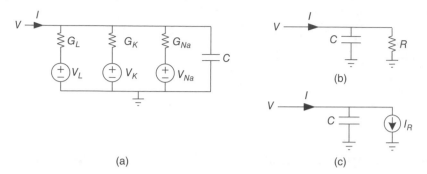

Figure 4.1 Schematic of the equivalent circuit of (a) the Hodgkin-Huxley model, (b) the conductance-based LIF model, and (c) the current-based LIF model. In the Hodgkin-Huxley model, three voltage sources model the reverse potentials and the three conductances model three leakage channels. For the LIF model, the conductance-based model is more biologically feasible, yet the current-based model is more computationally efficient.

4.1.1 Popular Spiking Neuron Models

To study SNNs, one needs to first understand how a single spiking neuron works. As mentioned previously, SNNs were initially created to model how the brain functions. Therefore, earlier spiking neuron models put more emphasis on reproducing the biological phenomenon observed in experiments. In other words, SNNs were mainly used for mapping biological neural networks. Later, more and more researchers in the artificial intelligence (AI) community started employing SNNs to conduct various brain-inspired computations. Neuron models used in these applications were made much simpler so that networks with a moderate size can be evaluated within a reasonable amount of time. Consequently, there have been many different neuron models developed in the literature. Some of the neuron models are more biologically feasible yet complex, whereas some of them are easier to implement and are computationally more efficient. In this section, we review three popular and representative neuron models that are widely used in the literature.

4.1.1.1 Hodgkin-Huxley Model

The Hodgkin-Huxley model is a mathematical model that describes the behavior of biological neurons. This model was proposed by Hodgkin and Huxley in 1952 [3]. They received the Nobel Prize in Physiology or Medicine in 1963 in recognition of this work.

The equivalent circuit for the Hodgkin-Huxley model is shown in Figure 4.1a. The current I flowing into a cell can be calculated as

$$I = C\frac{dV}{dt} + G_{Na}m^3h(V - V_{Na}) + G_K n^4(V - V_K) + G_L(V - V_L) \tag{4.1}$$

In Eq. (4.1), V_{Na}, V_K, and V_L are called reverse potentials, as shown in Figure 4.1a. G_{Na}, G_K, and G_L are parameters modeling conductances of sodium, potassium, and leakage channels, respectively, and m, n, and h are three gating variables. The dynamics of these three variables are described by Eqs. (4.2) to (4.4). The coefficients, m and h, control the sodium channel, whereas n controls the potassium channel [12]:

$$\frac{dm}{dt} = \alpha_m(V)(1 - m) - \beta_m(V)m \tag{4.2}$$

$$\frac{dn}{dt} = \alpha_n(V)(1-n) - \beta_n(V)n \tag{4.3}$$

$$\frac{dh}{dt} = \alpha_h(V)(1-h) - \beta_h(V)h \tag{4.4}$$

In Eqs. (4.2) to (4.4), $\alpha_m(V)$, $\alpha_n(V)$, $\alpha_h(V)$, and $\beta_m(V)$, $\beta_n(V)$, $\beta_h(V)$ are empirically-set functions.

4.1.1.2 Leaky Integrate-and-Fire Model

Even though the biologically feasible Hodgkin-Huxley model can be used to capture the dynamics of many real neurons accurately, its computational complexity is too high to be adopted in large-scale neural network simulations. Therefore, a simpler neuron model is needed. The leaky integrate-and-fire (LIF) model is one of the most popular spiking neuron models used to simulate SNNs efficiently.

The model for an LIF neuron can be described as

$$C\frac{dV(t)}{dt} = I - I_L \tag{4.5}$$

where I is the current flowing into the neuron and I_L is the leakage current. The definition for the leakage current term can vary. One way to compute the leakage current is through $V(t)/R$. We call this type of leakage current a conductance-based leakage. This is because the leakage current can be thought as the current that flows through a conductance, as shown in Figure 4.1b. This way of defining leakage current is more biological plausible, yet the computational overhead is somewhat high as the leakage at any moment is a function of the membrane voltage. Another form of leakage current, which is more popular in hardware implementation, is what we call a current-based leakage, as shown in Figure 4.1c. The leakage current, in this case, is independent of the membrane voltage.

Similarly, the injected current I in Eq. (4.5) is determined by the type of synapse used to interconnect neurons. There are, in general, two types of synapses: current-based synapses and conductance-based synapses [13]. Conductance-based synapses are more biologically accurate. When this type of synapse is used, the injected current is a function of the membrane voltage of the postsynaptic neuron. On the other hand, for a current-based synapse, the injected current from a presynaptic neuron is independent of the postsynaptic potential. Current-based synapse and leakage can help reduce the computational complexity significantly. Therefore, they are often employed by many SNN accelerators.

4.1.1.3 Izhikevich Model

The Hodgkin-Huxley model offers a good biological accuracy yet is computationally prohibitive for many real-life applications. On the other hand, the LIF model might be useful in many artificial SNNs due to its simplicity, yet it is deficient in mimicking biological neurons very well. An intermediate choice is the Izhikevich model proposed by Izhikevich in 2003 [14]. The model can be described mathematically as

$$\frac{dV}{dt} = 0.04V^2 + 5V + 140 - U + I \tag{4.6}$$

$$\frac{dU}{dt} = a(bV - U) \tag{4.7}$$

with the auxiliary after-spike resetting:

If $V \geq 30$ mV, then V is reset to c and U is reset to $U + d$

In Eqs. (4.6) and (4.7), V is the membrane voltage, U is a membrane recovery variable, and a, b, c, and d are model parameters. Using this simple model, many phenomena observed in biological neurons can be reproduced with a computational complexity similar to that of an LIF neuron.

4.1.2 Information Encoding

How to encode information in spikes is a popular research topic that both neuroscientists and computational artificial-intelligence researchers have studied for decades. There are two popular encoding methods: (i) rate coding and (ii) temporal coding.

Rate coding is the simplest way of encoding information on spikes. It assumes that the information is modulated on the mean firing rate of a neuron. This type of encoding method is similar to the frequency modulation in communication systems. Rate coding was demonstrated as early as in 1926 by Adrian [15]. This type of information encoding is actually the foundation for ANNs. Indeed, the activation level of a neuron in an ANN can be treated as an abstraction of the mean firing rate in an SNN. One of the biggest advantages of the rate coding is the low complexity in encoding and decoding. Therefore, they have been employed extensively for SNNs in the literature.

In recent years, many researchers have argued that rate coding might be too simple to represent complicated information in the brain [16, 17]. The main criticism comes from the fact that rate coding relies on averaging the information collected over a long period. Such a slow response time is not consistent with the response latency observed in the biological experiment [17]. To circumvent this difficulty, temporal coding is often used as an alternative. It was hypothesized that the information is encoded in the exact timing of the spikes. In recent years, more and more experimental results are obtained to support the feasibility of this type of information encoding [18, 19].

As an example, Figure 4.2 shows two spike patterns. For the part that is enclosed by the dashed-line window, two patterns are the same from the perspective of rate coding, as the number of spikes in the window are the same. However, from the perspective of temporal coding, these two patterns are different, as the timings of the spikes are different. Apparently, the temporal coding is able to encode more information with the same number of spikes. On the other hand, despite being less biologically feasible and less efficient in carrying information, rate coding does have some advantages in hardware implementation.

The first advantage of rate coding is the simplicity of the associated encoding and decoding circuitry, which is especially important for large-scale neuromorphic systems containing many encoders and decoders. The second advantage is the robustness. For

Figure 4.2 Illustration of two spike trains. Depending on the encoding and decoding method employed, these two spike trains might carry similar or different information.

example, let us imagine that due to some non-idealities, such as noise and process variations, the original spike pattern A becomes pattern B. Under this circumstance, a rate-based decoder is still able to recognize and decode the correct information from the distorted pattern B, whereas the temporal-based decoder is likely to fail. This contrast in robustness is similar to that between stochastic computing and binary computing. The error-tolerant feature provided by the rate coding is valuable in hardware implementation, as nanometer CMOS technology as well as other emerging nanotechnologies such as memristors behave less and less deterministically owing to a wide-scale device-to-device process, supply voltage, and temperature variations.

4.1.3 Spiking Neuron versus Non-Spiking Neuron

To summarize and highlight the differences between a spiking neuron and an artificial (non-spiking) neuron, Figure 4.3 compares these two neuron models. For both neuron models, there is an operation for summing up contributions from each synapse. In a non-spiking neuron, most likely a multiply-accumulate (MAC) operation is needed, whereas a masked addition is enough for a spiking neuron as spikes are binary in nature. Another key difference is that there is a memory element associated with each spiking neuron. For a simple LIF neuron, the memory element can be implemented as a linear capacitor. More complicated memory effects might exist in a more complex neuron model.

An analogy is given in the figure. The artificial neuron is like combinational logic. No matter how complicated the activation, it can still be implemented with combinational

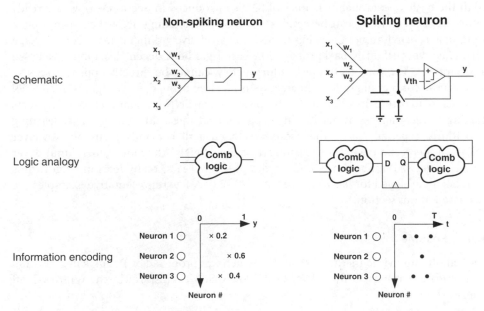

Figure 4.3 Comparison between spiking neurons and non-spiking neurons. A non-spiking neuron is memoryless. Its present output is solely dependent on its current input. On the other hand, a spiking neuron has inherent memory. The output at the present time depends not only on the current input but also the history of the input. Non-spiking neurons often use spatial patterns to represent information, whereas spiking neurons modulate information in spatiotemporal patterns.

logic gates. A spiking neuron, on the other hand, is more like a finite-state machine where the output at the present state is not only affected by the present input but also the past history of the input values.

It is worth noting that even though artificial neurons are often memoryless, it does not necessarily mean all ANNs are memoryless. In fact, as one of the most useful types of neural networks, the long short-term memory (LSTM), discussed in Chapter 2, obviously has memory, which is conveyed by its name. Indeed, memoryless artificial neurons can still be used to construct ANNs with memory through a careful selection of the network topology.

In contrast to ANNs, SNNs have inherent memory that does not completely rely on the network topology. As a result, the ways in which information is expressed in these two types of neural networks are also very different, as shown in Figure 4.3. For an artificial neuron, the output is a real number, i.e. a scalar quantity. The distribution of the activations of the output neurons indicates the inference result. For example, if each output neuron represents a class, then the activations of the output neurons could be the proof that the applied input belongs to this category. For an SNN, on the other hand, the outputs from the neurons are a binary vector. Therefore, the output of an SNN does not only have a spatial distribution but also a temporal distribution. This is partially the reason why SNNs are often trained to learn spatiotemporal patterns.

4.2 Learning in Shallow SNNs

With the basic operational principles of SNNs introduced in Section 4.1, we are ready to discuss how learning can be conducted in SNNs. Training SNNs has always been a popular research topic. Learning in SNNs is a relatively difficult task. For ANNs, a very effective method is backpropagation-based gradient decent learning, discussed in Chapter 2. The same technique, unfortunately, cannot be directly applied to SNNs because the concept of gradient in an SNN is not well defined, since signals in an SNN are discrete in nature. In addition, the outputs from an SNN is a spatiotemporal pattern, making the learning even harder. In recognition of these difficulties, many learning algorithms, especially those that were developed earlier, mainly focus on two-layer SNNs. In this section, we study learning in shallow SNNs. The more general multilayer case is discussed in Section 4.3. Over the past few years, many learning algorithms have been developed for learning in SNNs [20–29]. A few representative examples are discussed in this section.

4.2.1 ReSuMe

The remote supervised method (ReSuMe), which was proposed by Ponulak in 2005 [26], is a powerful learning rule for SNNs [23, 25, 26]. ReSuMe is inspired by the Widrow-Hoff rule, as shown below:

$$\Delta w_{oi} = \alpha x_i (y_d - y_o) \tag{4.8}$$

where Δw_{oi} is the synaptic weight change, x_i is the spike from the ith presynaptic neuron, y_o is the output spike, y_d is the target spike that needs to be learned, and α is the learning rate.

The ReSuMe learning rule can then be formulated as follows:

$$\frac{dw_{oi}}{dt} = (S_d(t) - S_o(t)) \left(a_d + \int_0^{\infty} a_{di}(s)S_i(t-s)ds \right) \tag{4.9}$$

where $S_i(t)$, $S_o(t)$, and $S_d(t)$ are the input, output, and desired spike train, a_d is the non-correlative factor, and $a_{di}(s)$ is a kernel function used to specify the learning window. A typical choice of $a_{di}(s)$ is an exponential window function. The goal of ReSuMe is to force the SNN to output desired spike trains. For each neuron, there is an implicit teacher neuron associated with it. The teacher neuron outputs the desired spike pattern and the training neuron learns from the teacher neuron. A learning example using ReSuMe is illustrated in Figure 4.4. The synapse is potentiated when there is a target spike, whereas the synapse is depressed when it observes an actual neuron fire. The amount of weight change is proportional to the time difference between the desired firing time $S_d(t)$ and the actual firing time $S_o(t)$. The non-correlative factor a_d is used to drive the mean firing rate of $S_o(t)$ toward that of $S_d(t)$. Clearly, when the learning converges, the synaptic weight does not change any more because the desired spike timing and the actual timing coincide with each other.

One of the striking features of the ReSuMe learning rule is that it works with many different neuron models. Such a property arises from the origin of the algorithm: the correlation between spikes. Since the neuron model is not used in deriving the learning algorithm, it is expected that the success of learning does not depend on the neuron model used. Such an assumption is indeed confirmed by many experiments and analyses in the literature [23, 25, 26]. In the experiment made by Ponulak and Kasiński [25], three popular neuron models covered in Section 4.1 were used: (i) the LIF model, (ii) the Hodgkin-Huxley model, and (iii) the Izhikevich model. Successful learning of target patterns was demonstrated for all these three neuron models.

4.2.2 Tempotron

Tempotron is a learning rule that was first formulated by Gütig and Sompolinsky in 2006 [24]. It trains the neural network to distinguish between two types of spike patterns. The neural network is trained such that it fires when one type of pattern is presented and stays quiet otherwise. The tempotron learning rule is a supervised learning rule. During the training phase, labeled data are fed into the neural network. When there is a misclassification, the synaptic weights are updated accordingly to increase or decrease the firing probability, depending on the error pattern.

The tempotron learning rule can be represented mathematically as follows:

$$\Delta w_i = \lambda \sum_{t_i < t_{max}} K(t_{max} - t_i) \tag{4.10}$$

where t_{max} denotes the time at which the postsynaptic potential reaches its maximal value, $K(\cdot)$ is the normalized postsynaptic potential resulting from the presynaptic spike, and λ specifies the maximum step of the synaptic update per input spike, which serves a role similar to that of the learning rate. Equation (4.10) is used in the case where a neuron should have fired but did not. A minus sign is included in Eq. (4.10) for the case where a neuron should not have fired but it did.

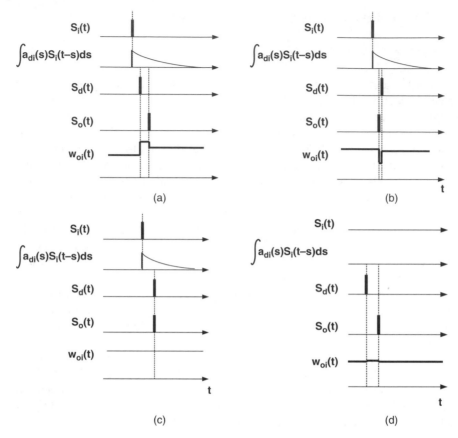

Figure 4.4 Illustration of the ReSuMe learning rule. The synaptic weight is updated based on the timings of the input spike $S_i(t)$, the actual output $S_o(t)$, and the desired output $S_d(t)$. (a) The synaptic weight is potentiated when the teacher neuron fires. (b) The synaptic weight is depressed when the output neuron fires. (c) When the actual output and the desired output coincide, the synaptic weight does not change. (d) When there is no input neuron, the synaptic weight is updated based on the value of the non-correlative factor, which helps control the mean firing rate of the output neuron. Source: adapted from [25].

Figure 4.5 illustrates the learning algorithm. In Figure 4.5a, two different spike patterns are shown. The black pattern is the target pattern for which the neuron should fire, whereas the gray one is the null pattern for which the neuron should stay silent. When these two patterns are presented to the network, the output neuron reacts differently by generating different postsynaptic membrane potentials, as shown in Figure 4.5b. The black curve does not exceed the firing threshold, whereas the gray curve does. Classification errors occur for both of these two patterns. Therefore, the synaptic weights are updated accordingly, as shown in Figure 4.5c.

It was shown in [28] that the tempotron learning rule was actually a particular form of the ReSuMe rule discussed in Section 4.2.1 under certain conditions. Due to this connection, a tempotron-like ReSuMe learning rule was proposed in [28]. Yu et al. compared these three related learning rules, (i) tempotron, (ii) ReSuMe, and (iii) tempotron-like ReSuMe, in [30] and it was observed that the tempotron learning rule was the fastest

Figure 4.5 Illustration of the tempotron learning rule. (a) Two spike patterns are presented to the neural network. The black pattern is the target pattern, whereas the gray one is not. (b) postsynaptic membrane voltages corresponding to the two patterns. (c) Synaptic change according to the tempotron rule [24]. Reproduced with permission of Springer.

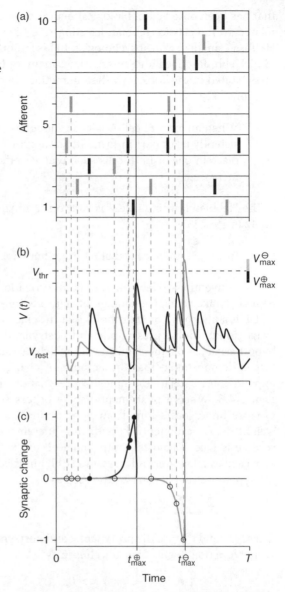

among these three. The tempotron rule was then employed in a pattern recognition task. The Modified National Institute of Standards and Technology (MNIST) dataset was utilized to verify the system equipped with this learning rule. A recognition rate that was comparable to a support vector machine, which is one of the most effective machine-learning models used for a classification task, was reported.

4.2.3 Spike-Timing-Dependent Plasticity

The biologically plausible spike-timing-dependent plasticity (STDP) learning rule has been widely employed by many researchers in recent years. STDP is a type of plasticity that has been observed in biological neural networks. STDP is generally believed to be related to the classic Hebbian learning rule [29, 31], which was named after Donald Hebb, a Canadian neuropsychologist. In his famous book, which was published in 1949 [32], Hebb described and summarized many of his important findings. For example, Hebb stated in the book how the interactions between neurons can decide their connections [32]:

> When an axon of cell A is near enough to excite a cell B and repeatedly or persistently takes part in firing it, some growth process or metabolic change takes place in one or both cells such that A's efficiency, as one of the cells firing B, is increased.

The Hebbian learning rule was later on summarized as a short and catchy phrase by Shatz in 1992 [33]:

> In a sense, then, cells that fire together wire together.

The conventional Hebbian rule is a correlation-based learning rule that does not explicitly take the spike timings of neurons into consideration. On the other hand, STDP is a phenomenon that shows that the change in synaptic strength is related to the exact timings of the presynaptic and postsynaptic neurons [34]. STDP has been long hypothesized as the mechanism with which mammals can learn.

It is demonstrated that the amount of change in synaptic strength for a synapse depends on the relative timings of presynaptic and postsynaptic spikes, as shown in Figure 4.6. When a postsynaptic spike occurs shortly after a presynaptic spike, the synapse undergoes a long-term potentiation (LTP) and the increase in the synaptic weight decays exponentially with the difference between the two spike timings. The reverse is true if a postsynaptic spike occurs before a presynaptic one. The synapse experiences a long-term depression (LTD) in this case. Mathematically, the change in the synaptic strength can be expressed as

$$\Delta w = \sum_n \sum_m K(t_{post}^m - t_{pre}^n) \tag{4.11}$$

where t_{pre}^n and t_{post}^m are the presynaptic and postsynaptic spike timings, respectively. The kernel function $K(x)$ is typically chosen as

$$K(x) = \begin{cases} A_+ \exp(-x/\tau_+), x > 0 \\ A_- \exp(x/\tau_-), x < 0 \end{cases} \tag{4.12}$$

In Eq. (4.12), τ_+ and τ_- are used to control the decay of the exponential-shaped window. A_+ and A_- are two quantities that determine the scale of the synaptic update. The STDP protocol shown in Eq. (4.11) is sometimes referred to as a pair-based STDP rule, which is a common choice in the literature. There also exist many variants of this basic STDP protocol, such as the triplet-based STDP rule [35], the spike-driven synaptic plasticity [36], and so on.

Clearly, the learning rule specified in Eq. (4.11) does not pose any restriction on the maximum or minimum synaptic weights. Such a learning rule might not be practical in reality since both the biological and artificial synaptic weights should have some upper

Figure 4.6 Illustration of a typical STDP protocol. The change in the synaptic weight depends on the causality between the presynaptic and postsynaptic spikes as well as the differences between the spike timings.

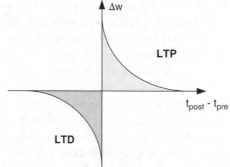

Figure 4.7 Illustration of a few examples of $A_+(w)$ that can be used to limit the maximum value that the synaptic weight can reach. In order to limit the maximum value of w to w_{max}, $A_+(w)$ needs to be 0 for $w > w_{max}$.

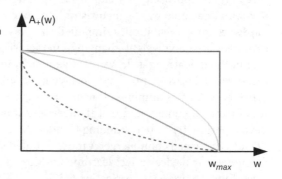

limit. One straightforward way is to put a hard limit on the maximum weight the synapse can achieve. Such a simple method is indeed effective in controlling the range of synaptic weights. However, it might deteriorate the learning performance.

More sophisticated solutions that can effectively limit the maximum weight exist. The two constants A_+ and A_- can be replaced by two functions of the weight: $A_+(w)$ and $A_-(w)$ [37]. This is conceptually illustrated in Figure 4.7 for $A_+(w)$. By doing so, the STDP protocol becomes a weight-dependent STDP protocol. In other words, the amount of change in the synaptic weight is not only a function of the spike timings but also of the value of the weight itself. Normally, the weight update becomes smaller and smaller as the synaptic weight approaches the maximum limit. This way of controlling the scaling factor in an STDP protocol is effective in preventing the synaptic weight from growing without limit. There are other elegant ways of achieving a similar effect and one example is demonstrated in Section 4.2.4.

An STDP protocol dictates how synaptic weights should change based on spike timings. This, however, does not explicitly specify how the learning is conducted. Therefore, the learning setup can vary from implementation to implementation. For example, Diehl and Cook demonstrated unsupervised learning based on STDP for a handwritten digit recognition task in [38]. The images from the MNIST dataset are first converted to Poisson spike trains. The spikes corresponding to all pixels of the images are fed to excitatory neurons in an all-to-all fashion. In other words, the input layer of the SNN is fully connected to the excitatory layer. Each excitatory neuron is then connected to its corresponding inhibitory neuron in a one-to-one fashion. The inhibitory neuron, on the other hand, inhibits all other neurons except its corresponding excitatory neuron once it spikes. Such a mechanism is similar to the winner-take-all approach in the sense that it introduces a competition among neurons.

The STDP learning presented in [38] is based on unsupervised learning. Therefore, in order to perform a classification task, a step of associating the learned structure with labels is needed. This was achieved by associating each excitatory neuron with the class that had the highest response in [38]. It was observed that, after training, the weight connecting the input layer and each excitatory neuron resembled the shape of handwritten digits in the dataset. It was also shown that by providing more neurons in the excitatory layer, the recognition rate can be gradually enhanced. Furthermore, several variants of STDP learning rules have been employed in [38] and the conclusion was that learning was effective regardless of the type of STDP rule used.

A similar approach was employed by Querlioz et al. in [39] where the platform used for learning was a memristor crossbar-based SNN. Since the target was for hardware implementation, a simplified rule was adopted. Instead of having the complex exponential-shape weight-changing characteristic, a rectangular one was used. Such a simplified learning rule was conveniently implemented by deliberately adjusting the overlap between pre- and postsynaptic spikes. A recognition rate of 93.5% was achieved with the simplified learning rule when 300 output neurons were involved. It was further demonstrated that the learning was rather insensitive to the variations in device parameters. Such a finding is indeed encouraging as the variations in nanoscale devices are usually quite large. A more detailed study on how the variations and noises in memristor devices affect learning and how to achieve reliable learning under large variations are presented in Section 5.3.5.

In addition to unsupervised learning, STDP has also been employed in conducting reinforcement-learning tasks [40–43]. Florian showed that reinforcement learning was possible through modulating the STDP term with a global reward signal [40]. The formulated learning algorithm was verified through an XOR problem for both the rate-coded input and the temporally coded input. An actor–critic structure was employed to study the possibility of reinforcement learning with reward-modulated STDP [43]. Frémaux et al. found that a temporal difference gradient decent learning could be achieved with an STDP-like learning rule modulated with the temporal-difference (TD) error. Through a sophisticated arrangement of neurons and synapse connections, the learning rule was verified on various tasks. Inspired by this work, conducting reinforcement learning with the help of STDP is studied in more detail in Section 4.2.4.

Learning using STDP rules has been studied by many researchers in recent years. Nevertheless, for most learning examples demonstrated in the literature, STDP was employed as a biologically plausible and empirically successful learning rule without its underlying principle being explained. The application of STDP-based learning was also mainly restricted to unsupervised learning. Despite being effective, the usefulness of unsupervised learning is rather limited, as most applications of neural networks nowadays are based on supervised learning or reinforcement learning, which requires neural networks to be able to approximate arbitrary functions. Furthermore, how to apply the learning in a multilayer neural network is not obvious with the conventional free-running STDP algorithm. Nevertheless, most successes in various machine-learning tasks were achieved using deep neural networks. In addition, it is unclear how closely one should follow the weight update curve shown in Figure 4.6

in order to achieve successful learning. Many of these limitations and questions are addressed in this and the following chapters.

4.2.4 Learning Through Modulating Weight-Dependent STDP in Two-Layer Neural Networks

4.2.4.1 Motivations
STDP has long been hypothesized to be an underlying mechanism of learning in mammal brains. With the target of mimicking how a brain learns, STDP is often used in SNNs as a biologically plausible and empirically successful learning algorithm. Numerous attempts have been made from both the neuroscience community and the AI community with the aim to explain the underlying principles of learning with STDP [27]. Hinton hypothesized that STDP-based learning might be a form of gradient descent learning [44]. Similar concepts have also been shown in different contexts using various forms [40–43].

The possibility of using the STDP learning rule as a hardware-friendly learning algorithm was explored in [45]. It was demonstrated that a term similar to the quantity measured in an STDP protocol could be utilized to estimate the gradients in a neural network. With the help of the estimated gradients, stochastic gradient descent (SGD) learning could be formed. In this section, the main results presented in [45] are reviewed. An actor–critic network-based reinforcement learning is studied as an example. The objective of this study is not to find out the exact role of STDP in a biological SNN. Rather, the main focus of this section is to develop a learning algorithm that can be efficiently implemented in hardware SNNs. To this end, various aspects of the algorithm are examined through numerical studies. This section serves as the first step toward a hardware-friendly learning algorithm in SNNs. Only two-layer neural networks as shown in Figure 4.8 are considered here. Multilayer neural networks that are suitable for deep learning are studied in Section 4.3.5.

4.2.4.2 Estimating Gradients with Spike Timings
To study the neural network shown in Figure 4.8, let us assume that the spike trains generated from the presynaptic and the postsynaptic neurons are

$$x_i(t) = \sum_{k_i} \delta(t - t_{k_i}) \tag{4.13}$$

$$y_j(t) = \sum_{k_j} \delta(t - t_{k_j}) \tag{4.14}$$

where a constant excitatory postsynaptic potential is assumed in the equations. This way of representing spikes is very popular in hardware implementation of SNNs, considering its ease of implementation.

Figure 4.8 Illustration of a two-layer neural network, where $x_i(t)$ is the presynaptic (input) spike train and $y_j(t)$ is the postsynaptic (output) spike train [45]. Reproduced with permission of IEEE.

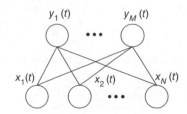

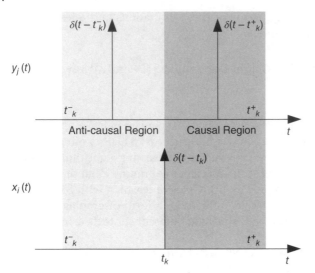

Figure 4.9 Illustration of the two regions divided by the spike timing of the input neuron. The causal and anti-causal regions are defined according to the causal relationship between the input and the output spikes [45]. Reproduced with permission of IEEE.

If we treat the presynaptic spike timing t_k as a boundary, two regions can be formed for the postsynaptic spikes, as shown in Figure 4.9. The causal region corresponds to the region after t_k, whereas the anti-causal region corresponds to the region before t_k.

Let us first focus on one pair of presynaptic and postsynaptic spike trains, $x_i(t)$ and $y_j(t)$. For ease of illustration, three variables are defined as

$$X_k = \int_{t_k^-}^{t_k^+} x(t)dt \tag{4.15}$$

$$Y_k^+ = \int_{t_k}^{t_k^+} y(t)dt \tag{4.16}$$

$$Y_k^- = \int_{t_k^-}^{t_k} y(t)dt \tag{4.17}$$

Subscripts for the neurons are understood and dropped in the equations. These three quantities are Bernoulli random variables that indicate whether certain events occur or not. For example, X_k represents whether there is a presynaptic spike at t_k, Y_k^+ indicates whether a postsynaptic causal spike occurs, and Y_k^- shows the existence of an anti-causal spike.

With these three variables, the probability that only an anti-causal spike occurs can be written as

$$Pr(Y_k^+ = 0 \cap Y_k^- = 1 | X_k = 1) = \frac{Pr(Y_k^+ = 0 \cap Y_k^- = 1 \cap X_k = 1)}{Pr(X_k = 1)} \tag{4.18}$$

The probability that only a causal spike occurs is

$$Pr(Y_k^+ = 1 \cap Y_k^- = 0 | X_k = 1) = \frac{Pr(Y_k^+ = 1 \cap Y_k^- = 0 \cap X_k = 1)}{Pr(X_k = 1)} \tag{4.19}$$

We define another quantity called *stdp* as

$$
stdp = \begin{cases} 1 & X_k = 1 \text{ and } Y_k^+ = 1 \text{ and } Y_k^- = 0 \\ -1 & X_k = 1 \text{ and } Y_k^+ = 0 \text{ and } Y_k^- = 1 \\ 0 & \text{otherwise} \end{cases} \tag{4.20}
$$

The reason that we name this variable as *stdp* is that it represents a quantity that is similar to what is measured in a typical STDP protocol. Our main objective is to find out a way to estimate $\partial\rho(y(t))/\partial\rho(x(t))$, where $\rho(\cdot)$ represents the normalized density of a spike train. Let us consider the quasi-stationary case where $\rho(x(t))$ and $\rho(y(t))$ vary much more slowly compared to the unit time step at which the neuron is allowed to spike. Under this circumstance, we have

$$
\rho(stdp(t)) = Pr(Y_k^+ = 1 \cap Y_k^- = 0 \cap X_k = 1) - Pr(Y_k^+ = 0 \cap Y_k^- = 1 \cap X_k = 1) \tag{4.21}
$$

With the help of Eqs. (4.15) to (4.21), an estimation of $\partial\rho(y(t))/\partial\rho(x(t))$ can be obtained as

$$
\frac{\partial\rho(y(t))}{\partial\rho(x(t))} \approx E(\Delta Y_k | X_k = 1) \approx \frac{\rho(stdp(t))}{\rho(x(t))} \tag{4.22}
$$

Intuitively, Eq. (4.22) indicates that $\partial\rho(y(t))/\partial\rho(x(t))$ can be estimated through observing how the output spike alters its statistical behavior upon an input spike that serves as a small perturbation to the network. Even though Eq. (4.22) is obtained by assuming there is only one pair of presynaptic and postsynaptic neurons, it can be extended to cover the more general cases where multiple presynaptic and postsynaptic neurons are involved, given that the spike timings of neurons are reasonably uncorrelated. For example, at any given time t_k, the postsynaptic neuron can spike at either side of t_k and the chances to spike at each side should be equal given that there does not exist any presynaptic spike. However, when the presynaptic spike is present, the probability for the postsynaptic neuron to spike at the causal region alters depending on whether the synapse connecting these two neurons is excitatory or inhibitory. With the mild assumption that spike timings for different presynaptic neurons are uncorrelated, the contributions from other presynaptic neurons appear as noises and thus can be filtered out easily. Even though the spike density of each presynaptic neuron might be highly correlated due to the spatial correlation in the input data, the spike timings of neurons can be largely uncorrelated, especially with some decorrelation techniques. Consequently, the gradients in the network can be estimated simultaneously, which can speed up the learning process significantly. Such a simultaneous gradient estimation method shares the same spirit as the simultaneous perturbation stochastic approximation (SPSA) method [46].

After obtaining $\partial\rho(y(t))/\partial\rho(x(t))$, the next step is to find a way to estimate $\partial\rho(y(t))/\partial w_{ij}$, which is needed in gradient descent learning. We assume that

$$
\rho(y_j(t)) = f_j\left(\sum_i w_{ij}\rho(x_i(t))\right) \tag{4.23}
$$

which is the firing density of the postsynaptic neuron, is related to that of the presynaptic neurons through a certain functional mapping.

From Eq. (4.23), we can obtain

$$f_j'\left(\sum_i w_{ij}\rho(x_i(t))\right) = \frac{\partial\rho(y_j(t))}{\partial\rho(x_i(t))}/w_{ij} \tag{4.24}$$

With the help of Eqs. (4.22) and (4.24), the following equation can be obtained:

$$\frac{\partial\rho(y_j(t))}{\partial w_{ij}} = \rho(x_i(t))f_j'\left(\sum_i w_{ij}\rho(x_i(t))\right) = \frac{\rho(stdp_{ij}(t))}{w_{ij}} \tag{4.25}$$

Close examination of Eq. (4.25) reveals that the gradient information $\partial\rho(y_j(t))/\partial w_{ij}$ can be estimated by a term that resembles the quantity measured in a common STDP protocol. The denominator w_{ij} in the equation is not present in previous literature [40, 43]. We believe this denominator plays a key role in learning. Mathematically speaking, including the weight provides the sign information. If negative weights are allowed, then it is necessary for a term to change the sign in Eq. (4.25), which would otherwise induce a wrong direction for gradient descent. In addition, the introduction of the weight denominator ensures an upper bound on w_{ij}, which serves a similar purpose as the weight decay technique that is widely used in ANNs [47].

In order to verify (4.25), two numerical simulations are conducted on a small neural network with five input neurons and one output neuron. In both of these two simulations, an LIF neuron model is used. A small amount of noise is injected into the neurons to decorrelate the spike timings of the neurons. In the first numerical study, the membrane voltage of one presynaptic neuron is swept, whereas the membrane voltage of other presynaptic neurons is kept at fixed values that are randomly chosen. The gradients estimated from the STDP information are compared with the results obtained from the numerical results in Figure 4.10. It is shown that the two sets of gradients match well

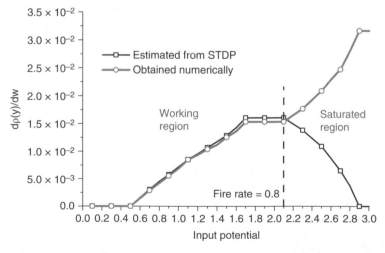

Figure 4.10 Gradients obtained from STDP and numerical simulation as the input membrane potential increases. Two sets of results match well until the firing rate of the input neuron reaches 0.8. The gradient estimation becomes inaccurate in the saturation region because it is difficult to distinguish a causal spike from an anti-causal spike when the input spike is too dense [45]. Reproduced with permission of IEEE.

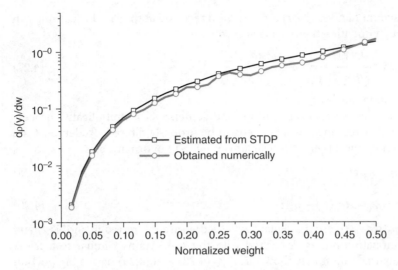

Figure 4.11 Gradients obtained from STDP and numerical simulation as the weight of the synapse increases. Two sets of results match well [45]. Reproduced with permission of IEEE.

until the firing rate of the presynaptic neuron reaches a certain value. The main reason for such a degradation in estimation accuracy is that when the firing rate is too large it is difficult to determine whether a postsynaptic spike is a causal or an anti-causal spike as the input spikes occupy almost every possible slot on the time axis. Such a saturation region, as shown in Figure 4.10, can be avoided by keeping the time resolution of the discrete-time neuron fine enough. Another technique that can avoid this saturation is presented in Section 4.3.5.

In the second simulation, spike densities of all five input neurons are fixed at randomly chosen levels, while the weight of one synapse is swept. Two sets of gradients are compared in Figure 4.11. Good matching is achieved between these two groups of results, which demonstrates the effectiveness of Eq. (4.25) in estimating the gradients. It is worth noting that the accuracy of Eq. (4.25) depends largely on how well the model in Eq. (4.23) can describe the actual dynamics of the chosen neuron model. In spite of being effective, some limitations in Eq. (4.23) do indeed exist. For example, Eq. (4.23) assumes that only the product of the weights and the presynaptic spike density matters. Even though such an assumption is indeed valid when the weights and the spike density are not too large, it might not hold in some extreme cases. Nevertheless, one can always avoid these cases through choosing an operating frequency that is high enough.

4.2.4.3 Reinforcement Learning Example

With the estimated gradient information available, the TD learning that is discussed in Section 2.2.4 can be readily conducted in order to minimize the absolute value of the TD error, which is defined as

$$\delta(t) = \gamma V(t) - V(t - \Delta t) + r(t) \tag{4.26}$$

where γ is the discount factor and $r(t)$ is the reward received at time t. The learning can be achieved through updating the weights according to

$$\frac{\partial w_{ij}(t)}{\partial t} = \alpha\delta(t)\frac{\partial\rho(y_j(t))}{\partial w_{ij}(t)} = \alpha\delta(t)\frac{\rho(stdp_{ij}(t))}{w_{ij}(t)} \qquad (4.27)$$

where α is the learning rate.

In Eq. (4.27), two time sequences are multiplied together. Such a multiplicative process may produce down-mixed noise, which needs to be properly filtered. To demonstrate this idea, let us denote the TD error and spike-timing (ST) information as

$$\delta(t) = \delta_0(t) + n_\delta(t) \qquad (4.28)$$

$$\rho(stdp_{ij}(t)) = \rho_0(stdp_{ij}(t)) + n_\rho(t) \qquad (4.29)$$

where $\delta_0(t)$ and $\rho_0(stdp_{ij}(t))$ are the signal parts that we are interested in and $n_\delta(t)$ and $n_\rho(t)$ are the quantization noises associated with the spike trains. Similar to a $\Sigma-\Delta$ modulator, the quantization noises shown in Eqs. (4.28) and (4.29) have a big portion of energy located at high frequency. This is particularly true for $n_\delta(t)$, to which a difference-like filter is applied.

By substituting Eqs. (4.28) and (4.29) into Eq. (4.27), we obtain

$$\frac{\partial w_{ij}}{\partial t} = \frac{\alpha}{w_{ij}}(\delta_0(t)\rho_0(stdp_{ij}) + \delta_0(t)n_\rho(t) + \rho_0(stdp_{ij})n_\delta(t) + n_\rho(t)n_\delta(t)) \qquad (4.30)$$

In this equation, the terms $\delta_0(t)n_\rho(t)$ and $\rho_0(stdp_{ij})n_\delta(t)$ do not cause much trouble because the weight update is accumulated over time. Such an accumulation behaves as a low-pass filter that can filter out the high-frequency noises in the scaled versions of the noises $n_\rho(t)$ and $n_\delta(t)$ effectively. The term $n_\rho(t)n_\delta(t)$, however, is more problematic because it produces a down-mixed noise through multiplication. This is illustrated in Eq. (4.31) and Figure 4.12:

$$n_\rho(t)n_\delta(t) = \mathcal{F}^{-1}(\mathcal{F}(\Phi(n_\rho(t))) * \mathcal{F}(\Phi(n_\delta(t)))) \qquad (4.31)$$

Here, $\Phi(\cdot)$ stands for the autocorrelation and $\mathcal{F}(\cdot)$ and $\mathcal{F}^{-1}(\cdot)$ represent Fourier and inverse Fourier transforms, respectively. It is difficult to evaluate Eq. (4.31) analytically, as the phase information of the noise spectrum is generally unknown. Nevertheless, some useful insights can be provided with the help of numerical simulations. The down-mixed noise, which is generated by the product of $n_\rho(t)$ and $n_\delta(t)$, can be several orders of magnitude more significant compared to the desired signal part. To make

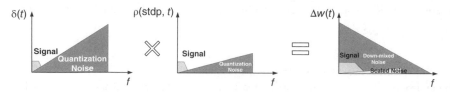

Figure 4.12 Illustration of the noise down-mixing process. Quantization noises in $\delta(t)$ and $\rho(stdp_{ij}(t))$ are mostly at high frequency. After multiplication, however, the noise is down-mixed to the baseband, which saturates the desired signal. In addition, because the two quantization noises are correlated, the resultant down-mixed noise is severely biased [45]. Reproduced with permission of IEEE.

things worse, the quantization noise $n_\delta(t)$ and $n_\rho(t)$ are correlated because they are essentially generated from the same signal. Such a correlation makes the down-mixed noise severely biased. The resultant noise buries the desired signal if the noise is not being effectively filtered before mixing.

From the perspective of a circuit designer, a high-performance filter, e.g. a sophisticated finite impulse response (FIR) filter should be used for $V(t)$ whereas a simple low-power filter, e.g. a moving-average filter, should be used for $stdp_{i,j}$. The reason associated with this choice is twofold. The first reason is that we only have one $V(t)$ signal in the system, yet every synapse has an $stdp_{ij}$ term that is associated with it. In other words, the $V(t)$ signal is shared by all the synapses. Therefore, it is wise for us to spend most effort in reducing the quantization noise associated with the signal that is reused repeatedly by different synapses. The second reason is that $n_\delta(t)$ is shaped twice: once by the neuron and the second time by Eq. (4.26). Consequently, $\delta(t)$ has a much stronger high-frequency noise associated with it. Following the filtration, decimation can be employed to convert the over-sampled low-resolution data into high-resolution data with a reduced sampling rate in order to save energy. The process of down-sampling is illustrated in Figure 4.13.

Two examples are employed to examine the learning algorithm in two-layer neural networks. The first example is a one-dimensional state-value learning problem similar to the one used in [43]. The configuration of the problem is illustrated in Figure 4.14. An agent moves toward the target with a fixed policy. The location of the agent is encoded through the input neuron whose membrane potential is computed according to

$$p_i(t) = k(x - c_i) \tag{4.32}$$

where c_i is the center of the input neuron i, x is the current location of the agent, and $k(\cdot)$ is the kernel function. One example of the kernel function is the Gaussian kernel. The input neuron, whose center is near the current location of the agent, fires intensively, whereas neurons with centers far away from the current agent location stay quiet, as illustrated in Figure 4.14. Such an encoding scheme is similar to those used in radial basis function (RBF) networks [48, 49].

In this learning task, the agent receives a reward if it reaches the target located at $x = 300$. The whole training process includes multiple iterations. In each iteration, the agent starts from a fixed point and moves toward the target at a fixed speed. The

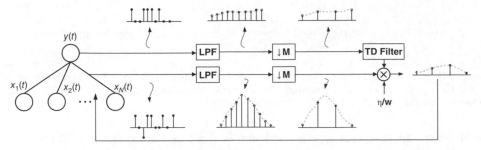

Figure 4.13 Illustration of the decimation process. Down-sampling is taken place after filtration. This process effectively reduces the operating frequency of corresponding blocks. In addition, the associated decimation gain is helpful in reducing the number of bits needed by the synaptic weights [45]. Reproduced with permission of IEEE.

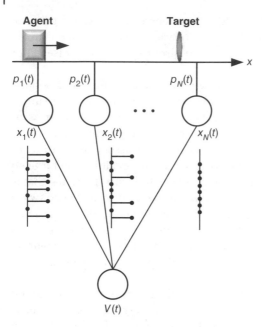

Agent   **Target**

$p_1(t)$ $p_2(t)$ $p_N(t)$

$x_1(t)$ $x_2(t)$ $x_N(t)$

$V(t)$

Figure 4.14 Configuration of the one-dimensional test case. An agent is moving toward a target under a fixed policy. Each input neuron is a place cell that fires intensively when the agent is close to its center [45]. Reproduced with permission of IEEE.

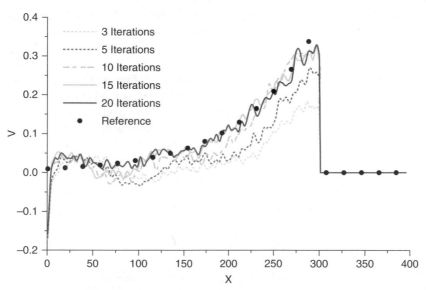

Legend:
- 3 Iterations
- 5 Iterations
- 10 Iterations
- 15 Iterations
- 20 Iterations
- ● Reference

Figure 4.15 Comparison between the output of the critic and the analytical result. Backpropagation of the TD error is clearly shown in the figure. The results obtained after 15 iterations start matching the reference result well [45]. Reproduced with permission of IEEE.

agent learns the correct state-value function in this process through the TD learning procedure.

The learned state-value function is shown in Figure 4.15. For the purpose of comparison, the correct state-value function obtained analytically is also plotted. As expected, the estimation made by the agent becomes more and more accurate as the learning goes

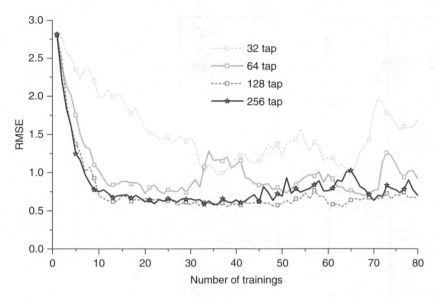

Figure 4.16 Comparison of the output of the critic network when filters of different orders are used. Learning rates are lowered for cases where only a few taps are used. Moving average FIR filters are used for a fair comparison. The result obtained with a 32-order filter barely features a learning. The learning becomes more effective as the order of the filter increases. The improvement starts saturating when the order of the filter reaches 128 [45]. Reproduced with permission of IEEE.

on. After approximately 10 learning iterations, the estimated state-value function starts matching with the analytical result.

Figure 4.16 demonstrates the importance of proper filtration through showing the root-mean-square error (RMSE) of the estimation from the critic network compared to the analytical reference. The results in the figure are obtained with different levels of filtration. More specifically, FIR filters with different taps, ranging from 32 to 256, are used for the purposes of comparison. Learning rates for the cases with fewer taps are reduced, which would otherwise diverge. Clearly, the RMSE reduces smoothly and quickly for cases where higher-order FIR filters are used.

In the derivation of the learning rule, we assume that spike timings of presynaptic neurons are uncorrelated. Such an assumption is indeed reasonable in many practical cases. Even for a deterministic LIF neuron model, the spike trains outputted by the input neurons are reasonably uncorrelated when a non-linear input kernel is used and the input is varying. To further decorrelate the spike timings, noise can be injected into the LIF neuron model to form a stochastic LIF model. The randomness involved in this process, however, is expensive because pseudo-random or true-random number generators are needed, which increase the area and power consumption of the system.

To circumvent this difficulty, we proposed a quantization residue injection method in [45]. Figure 4.17 illustrates this idea. The LIF model with a noisy threshold is shown in Figure 4.17a. By using a noisy threshold, randomness is introduced in the neuron model as the firing of the neuron is somewhat stochastic. Figure 4.17b demonstrates a similar stochastic model where a noisy residue is employed. Upon emitting a spike, instead of resetting the membrane voltage to a fixed resetting potential, a random residue is left. Figure 4.17c shows the proposed neuron model. Instead of injecting conventional

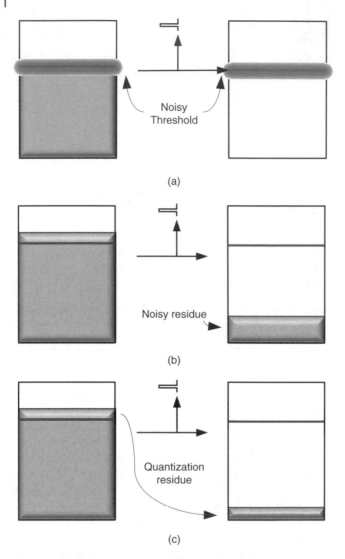

Figure 4.17 Illustration of three different ways of injecting noise into neuron models: (a) noisy threshold, (b) noisy residue where the noise is white, and (c) quantization residue [45]. Reproduced with permission of IEEE.

noise into the neuron as the residue, quantization residue associated with the membrane voltage is exploited. In other words, the excessive potential exceeding the threshold is kept as the residue in the neuron after firing, as shown in Figure 4.17c. This neuron model is actually one of the supported models in the IBM TrueNorth chips [50], which is discussed in more detail in Chapter 5.

With the quantization residue injection technique, the neuron model behaves similarly to a first-order $\Sigma - \Delta$ modulator. Different from the conventional noise, the quantization residue injected into each neuron depends on the input of that neuron. However, the injected quantization residue becomes reasonably uncorrelated in practical cases where inputs are varying.

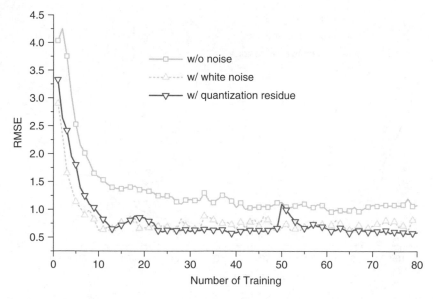

Figure 4.18 Comparison of the RMSEs obtained with a Gaussian kernel and three different neuron models: the deterministic model, the stochastic model with white noise injected as the membrane potential residue, and the model with the quantization residue [45]. Reproduced with permission of IEEE.

To demonstrate the efficacy of the quantization residue injection technique, three different neuron models, a deterministic LIF model, a stochastic LIF mode, and the modified LIF model with the quantization residue injection, are compared in Figures 4.18 and 4.19. Two different kernels are used: a Gaussian kernel is used in Figure 4.18, whereas a triangular kernel is used in Figure 4.19. In Figure 4.18, it is observed that all three neuron models can achieve impressive results, albeit the RMSE obtained from the deterministic neuron model is slightly worse. When a triangular kernel is used, however, the performance of the deterministic neuron model deteriorates significantly. To show that this degradation in performance is mainly attributed to the correlation in spike timings, Figure 4.20 compares the average magnitudes of the correlations between the spike timings of adjacent input neurons in the network. As shown in Figure 4.20, the spike timings are highly correlated when a triangular kernel is used for the poorly performing deterministic neuron model. The correlations of spike timings among different neurons can be remarkably reduced due to the proposed quantization residue injection. Such a reduction in correlations then improves the RMSE shown in Figure 4.19.

In the second example, a two-dimensional maze-searching task is demonstrated. The agent is placed at a random starting point in a maze with a size of 120×120. The objective of the agent is to reach the target and avoid walls and obstacles. It moves toward the direction that is determined by the actor network at a constant speed. Four actor neurons that represent four moving directions are used, as shown in Figure 4.21. At every time unit, neuron N and neuron S compare their membrane potentials. The neuron with a larger potential spikes and the potentials of both neurons are reset. Similar operations are also conducted for neuron E and neuron W. This method is similar to the winner-take-all scheme. Nevertheless, this arrangement avoids the cumbersome

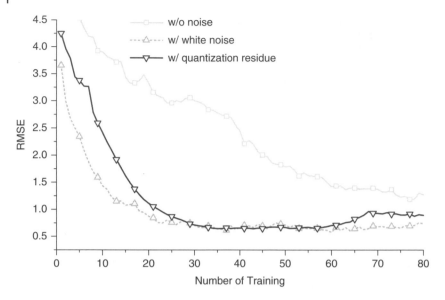

Figure 4.19 Comparison of the RMSEs obtained with a triangular kernel and three different neuron models: the deterministic model, the stochastic model with white noise injected as the membrane potential residue, and the model with the quantization residue [45]. Reproduced with permission of IEEE.

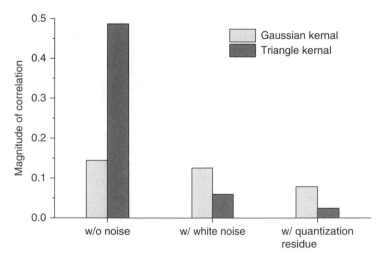

Figure 4.20 Magnitudes of correlations obtained from the simulation. The correlations of spike timings of adjacent neurons are measured. The deterministic neuron model with a Gaussian kernel and a varying input can achieve reasonably low correlations, whereas a stochastic model is needed to ensure a low correlation when a simple kernel is used. The proposed quantization residue injection is effective in randomizing the spike timing of each neuron [45]. Reproduced with permission of IEEE.

tuning for the inhabitation weights used in [43]. The moving direction of the agent is then determined by

$$\theta(t) = \tan^{-1}\left(\frac{\rho(E(t)) - \rho(W(t))}{\rho(N(t)) - \rho(S(t))}\right) \qquad (4.33)$$

Figure 4.21 Actor neurons employed in the maze problem. Action a_k and $\overline{a}_k$ are mutually exclusive in nature. The actual moving direction is interpolated according to the relative firing rate of each neuron [45]. Reproduced with permission of IEEE.

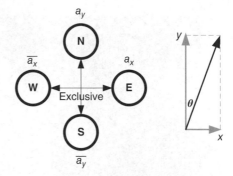

For the critic network, the same critic used in the one-dimensional learning problem is employed in this two-dimensional example. Initially, the agent knows nothing about the location of the target, the wall, and the obstacle. It does not know what will happen if it runs into those objects. All information is obtained through reinforcement in each trial. To help the agent explore the problem, noise is injected into the actor neurons. In this maze problem, the mutual exclusiveness and the orthogonality of actions can be exploited, which leads to a policy of

$$\pi(s, a_k) = Pr(a(t), s(t)) = Pr\left(\sum_i (w_{ij}^{a_k} - w_{ij}^{\overline{a}_k})\rho(x_i(t)) + n_k > 0 \right) \tag{4.34}$$

where $\pi(s, a)$ is the policy, $a(t)$ is the chosen action, $s(t)$ is the current state, a_k is the action k, $\overline{a}_k$ is the opposite of a_k, and n_k is the injected noise. Equation (4.34) behaves similarly to a Gibbs softmax method [51]. The preferred actions are chosen with a higher probability, whereas other actions are picked occasionally for the purpose of exploration. Synaptic weights in the actor network are updated according to the learning rule in Eq. (4.27). The actions undertaken are reinforced if a positive TD error is received. Since the selected actions result in states that are better than expected, the corresponding actions receive credits for that.

For problems where mutual exclusiveness is not present, a more general policy as shown in Eq. (4.35) can be applied, where the most preferred action is selected with the highest probability. Again, noise is added to encourage exploration.

$$\pi(s, a_k) = Pr(a(t), s(t)) = Pr\left(\underset{k}{\text{argmax}} \left(\sum_i w_{ij}^{a_k} \rho(x_i(t)) + n_k \right) = k \right) \tag{4.35}$$

To examine the effectiveness of the learning, two sets of simulations are conducted. The configurations of the maze in these two sets of experiments are illustrated in Figures 4.22 and 4.23. For the first set of simulations, there is no obstacle placed in the maze. To increase the difficulty of the test, obstacles are added in the maze for the second set of simulations. Eighty-one place cells are used in the simulations. The agent is placed at a random starting point in the maze at the beginning of each learning iteration. The agent then walks in the maze with the direction determined by Eq. (4.33). Once the agent reaches the target region, which is less than 3% of the total area of the maze, a reward is given to the agent and the current learning iteration is ended. If the agent hits the wall or the obstacle, a negative reward is delivered to the agent. There is a

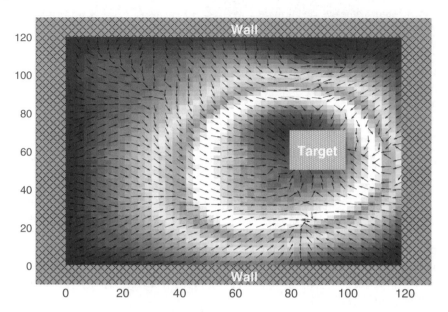

Figure 4.22 State-value function and preferred actions outputted by the critic and actor networks. The brighter the color is, the higher the chance that the agent "thinks" there is a reward. Arrows in the figure represent the preferred moving directions [45]. Reproduced with permission of IEEE.

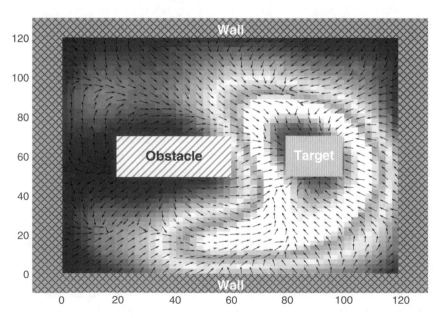

Figure 4.23 State-value function and preferred actions outputted by the critic and actor networks. The brighter the color is, the higher the chance that the agent "thinks" there is a reward. Arrows in the figure represent the preferred moving directions. Compared to Figure 4.22, the color around the obstacle is darker, indicating a possible punishment. Actions are changed accordingly to avoid the obstacle [45]. Reproduced with permission of IEEE.

time limit for the agent to reach the target. If the agent cannot succeed within the time limit, the current learning trial is ended and a new learning iteration is started.

To examine how well the agent learns about walking in the maze, the accumulated rewards received by the agent and the time it takes for the agent to reach the target area during each learning iteration are plotted in Figures 4.24 and 4.25. The results shown in these two figures are filtered to show the trend of the learning. The agent has no knowledge about its surroundings at the beginning, so it is not able to reach the target

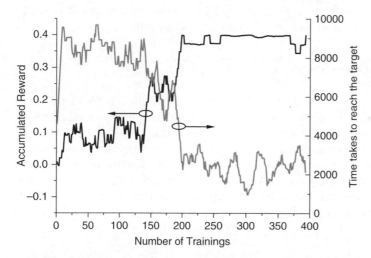

Figure 4.24 Accumulated reward obtained and time it takes for the agent to reach the target versus the number of training conducted. The maze configuration is shown in Figure 4.22. As the training goes on, the agent gradually learns about its environment and starts receiving more rewards [45]. Reproduced with permission of IEEE.

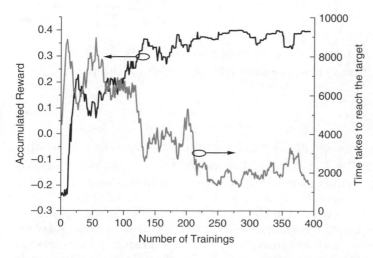

Figure 4.25 Accumulated reward obtained and time it takes for the agent to reach the target versus the number of trainings conducted. The maze configuration is shown in Figure 4.23 [45]. Reproduced with permission of IEEE.

within the time limit, which is 10 000 in these experiments. As the learning continues, the agent starts to learn how to avoid obstacles and walls and even how to move toward the target region. This is demonstrated by the growing accumulated reward and the reduced time-to-destination shown in Figures 4.24 and 4.25. To provide more insights into how the agent learns to perform the task. The outputs from the critic and the actor networks are plotted in Figures 4.22 and 4.23. Color maps in the figures represent the output from the critic network. A bright color in the figure translates to a high reward probability. The arrows in the figures indicate the preferred moving direction outputted by the actor network. Clearly, the agent has acquired the information that it needs to maximize the reward that it can receive in the process of learning.

4.3 Learning in Deep SNNs

In the previous section, various ways of conducting learning in shallow SNNs are introduced. In the literature, it has been shown that deep ANNs can achieve a much better performance through leveraging a hierarchical representation [52–54]. Therefore, in recent years numerous research efforts have been spent developing learning algorithms that can be applied to multilayer SNNs. Intuitively, training a shallow neural network can be thought as a subtask of training a deep network. One key component that is still missing is how to assign the credit to each layer and then to each synapse. In this section, several popular ways of training multilayer SNNs are discussed.

4.3.1 SpikeProp

SpikeProp is one of the earliest methods used for supervised learning in SNNs. This method was originally proposed by Bohte et al. in 2002 [55]. The algorithm is analogous to the backpropagation employed in conventional ANNs. The target is to minimize the difference between the actual spike timing and the desired spike timing

$$E = \frac{1}{2} \sum_j (t_j^a - t_j^d)^2 \tag{4.36}$$

where t_j^a and t_j^d are the actual and designed firing timings.

Similar to the gradient descent learning in ANNs, the update of synaptic weights can be shown to follow

$$\Delta w_{ij}^k = -\alpha \frac{\partial E}{\partial w_{ij}^k} \tag{4.37}$$

where α is the learning rate and $\partial E/\partial w_{ij}^k$ can be derived using the chain rule, which is similar to that used in ANNs. A more detailed derivation of $\partial E/\partial w_{ij}^k$ can be found in [55].

Since the SpikeProp algorithm was introduced, many researchers have continued to develop and improve the algorithm over the past years. The original SpikeProp algorithm outlined in [55] was extended in [56] so that not only the synaptic weights but also the delay and time constant associated with each synapse and the threshold of the neurons could be learned. Such an extension was effective in reducing the number of synapses needed for the same task, which made the neural network more compact. In [57], RProp and QuickProp, which are used to accelerate training in ANNs, have been adapted and applied to SNNs. RProp is an algorithm for adjusting the learning rate, whereas QuickProp is based on Newton's method. McKennoch et al. reported that both of these two methods could help improve the learning speed significantly.

When using the SpikeProp algorithm for learning, a problem one may encounter is the sudden rise of errors in the training process, which is called surge [58]. An in-depth study was carried out by Haruhiko et al. to study the origin of this surge phenomenon. It was concluded that the non-monotonic behavior caused by the spike-response model was the underlying reason for surge. To alleviate this problem, a modified algorithm with an adaptive learning rate technique, called SpikePropAd, was proposed in [59]. It was reported that the proposed algorithm could achieve a faster learning speed compared to the original SpikeProp while suffering less from the surge problem.

4.3.2 Stack of Shallow Networks

Even though many algorithms outlined in Section 4.2 were initially designed for training shallow SNNs, some of them have the potential to be applied to a deep network through stacking [60–63]. With this method, even though the SNN can be deep, the learning is often only conducted on a portion of the network. By doing so, learning in the shallow network can be used while a hierarchical representation available in a deep network can also be exploited.

Many studies on this topic utilize an architecture inspired by HMAX [64]. The idea of HMAX is conceptually illustrated in Figure 4.26. HMAX is inspired by the human visual system. The input image is convolved to form S1 cell responses, which normally detect simple features such as edges. Max pooling operations are conducted on S1 cells to obtain C1 cell responses. One more convolution and max pooling operations are conducted to obtain the responses from S2 cells and C2 cells, respectively. These cells contain features that are of intermediate complexity. A classifier at the end of the pipe can be used for classification purposes. One may notice that this architecture highly resembles an artificial convolutional neural network (CNN) structure discussed in Chapter 2. This should come as no surprise, considering that both HMAX and CNN are inspired by human visual systems.

Many researchers utilize a HMAX-like architecture for feature extraction and image classifications, where shallow learning is often performed on one layer in the network. For example, in [60], a shallow fully connected SNN was combined with a multilayer front-end processing network that consisted of a convolution layer, a competition

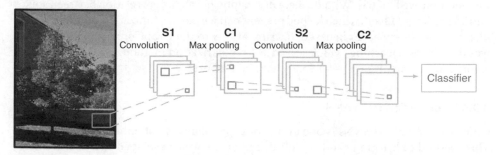

Figure 4.26 The configuration of an HMAX-like architecture. Two alternate convolution and max pooling operations are conducted to form S1, C1, S2, and C2 cells, respectively. At the end of the pipe, a classifier can be used to form the final inference.

layer, a feature-spike conversion layer, etc., to form a deep network. The last layer of the neural network, which served as a classifier, was trained using the tempotron rule. The developed system was able to conduct inference on data provided from an address-event representation (AER) sensor. Information in the input AER signals was extracted layer by layer and the formed high-level features were then classified by the last classification layer.

Another example uses STDP learning in a deep neural network. It was shown in [63] and later on in [61] that unsupervised STDP learning was very effective in learning features in images. In [61], a multilayer SNN was used to conduct image classification, even though the STDP learning was only conducted in the layer connecting C1 and S2 cells. S1 cells detected edges in the input images through convolution, whereas C1 cells conducted the max-pooling operation to reduce the dimension of the input. S2 features were the weighted combination of C1 features and contained the intermediate-complexity features that were suitable for conducting classification tasks. C2 cells performed the global-max operation by taking the maximum response from S2 cells over all positions and scales. The final classification was done through a simple classifier with the C2 features as its input. The synapses connecting the C1 layer and the S2 layer were trained with an unsupervised STDP learning rule in order to learn good features for the purpose of classification. With such an arrangement, it was reported in [61] that the classifier implemented with the unsupervised STDP learning was able to outperform both the biologically inspired HMAX [65] and AlexNet [66].

Contrast divergence (CD) is the driving horse for learning with restricted Boltzmann machines. It works quite well in ANNs through stacking multiple layers of layer-wise trained neural networks to form a deep belief network. Neftci et al. extended the CD algorithm into SNNs with the help of STDP [67]. An event-driven CD rule was developed. In the event-driven CD algorithm, the two phases needed in a CD algorithm are conducted with the help of a global gating signal $g(t)$. The synaptic weight update, in this case, is conducted based on a symmetrical STDP learning rule modulated by the global gating signal. By utilizing this scheme, effects similar to the conventional CD algorithm can be achieved.

The algorithm was tested on the MNIST dataset, where 784 visible units corresponding to the 784 pixels from the images were employed. Forty output neurons were used to reduce the spike-to-spike correlation. With the event-driven CD, a 91.9% recognition rate was achieved on the MNIST dataset. The performance of the event-driven CD was further improved in [68]. With the help of synaptic noise, the correlation between spikes and spikes was reduced, which helped prevent pairwise synchronization. In addition, the blank-out in the synapses behaved in a similar way to the DropConnect technique that was used in deep learning, which served as a regularization method. With the upgraded algorithm, a recognition rate of 95.6% has been reported.

4.3.3 Conversion from ANNs

Converting ANNs to SNNs is one of the most popular ways of obtaining useful SNNs. This method of learning can take advantage of the well-developed network architectures and theories in ANNs. The first step in converting is to establish the relationship between an SNN neuron and an ANN neuron. This can be done empirically [69, 70] or mathematically [71, 72]. Cao et al. assumed that the dynamics of a spiking neuron can be approximated using a linear neuron followed by a rectified linear unit (ReLU) activation function [69]. Three design strategies have been used to restrict the ANN in order to achieve a successful mapping. (i) Ensure that the outputs from all neurons are positive. This is mainly because it is more natural for SNNs to represent positive numbers with the spiking rate. The positivity can be ensured by feeding only positive inputs to the network and using ReLU as the activation function. (ii) Set all the bias to zero. (iii) Use average pooling instead of max pooling. This is mainly because an average pooling can be naturally implemented in SNNs.

With these restrictions on the ANN, the methodology of converting from ANNs to SNNs has been verified with the Neovision2 Tower dataset and the CIFAR-10 dataset in [69]. A similar performance was reported for the converted SNN when compared to the ANN baseline.

The conversion method outlined in [69] was further refined in [70]. Diehl et al. discussed three reasons why the converted SNNs suffered from the performance degradation. The first reason was that some neurons did not receive sufficient input to spike. The second reason was that some neurons received too much input current, but it could only spike once, which might introduce errors in the mapping. The third reason was that due to statistical fluctuation of spiking input, certain features might be over- or underactivated. To address these issues, a weight normalization procedure was introduced in [70]. Two normalization methods were proposed. The normalization techniques were tested on the standard MNIST dataset and recognition rates of 98.6% and 99.1% were achieved by a spiking fully connected neural network (FCNN) and a spiking CNN, respectively. Such good classification accuracy is one of the highest recognition rates that has been achieved for SNNs in the literature.

In addition to the empirically derived activation functions, an activation function for the artificial neuron was derived from the LIF model in [71]. A Siegert neuron model was employed to predict the average firing rate of the LIF model when the input obeyed the Poisson process. With the proposed converting method, a deep belief networks that was trained offline was mapped to an efficient event-driven SNN. It was demonstrated that the degradation in the recognition rate associated with the conversion was less than 1%.

Even though many operations in ANNs can be naturally converted in SNNs with the help of a good neuron model, it is still not trivial to implement some common operations that are often used in ANNs, such as softmax and max pooling, in SNNs. In [73], a complete methodology of converting an ANN into an SNN was introduced, which allowed the conversion of nearly all CNN structures. Many ANN common operations that were not previously being mapped to an SNN were studied, including batch normalization, max pooling, softmax, and inception modules. Thanks to the mapping methodology, well-trained ANNs could be converted, and state-of-the-art classification accuracies have been reported [73].

Converting ANNs to SNNs indeed provides a convenient and flexible way of obtaining SNNs that can be used for energy-efficient inference. The biggest advantage of this approach is that there is no training algorithm needed for SNNs. Rather, well-developed learning methodologies for ANNs can be used instead. Nevertheless, the main drawback of this method is that it fails to provide an online (on-chip) learning capability since this method relies on an off-line conversion. There might be some performance degradations associated with the conversion, especially when there are certain restrictions on the SNN hardware. Online learning is very essential for many applications where learning is needed in-situ. One example is when implementing SNNs with nanoscale devices where variations of the devices are substantial. On-chip learning, in this case, is needed to compensate for the variations. This is discussed in depth in Chapter 5.

4.3.4 Recent Advances in Backpropagation for Deep SNNs

As discussed in Chapter 2, deep ANNs have achieved tremendous successes in recent years. Inspired by the backpropagation-based SGD, the driving horse of training deep ANNs, many researchers started looking into backpropagation in deep SNNs. The main difficulty of using SGD directly in SNN learning is that outputs from SNNs are discrete events. In contrast to an artificial neuron where the output is a differentiable continuous variable, the discrete nature of a spike does not lead to a natural definition of the gradient. As a consequence, it is not a straightforward task to apply a conventional SGD learning algorithm to an SNN directly.

To circumvent this difficulty, one approach that many researchers have used recently is to define a differentiable auxiliary quantity that is related to the discrete spike outputs. The hope is that by controlling these differentiable quantities, the output from neurons can be controlled as well. One example is to abstract the output from a spiking neuron as a random variable [45, 74, 75]. The abstracted probability is differentiable and can be used for backpropagation and SGD. Another example is to use the membrane voltage of a spiking neuron instead of its output as the differentiable signal for training [76]. Even though the membrane voltage is not differentiable in a strict sense as it has discontinuity when spikes occur, Lee et al. showed that this discontinuity could be treated as noise and the resultant learning performance was still quite impressive. One more way of tackling the non-differentiable problem was presented in [77], where an approximation of the derivative was proposed to estimate the gradients. Such a method shares a similar spirit to that of the straight-through estimator used in binarized networks [78]. Indeed, a binarized network faces a similar problem where the derivative of a hard-threshold activation function is ill-defined.

Another concern with using backpropagation in SNNs is the so-called weight transport problem [79–81]. In a conventional backpropagation algorithm, the weights used for a forward pass and a backward pass should be identical, which appears to be biologically implausible. In addition to the biological requirement, the requirement of identical weights might induce some difficulties when implementing certain analog SNNs where synaptic weights are not directly accessible. To address this problem, random feedback has been proposed where instead of scaling the errors at the outputs by the synaptic weights, the errors were scaled randomly [80, 81]. It was demonstrated that with such a simplified processing, satisfactory learning performance could still be achieved. An alternative way of avoiding using weights in the backward phase is to use spike timings for estimating gradients. This is discussed in great detail in Section 5.3.5.

Looking for the possibility of conducting backpropagation in SNNs is not a new topic yet it has started to receive more and more attention in recent years, partially driven by the success achieved by backpropagation in deep ANNs. It is expected that more and more efficient and high-performance backpropagation algorithms will be developed in the near future. In the next section, we discuss one type of backpropagation algorithm in SNNs in detail.

4.3.5   Learning Through Modulating Weight-Dependent STDP in Multilayer Neural Networks

4.3.5.1   Motivations

An enormous amount of effort was made from both the computational intelligence community and the neuroscience community to develop learning algorithms for SNNs. Many of them have been reviewed in previous sections. For example, the STDP learning rule presented in Section 4.2.3 has shown some promising results in many unsupervised learning tasks. However, how to employ an STDP learning rule in a deep neural network for supervised learning and reinforcement learning is still not obvious. Many learning algorithms such as SpikeProp, ReSuMe, etc., attempt to learn the precise firing timings of neurons. In spite of being effective, it is still debatable what the optimum way is to encode information on spikes.

A learning algorithm for training SNNs with a backpropagation-like scheme was proposed in [75]. It was a natural extension of the learning algorithm outlined in Section 4.2.4. The learning rule was formulated with two objectives in mind. The first objective is to provide a hardware-friendly learning algorithm that is suitable for implementing in very large-scale integration (VLSI) circuits. The second objective is to stay relatively compatible with the conventional ANN-based learning so that various design methodologies, techniques, and network architectures can be readily adapted and used in SNNs. In this section, the main results from [75] are presented to study backpropagation-based learning in multilayer SNNs in depth.

4.3.5.2   Learning Through Modulating Weight-Dependent STDP

The configuration of the multilayer neural network studied in this section is illustrated in Figure 4.27. In the figure, there are N_l neurons at the lth layer of the neural network and x_i^l denotes the *ith* neuron at that layer. In this section, we restrict ourselves to a discrete-time system, as this is the most popular choice in hardware SNNs.

Following the convention used in TrueNorth, we assume spikes in our system occur synchronously with a time unit called a tick [82]. In the remainder of this section, a tick is used as the minimum temporal resolution as well as the unit for time-related quantities. The spike trains from a presynaptic neuron x_i^l and a postsynaptic neuron x_j^{l+1}, which can

Figure 4.27 Illustration of a multilayer neural network. A neuron located at the lth layer is denoted as $x_{j,}^l$ where i represents the index of that neuron [75]. Reproduced with permission of IEEE.

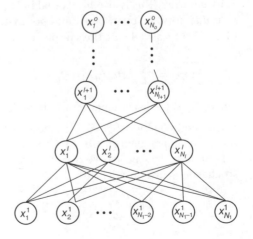

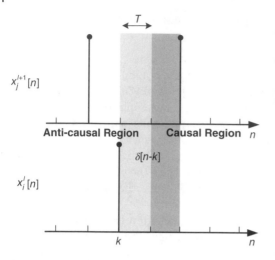

Figure 4.28 Illustration of the two regions divided by the spike timing of a presynaptic neuron. The causal and anti-causal regions are defined according to the causal relationship between the presynaptic and postsynaptic spikes [75]. Reproduced with permission of IEEE.

be represented as

$$x_i^l[n] = \sum_m \delta[n - n_{i,m}^l] \tag{4.38}$$

$$x_j^{l+1}[n] = \sum_m \delta[n - n_{j,m}^{l+1}] \tag{4.39}$$

where $\delta[n]$ is the unit sample sequence and $n_{i,m}^l$ and $n_{j,m}^{l+1}$ are spike timings for the mth spikes from neuron x_i^l and neuron x_j^{l+1}, respectively. Note that since we focus on discrete-time systems in this section, the notations used here are slightly different from those used in Section 4.2.4 in order to emphasize the discrete-time nature. Figure 4.9 is also replotted in Figure 4.28 to reflect the changes in the notation. In Eqs. (4.38) and (4.39), a constant postsynaptic potential is used, considering its ease of implementation in hardware SNNs.

Similar to the derivation presented in Section 4.2.4, let us define a quantity as

$$stdp_{ij}^l[n] = x_i^l[n - T](1 - x_i^l[n - T - 1])(x_j^{l+1}[n] - x_j^{l+1}[n - 1]) \tag{4.40}$$

Here, we explicitly take the time delay of the employed neuron model, T, into the consideration. This is also illustrated in Figure 4.28. The quantity $stdp_{ij}^l[n]$ measures the causality between presynaptic and postsynaptic spikes, similar to the one defined in Eq. (4.20). Let us also define the sample mean of $stdp_{ij}^l[n]$ as

$$\overline{stdp_{ij}^l} = \sum_{n=T+1}^{D_L} stdp_{ij}^l[n]/(D_L - T) \tag{4.41}$$

where D_L is the learning duration. It serves as a design parameter whose implication is discussed in Section 4.3.5.3. It is shown shortly that the quantity, $\overline{stdp_{ij}^l}$, measures how the postsynaptic neuron alters its behavior when a presynaptic spike is observed.

For analysis purposes, we treat spikes as stochastic processes. We consider a class of stochastic neuron model with the dynamics of

$$X_j^{l+1}[n] = H\left(\sum_{i=1}^{N_l} w_{ij}^l X_i^l[n - T] + S_j^{l+1}[n - T]\right) \tag{4.42}$$

where $S_j^{l+1}[n]$ is a random process used to model the internal state of neuron x_j^{l+1} and $H(\cdot)$ is the Heaviside function. The spike trains $x_i^l[n]$ and $x_j^{l+1}[n]$ in Eqs. (4.38) and (4.39) are particular realizations of the random processes $X_i^l[n]$ and $X_j^{l+1}[n]$ in (4.42). Let us make two assumptions on this neuron model:

A1. $X_i^l[n]$ and $X_k^l[n]$ are independent for $k = 1, 2, \ldots, N_l$, and $k \neq i$.

A2. $X_i^l[n]$ and $X_j^{l+1}[n]$ are strictly stationary processes, and $C_{X_i^l X_j^{l+1}}(n, m) = 0$ for $n \neq m - T$. $C_{X, Y}(\cdot)$ here stands for the cross-covariance function.

The goal here is to derive how the mean firing rate of neuron x_j^{l+1}, which is denoted as μ_j^{l+1}, is related to μ_i^l, which is the mean firing rate of neuron x_i^l. The first step is to show that μ_j^{l+1} is a function of $\boldsymbol{\mu}^l$, where $\boldsymbol{\mu}^l = [\mu_1^l, \mu_2^l, \cdots, \mu_{N_l}^l]^T$ is a vector containing all the mean firing rates of the neurons in the lth layer.

The mean firing rate of neuron x_j^{l+1} can be written as

$$\mu_j^{l+1} = Pr(X_j^{l+1} = 1) = \sum_{b_1^l=0}^{1} \cdots \sum_{b_{N_l}^l=0}^{1} \left[Pr(X_j^{l+1} = 1 \mid X_1^l = b_1^l, \ldots, X_{N_l}^l = b_{N_l}^l) \prod_{i=1}^{N_l} Pr(X_i^l = b_i^l) \right]$$

$$= \sum_{b_1^l=0}^{1} \cdots \sum_{b_{N_l}^l=0}^{1} \left\{ Pr(X_j^{l+1} = 1 \mid X_1^l = b_1^l, \ldots, X_{N_l}^l = b_{N_l}^l) \prod_{i=1}^{N_l} [\mu_i^l(2b_i^l - 1) - b_i^l + 1] \right\}$$

$$= g(\boldsymbol{\mu}^l) \tag{4.43}$$

In Eq. (4.43), b_i^l is a binary auxiliary variable for the ease of derivation. Note that the index n is omitted for a cleaner notation, considering that we deal with strictly stationary processes. Next, we show that $g(\cdot)$ is differentiable with respect to $\boldsymbol{\mu}^l$ and its mth derivative $\partial^m g/\partial(\mu_i^l)^m = 0$ for $m > 1$.

The first derivative of $g(\boldsymbol{\mu}^l)$ with respect to μ_i^l can be written as

$$\frac{\partial g}{\partial \mu_i^l} = \sum_{b_1^l=0}^{1} \cdots \sum_{b_{N_l}^l=0}^{1} \left[Pr(X_j^{l+1} = 1 \mid X_1^l = b_1^l, \ldots, X_{N_l}^l = b_{N_l}^l) \prod_{\substack{k=1 \\ k \neq i}}^{N_l} [\mu_k^l(2b_k^l - 1) - b_k^l + 1] (2b_i^l - 1) \right] \tag{4.44}$$

Clearly, $\partial g/\partial \mu_i^l$ is no longer a function of μ_i^l. Therefore, we have $\partial^m g/\partial(\mu_i^l)^m = 0$ for $m > 1$.

We then show that $\overline{stdp_{ij}^l}$ defined in (4.41) is an unbiased estimator of the term $(\partial \mu_j^{l+1}/\partial \mu_i^l)\mu_i^l(1 - \mu_i^l)$. According to the law of large numbers, we have

$$E[\overline{stdp_{ij}^l}] = Pr(X_j^{l+1} = 1, X_i^l = 1, X_i^{l'} = 0) - Pr(X_j^{l+1} = 1, X_i^l = 0, X_i^{l'} = 1)$$

$$= [Pr(X_j^{l+1} = 1 \mid X_i^l = 1) - Pr(X_j^{l+1} = 1 \mid X_i^l = 0)]\mu_i^l(1 - \mu_i^l) \tag{4.45}$$

where X_i^l and $X_i^{l'}$ are used to denote the random variables that are dependent and independent on X_j^{l+1}, respectively.

From Eq. (4.45), one can obtain

$$\frac{E[stdp_{ij}^l]}{\mu_i^l(1 - \mu_i^l)} = Pr(X_j^{l+1} = 1 \mid X_i^l = 1) - Pr(X_j^{l+1} = 1 \mid X_i^l = 0)$$

$$= g(\boldsymbol{\mu^l} + (1 - \mu_i^l)\mathbf{u_i^l}) - g(\boldsymbol{\mu^l} - \mu_i^l\mathbf{u_i^l}) = g(\boldsymbol{\mu^l}) + \frac{\partial g}{\partial \mu_i^l}(1 - \mu_i^l) - g(\boldsymbol{\mu^l}) - \frac{\partial g}{\partial \mu_i^l}(-\mu_i^l)$$

$$= \frac{\partial g}{\partial \mu_i^l} = \frac{\partial \mu_j^{l+1}}{\partial \mu_i^l} \tag{4.46}$$

where $\mathbf{u_i^l}$ is a unit vector in which all elements are zero except that the ith element is one. In Eq. (4.46), we have used the fact that $\partial^m g/\partial(\mu_i^l)^m = 0$ for $m > 1$ and the assumption that X_i^l and X_k^l are independent for $i \neq k$. The implication of Eq. (4.46) is that the quantity $\underline{stdp_{i,j}^l}$ defined in Eq. (4.41) is an unbiased estimator of the term $(\partial\mu_j^{l+1}/\partial\mu_i^l)\mu_i^l(1 - \mu_i^l)$.

Clearly, in deriving Eq. (4.46), we gain an effective tool to estimate the gradient information needed for stochastic gradient decent learning. The derivation of Eq. (4.46) utilizes the two assumptions A1 and A2, which might be too strong to be held for all the neuron models. As discussed in Section 4.2.4, despite the fact that the firing rate of each presynaptic neurons can be correlated, the spike timing of each neuron may remain largely uncorrelated. Therefore, A1 is a reasonable assumption that can be satisfied by many neuron models. The assumption A2 is a much stronger assumption that requires X_i^l and X_j^{l+1} to be strictly stationary random processes. A2 also implies that $S_j^{l+1}[n]$ needs to be strictly stationary as well. Such a requirement is hard to satisfy for neuron models with memory. Nevertheless, the dependence of $S_j^{l+1}[n]$ on the previous presynaptic and postsynaptic spikes can be diluted through the deliberate injection of noise. More conveniently, the quantization residue injection technique illustrated in Section 4.2.4 can also be employed for this purpose. This arrangement is elaborated in Section 4.3.5.3.1. A natural extension of Eq. (4.40) is to include more samples into the time sequence $stdp_{ij}^l[n]$ as

$$stdp_{ij}^l[n] = x_i^l[n - T](1 - x_i^l[n - T - 1]) \left(\sum_{m=1}^{WIN_{STDP}} x_j^{l+1}[n + m - 1] - \sum_{m=1}^{WIN_{STDP}} x_j^{l+1}[n - m] \right) \tag{4.47}$$

where WIN_{STDP} is a design parameter used to specify the window size of the summation. The purpose of introducing this parameter is to incorporate the effects of delayed perturbated outputs. Such an extension in a summation window is inspired by the biological STDP protocol where an exponential integration window is observed.

The intuition behind Eq. (4.46) is that the gradient of the postsynaptic mean firing rate with respect to the presynaptic firing rate can be estimated by observing how the postsynaptic neuron changes its behavior when a small perturbation in the form of a presynaptic spike is applied to the network. This is similar to Eq. (4.22) in Section 4.2.4, even though they are derived in different ways. Compared to the learning rule formulated in Section 4.2.4, the definition of the quantity $stdp_{ij}^l[n]$ and the way to estimate the gradient are slightly different. Nevertheless, when the spike trains are sparse, the two methods are very similar.

Once $\partial\mu_j^{l+1}/\partial\mu_i^l$ is obtained, a deriving procedure similar to the one presented in Section 4.2.4 can be followed. Let us assume the input–output relationship in the neuron model can be represented as

$$\mu_j^{l+1} \approx f_j^{l+1}\left(\sum_i w_{ij}^l \mu_i^l\right) \tag{4.48}$$

Taking the derivative with respect to μ_i^l in Eq. (4.48), one can obtain

$$f_j^{l+1'}\left(\sum_i w_{ij}^l \mu_i^l\right) = \frac{\partial\mu_j^{l+1}}{\partial\mu_i^l} / w_{ij}^l \tag{4.49}$$

Then, using Eqs. (4.46) and (4.49), we arrive at

$$\frac{\partial\mu_j^{l+1}}{\partial w_{ij}^l} = \mu_i^l f_j^{l+1'}\left(\sum_i w_{ij}^l \mu_i^l\right) = \frac{E[\overline{stdp_{ij}^l}]}{w_{ij}^l(1-\mu_i^l)} \tag{4.50}$$

It is worth noting that even though Eqs. (4.46) and (4.50) provide theoretical guidelines on how to estimate the gradients in an SNN in order to perform gradient descent learning, in practice we use the following equations to approximate the gradient information needed:

$$\frac{\partial\mu_j^{l+1}}{\partial\mu_i^l} \approx \overline{stdp_{ij}^l} / [\overline{x_i^l}(1-\overline{x_i^l})] \tag{4.51}$$

$$\frac{\partial\mu_j^{l+1}}{\partial w_{ij}^l} \approx \overline{stdp_{ij}^l} / [w_{ij}^l(1-\overline{x_i^l})] \tag{4.52}$$

In these equations, $\overline{x_i^l} = \sum\limits_{n=T+1}^{D_L} x_i^l[n]/(D_L - T)$.

In order to employ the learning algorithm in a deep neural network, we need a way to propagate the error from observable locations (typically at output neurons) back to each synapse in the network. With the gradient information estimated from spike timings, the backpropagation can be readily achieved through

$$\frac{\partial\mu_k^o}{\partial w_{ij}^l} = \frac{\partial\mu_k^o}{\partial\mu_j^{l+1}} \cdot \frac{\partial\mu_j^{l+1}}{\partial w_{ij}^l} \tag{4.53}$$

where the term $\partial\mu_j^{l+1}/\partial w_{ij}^l$ can be estimated according to Eq. (4.52), whereas the term $\partial\mu_k^o/\partial\mu_j^{l+1}$ can be obtained by

$$\frac{\partial\mu_k^o}{\partial\mu_j^{l+1}} = \sum_{i_o=k}^{k}\sum_{i_{o-1}=1}^{N_{o-2}}\cdots\sum_{i_{l+2}=1}^{N_{l+2}}\sum_{i_{l+1}=j}^{j}\prod_{p=l+1}^{o-1}\frac{\partial\mu_{i_{p+1}}^{p+1}}{\partial\mu_{i_p}^p} \tag{4.54}$$

Such a method of propagating errors from the output neurons to each layer is similar to the backpropagation process in conventional ANNs. Alternatively, the gradient can be directly propagated as

$$\frac{\partial \mu_k^o}{\partial \mu_j^{l+1}} = \frac{\overline{E[cstdp_{jk}^{l+1}]}}{\mu_j^{l+1}(1 - \mu_j^{l+1})} \tag{4.55}$$

where $cstdp_{jk}^{l+1}[n]$ is a quantity that is similar to the one defined in Eq. (4.40). The difference between $cstdp_{jk}^{l+1}[n]$ and $stdp_{jk}^{l+1}[n]$ is that $cstdp_{jk}^{l+1}[n]$ measures the causality between the $(l+1)$th layer and the output layer whereas $stdp_{jk}^{l+1}[n]$ only measures the causality in adjacent layers. Intuitively, similar to $stdp_{jk}^{l+1}[n]$, which measures how the postsynaptic neuron changes its firing probability upon a presynaptic spike, $cstdp_{jk}^{l+1}[n]$ measures how the output neuron behaves when the same presynaptic neuron spikes.

Numerical simulations are conducted in order to verify the proposed method of estimating gradients in neural networks. The modified LIF neuron model employed in Section 4.2.4 is used. The dynamics of the neuron model is shown as

$$x_j^{l+1}[n] = \begin{cases} 0, & V_j^{l+1}[n] < th_j^{l+1} \\ 1, & V_j^{l+1}[n] \geq th_j^{l+1} \end{cases} \tag{4.56}$$

$$V_j^{l+1}[n] = \max\left(0, V_j^{l+1}[n-1] + \sum_i w_{ij}^l x_i^l[n-1] - L_j^{l+1} - x_j^{l+1}[n-1] \cdot th_j^{l+1}\right) \tag{4.57}$$

where $V_j^{l+1}[n]$ is the membrane voltage associated with neuron x_j^{l+1} at tick n, L_j^{l+1} is the leakage, and th_j^{l+1} is the threshold to fire. It is shown in Section 4.2.4 that this neuron model leverages the readily available quantization residue to decorrelate spike timings, which helps achieve a better learning. We use this neuron model throughout this section unless otherwise stated. Nevertheless, this neuron model is not the only model that our learning algorithm can be applied to. It will shortly be demonstrated in Section 4.3.5.3.1 that, with a proper noise injection, the learning algorithm can be applied to a conventional LIF model as well.

The neural network employed in the numerical simulations is a four-layer neural network with 80 input neurons, 30 neurons in the first hidden layer, 100 neurons in the second hidden layer and 1 output neuron. The input neurons in the network are injected with fixed excitatory currents at every tick. The injected currents are randomly picked at the beginning of the simulations. Ten sets of experiments were performed. For each set of experiments, 100 synapses in each layer of the neural network were randomly picked for observation. For each synapse, its associated gradient information was estimated using two methods: estimation based on spike timings and the finite-difference (FD) method. The FD method relies on applying a small perturbation to each synapse, and then observing how the firing rate of the output neuron changes. The numerical value for the gradient is then obtained by dividing the change in firing density by the change in the synaptic weight. It is worth noting that the gradient obtained from this numerical method is only a noisy estimation of the true gradient due to the complicated dynamics of SNNs. Nevertheless, comparing with the numerical gradients provides some useful insights into how well we can estimate gradients from spike timings.

Table 4.1 Information for the limiting operations used to obtain data in Figures 4.29 to 4.33.

Layer number	Clamping threshold	Number of outliers
1	±0.05	1
2	±0.05	4
3	±0.5	16

Source: data are from [75]. Reproduced with permission of IEEE.

Even though the weight denominator shown in Eq. (4.52) introduces some benefits, such as soft-limiting the maximum synaptic weight, it has the side effect that the quantization noise in the density of the spike might cause a large estimated gradient when the synaptic weight is small. A solution to this problem is to use clamping operations to limit the minimum and maximum gradients. The detailed information on the limiting operation, such as the allowed gradient range and the number of points being clamped, is illustrated in Table 4.1. The gradients associated with each layer are compared in Figures 4.29 to 4.31. In total, 1000 data points are collected for each layer in all the experiments. The gradients estimated from spike timings match well with the gradients obtained from the FD method. It is worth noting that the relatively low correlation for w_{ij}^3 is due to a few negative outliers, shown in Figure 4.31. These outliers can be easily filtered out if the fact that there is no negative gradient for the last layer is considered.

Two sets of simulations are conducted to evaluate the implication of two design parameters: WIN_{STDP} and D_L. The correlation between the estimated gradients and the numerical gradients are shown in Figure 4.32 for different WIN_{STDP}. There is no noticeable correlation between the accuracy in estimating gradients and the choice of WIN_{STDP}. Some preliminary numerical studies on the effect of WIN_{STDP} on learning also show that different window sizes do not have much effect on the

Figure 4.29 Comparison of the gradients obtained from the numerical simulations and the gradients estimated based on spike timings. The gradients are associated with the first-layer synapses w_{ij}^1 [75]. Reproduced with permission of IEEE.

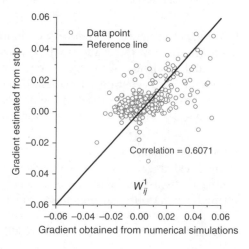

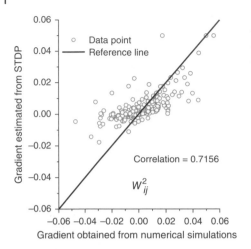

Figure 4.30 Comparison of the gradients obtained from the numerical simulations and the gradients estimated based on spike timings. The gradients are associated with the second-layer synapses w_{ij}^2 [75]. Reproduced with permission of IEEE.

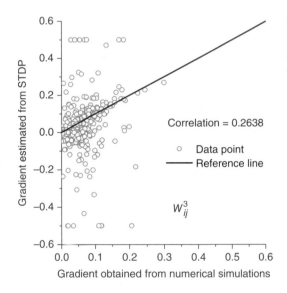

Figure 4.31 Comparison of the gradients obtained from the numerical simulations and the gradients estimated based on spike timings. The gradients are associated with the third-layer synapses w_{ij}^3 [75]. Reproduced with permission of IEEE.

learning performance. Therefore, the window size is set to one in this study. Figure 4.33 demonstrates how the change in the evaluation duration influences the accuracy of the estimated gradients. In the figure, one can observe that a longer evaluation duration leads to a more accurate estimation. Indeed, for a stochastic approximation method, any unbiased noise observed in the process of evaluation can be filtered out through averaging. The longer the averaging window, the more accurate the results are.

4.3.5.3 Simulation Results

In the previous section, how the gradient information can be estimated from the spike timing is discussed. With the estimated gradients, an SGD method can be readily formed for various learning tasks. Since supervised learning is the most popular learning method, we use it here as an example for demonstration. However, the proposed learning algorithm can be readily extended to other learning schemes, such

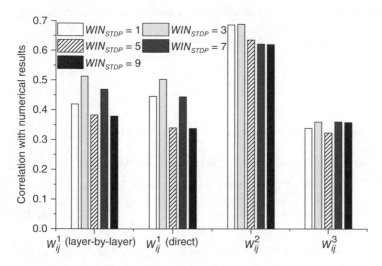

Figure 4.32 Correlations between the estimated gradients and the gradients obtained with the FD numerical method. The results obtained for all three layers of synaptic weights (w_{ij}^1, w_{ij}^2, and w_{ij}^3) are compared. Two different backpropagation methods (layer-by-layer and direct) are also compared. Different window sizes for evaluating STDP are compared. The estimation accuracy does not show significant dependency on the size of the STDP window [75]. Reproduced with permission of IEEE.

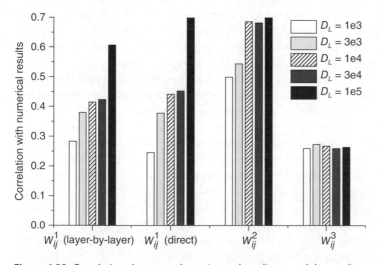

Figure 4.33 Correlations between the estimated gradients and the gradients obtained with the FD numerical method. The results obtained for all three layers of synaptic weights (w_{ij}^1, w_{ij}^2, and w_{ij}^3) are compared. Two different backpropagation methods (layer-by-layer and direct) are also compared. Different evaluation durations are used. The longer the evaluation duration, the more accurate the estimated gradients are [75]. Reproduced with permission of IEEE.

as reinforcement learning, which is demonstrated with a similar algorithm in Section 4.2.4.

As discussed in Chapter 2, in supervised learning, the target is to minimize the error function

$$E = \frac{1}{2} \sum_{k=1}^{N_o} (e_k^o)^2 \tag{4.58}$$

where $e_k^o = \overline{x_k^o} - t_k^o$ is the error at each output neuron and t_k^o is the expected mean firing rate of neuron x_k^o.

The error function can be minimized by updating the synaptic weights of the neural network according to

$$\Delta w_{ij}^l = -\alpha \cdot \sum_{k=1}^{N_o} \frac{\partial E}{\partial \mu_k^o} \cdot \frac{\partial \mu_k^o}{\partial w_{ij}^l} = -\alpha \cdot \sum_{k=1}^{N_o} e_k^o \cdot \frac{\partial \mu_k^o}{\partial w_{ij}^l} \tag{4.59}$$

where α is the learning rate. The term $\partial \mu_k^o / \partial w_{ij}^l$ can be computed according to Eq. (4.53).

To demonstrate the efficacy of the proposed learning algorithm in performing supervised-learning tasks, simulations are conducted on two neural networks. The configuration of the networks are selected based on the two examples presented in [83]. The MNIST benchmark task introduced in Section 2.4 is employed for the simulations. In order to input the intensities of the grayscale images in the MNIST data set into the SNNs, a proper encoding scheme is needed. In this study, we encode the floating-point numbers into spike trains with the modified LIF neuron model. The intensity is injected into the input neuron as excitatory current. Such an encoding scheme behaves similarly to a $\Sigma - \Delta$ modulator, which is famous for converting high-precision data into low-precision ones. This encoding method is also convenient in hardware implementation. The firing densities of the input neurons are proportional to the intensities of the image with the proposed encoding scheme.

The objective of the learning is to correctly classify the digit that is presented to the neural network. In both of these two neural networks, there are 10 output neurons and each of these neurons represents one output label, ranging from 0 to 9. During the learning phase, when an image of the digit i is presented to the network, the output neuron x_i^o is expected to fire with a high firing density, μ_H, whereas all other output neurons should fire with a low density, μ_L. The firing density of the neuron x_i^o is measured by $\overline{x_i^o}$ over a learning duration of D_L. The error function is then calculated according to (4.58) and the consequent weight updates are conducted to minimize this error function. During the test phase, the test image is presented to the neural network and an inference duration of D_I is spent for the neural network to conduct inference. At the end of D_I ticks, the neuron that fires most frequently is chosen as the wining neuron and its corresponding digit is read out as the inferred result.

4.3.5.3.1 *Three-Layer Neural Network*
The first neural network we consider is a three-layer SNN with 784 input neurons, 300 hidden-layer neurons, and 10 output neurons. To accelerate the simulations, the three-layer neural network is trained with the first 500 training images in the training set. The test process is conducted for all 10 000 images in the test set. All the results presented in this section are obtained from

10 independent runs. Error bars that correspond to the 95% confidence interval are plotted together with the simulation data.

The proposed method of estimating gradient starts losing accuracy as the spike trains become dense, which is implied by the term $(1 - \mu_j^{l+1})$ in the denominator of Eq. (4.50). This degradation in accuracy is mainly because when the spike train is dense, it is hard to distinguish a causal spike from an anti-causal spike. In Section 4.2.4, it is suggested that the density of the spike trains can be lowered through making the temporal resolution of the system finer, which prevents the spike density of any neuron being too high. This is analogous to avoiding the neuron activations to be close to 0 or 1 in a conventional ANN, which would otherwise slow down the learning process significantly. A more convenient alternative solution to this problem was proposed in [75], which utilized a bio-inspired refractory mechanism to avoid dense spike trains. With this method, we impose a refractory period for each neuron in the SNN. The neuron is not allowed to spike when it is in the refractory period. Clearly, the sparsity of spikes can be adjusted conveniently by choosing a proper refractory period.

With the introduction of the refractory mechanism, the dynamics of the neuron shown in Eq. (4.56) is then changed to

$$
x_j^{l+1}[n] = \begin{cases} 0, & V_j^{l+1}[n] < th_j^{l+1} \\ 1, & V_j^{l+1}[n] \geq th_j^{l+1} \text{ and } x_j^{l+1}[n-1] = 0 \\ 1 - R, & V_j^{l+1}[n] \geq th_j^{l+1} \text{ and } x_j^{l+1}[n-1] = 1 \end{cases} \tag{4.60}
$$

where R is a random variable with a Bernoulli distribution $B[1, p_r]$. Here, p_r is a design parameter that is used for controlling the sparsity. A larger p_r can lead to sparser spike trains. The reason to utilize a stochastic refractoriness instead of a deterministic refractoriness is that all the saturated neurons would have highly correlated spike timings if a fixed refractory period were used for all the neurons. Such a subtlety is demonstrated soon with numerical simulation results. Another point that is worth noting is that for the purpose of illustration, the maximum refractory period that can be implemented with Eq. (4.60) is only one tick. Nevertheless, refractory periods of more than one tick can be implemented readily by slightly changing the neuron dynamics shown in Eq. (4.60).

Numerical simulations are conducted to examine the effectiveness of the stochastic refractory scheme. The obtained training correct rate and the test correct rate are shown in Figure 4.34 for cases where different p_r are used. To demonstrate the advantage of a random refractory period over a deterministic refractory period, two sets of initial weights are employed in the simulations. One set of synaptic weights is initialized uniformly from the interval [0, 2]. In other words, $w_{ij}^l \sim U[0, 2]$, where $U[0, 2]$ stands for a uniform distribution between 0 and 2. The other set of synaptic weights is initialized from the interval [0, 8]. The results corresponding to these two sets of initializations are labeled as "2x" and "8x" in Figure 4.34.

When the small initial weights are used, successfull learning is achieved even for the case in which $p_r = 0$. This is because the dense spike trains are avoided through the proper choice of small initial weights. However, when the initial weights are large, the learning performance deteriorates when the non-refractory scheme ($p_r = 0$) is used. In addition, one can observe that when a deterministic refractory period is used ($p_r = 1$), the correct rates for both sets of initial weights are low. Such a degradation in learning performance is caused by the correlated spike timings in saturated neurons. With the

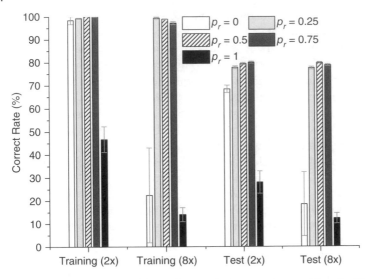

Figure 4.34 Comparison of the training and test correct rates achieved with different levels of refractoriness and different initial weights. The refractory mechanism is helpful in avoiding dense spike trains, which can improve the learning results. Two sets of initial weights are used. One set of weights is uniformly initialized between 0 and 2 (labeled with 2x), whereas another set of weights is uniformly initialized between 0 and 8 (labeled with 8x). Learning performance severely deteriorates when a deterministic refractory period ($p_r = 1$) is used, as all saturated neurons have highly correlated spike timings [75]. Reproduced with permission of IEEE.

stochastic refractory period technique, the learning performance can be improved when a proper p_r is employed, as shown in the figure.

One design consideration in our learning algorithm is the choice of initial conditions for the neurons. In the classification problem presented in this section, the learning is conducted independently for each image. Therefore, a proper set of initial conditions needs to be set up before an image is fed into the network. A pseudorandom initial condition is employed here to address this concern. A convenient choice is a uniform distribution in the interval of $[0, th_i^l]$, i.e. $x_i^l[0] \sim U[0, th_i^l]$. This choice of initial conditions is inspired by the observation that the run-time membrane voltages of an SNN approximately follows such a distribution. Consequently, the use of this initial condition can provide a warm start for the neural network.

Another motivation for the pseudorandom initial condition is that it can help achieve lower correlations among spike timings. For the MNIST dataset we employ in this study, many pixels associated with the strokes in the digit have an intensity of one. This may lead to correlated input spikes even when the modified LIF neuron model is utilized. Through the use of the proposed pseudorandom initial condition, the correlation can be lowered. Similarly, a pseudorandom leakage is also introduced at the input layer to further reduce the correlations among spike timings. From another perspective, the use of a pseudorandom initial condition and leakage is helpful in breaking the possible symmetry in the network, increasing the capacity of the network. Such a symmetry-breaking strategy is widely used by many researchers. Examples are the asymmetric connections in CNNs [83], the random weight initialization [47] in neural networks, and so on. The use of a pseudorandom initial condition and leakage is also easy and hardware-friendly.

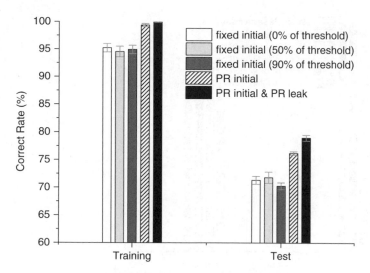

Figure 4.35 Comparison of the training and test correct rates achieved with different initial conditions. The case with pseudorandom initial membrane voltages outperforms the cases with fixed initial membrane voltages. A pseudorandom leakage technique is also employed to further improve the learning performance [75]. Reproduced with permission of IEEE.

No random number generators are actually needed. The values of the initial membrane voltages and leakages can be precomputed and stored conveniently in a static random-access memory (SRAM) array or can be hard-coded in the logic.

Figure 4.35 compares the results obtained from different fixed initial conditions. In the figure, "50% of threshold" means that the initial condition for the neuron is set as $x_i^l[0] \sim U[0, 0.5 \times th_i^l]$. Similar notations apply for other results. Through introducing the pseudorandom initial condition, the correct rate in classifying digits is boosted significantly compared to the fixed initial conditions. In addition, it is observed that the pseudorandom leakage is also helpful in improving the performance remarkably.

As mentioned in Section 4.3.5.2, even though a modified LIF neuron model is more suitable for the learning algorithm presented in this section, a conventional LIF mode can be used as well if proper noise injection is employed. To demonstrate this, Figure 4.36 compares the training and test correct rate obtained from different neuron models. For the conventional model, different levels of noises are injected into the neuron. For example, the "20% white noise" indicates that the noise added to the neuron follows a uniform distribution, $U[0, 0.2 \times th_i^l]$. The results obtained from the modified LIF neuron, which is labeled as "LIF w/quant.residue," is also shown for the purpose of comparison. From the results, one observation is that with the amount of added noise increasing, the obtained results become better. Such an observation is expected as the proposed learning algorithm relies on uncorrelated spike timings. The technique of quantization residue injection can effectively satisfy the need for weakly correlated spike timings without requiring random number generators.

As demonstrated in Figure 4.33, a long evaluation duration can improve the estimation accuracy on the gradients. A natural inference on this is that lengthening the learning duration is helpful in achieving a better classification error. To prove such a hypothesis, Figure 4.37 compares the test correct rate achieved with five different

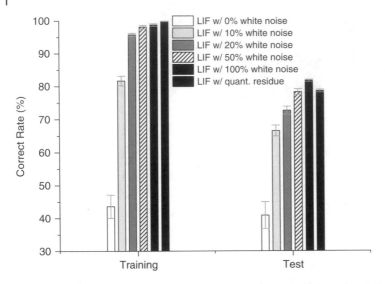

Figure 4.36 Comparison of the training and test correct rates achieved with the LIF neuron model and the modified LIF model. The results obtained with the conventional LIF model with white noise residue injection are labeled as "LIF w/white noise," whereas the results obtained with the modified LIF model is labeled as "LIF w/quant. residue." Learning with the conventional LIF model is effective when enough noise is injected [75]. Reproduced with permission of IEEE.

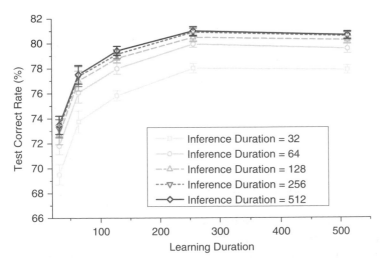

Figure 4.37 Comparison of the test correct rates achieved with different learning and inference durations. The longer the learning or inference duration is, the higher the correct rate [75]. Reproduced with permission of IEEE.

learning and inference durations. The trend that can be observed from this figure is that the longer the learning/inference duration, the higher the recognition rate. Again, such an improvement in performance as the evaluation duration increases is due to the stochastic nature of the algorithm. Any unbiased noise can be filtered out when the number of available samples is enough.

In Figure 4.37, the learning performance ramps up quickly when the learning duration increases from 64 to 128 and the improvement in performance starts saturating when the learning duration reaches approximately 256. With the curves shown in Figure 4.37, one can pick a proper learning duration for training the neural network. A natural strategy is to start the learning with a short learning duration in order to speed up the training. The noise associated with a short evaluation duration might even be helpful in escaping from bad local minima. In addition, it has been demonstrated recently that a noisy gradient is actually beneficial for training deep neural networks [84]. As the learning continues, the learning duration can be gradually lengthened to allow a better convergence. Besides the choice of the learning duration, inference duration also plays an important role in optimizing the system performance. The strategy in picking the inference duration is presented in Chapter 5.

4.3.5.3.2 *Four-Layer Neural Network* To evaluate how well the learning algorithm presented in Section 4.3.5.2 can conduct backpropagation in a deeper neural network, a four-layer neural network is examined. The configuration of the neural network is 784-300-100-10, where each number represents the number of neurons at each layer, from the input layer to the output layer. Similarly, the first 500 training images in the training set is employed for training and the test is conducted with all the 10 000 images in the test set. Figure 4.38 compares the test correct rate achieved by the three-layer neural network and the four-layer neural network. Two backpropagation schemes introduced in Section 4.3.5.2, the direct backpropagation and the layer-by-layer backpropagation, are also compared. The deeper network outperforms the three-layer neural network when a learning duration that is long enough is used for training. Besides, it is observed in the figure that the deep neural network requires a longer learning duration to achieve a reasonable classification accuracy. Such an observation

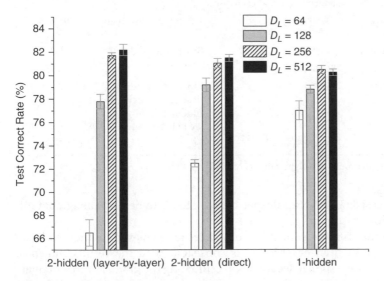

Figure 4.38 Comparison of the test correct rates achieved with the two different backpropagation schemes. The two methods achieve similar performances. The four-layer neural network can yield better performance, but it requires a longer learning duration [75]. Reproduced with permission of IEEE.

Table 4.2 Comparison of the recognition accuracy on the MNIST benchmark test.

Reference	Network type	Network configuration	Learning algorithm	Classification accuracy (%)
[38]	Spiking neural network	784 input neurons + 6400 excitatory neurons + 6400 inhibitory neurons + readout circuit	Unsupervised STDP	95.0
[39]	Spiking neural network	784 input neurons + 300 output neurons + readout circuit	Unsupervised STDP	93.5
[71]	Spiking deep belief network	784-500-500-10	ANN to SNN conversion	94.09
[60]	Spiking convolutional neural network	—	Tempotron	91.29
[70]	Spiking neural network	784-1200-1200-10	ANN to SNN conversion	98.6
	Spiking convolutional neural network	28x28-12c5-2s-64c5-2s-10o		99.1
[85]	Spiking deep belief network	484-256 + linear classifier	ANN to SNN conversion	89
[67]	Spiking deep belief network	784-500-40	Contrastive divergence	91.9
[86]	Spiking neural network	784 input + 10 output neurons each with 200 dendrites and 5000 synapses	Morphological learning	90.26
[68]	Spiking neural network	784-500-10	Contrastive divergence	95.6
[83]	Artificial neural network	784-300-10	Stochastic gradient descent	95.3
		784-300-100-10		96.95
[75]	Spiking neural network	784-300-10	Stochastic gradient descent through modulation of weight-dependent STDP	97.2
		784-300-100-10		97.8

Source: data are from [75]. Reproduced with permission of IEEE.

is similar to what was seen in ANNs: a deeper network tends to perform better, but it is more difficult and slower to train.

It is shown in Figure 4.38 that similar performances are achieved by the two back-propagation methods. Such an observation is consistent with the results shown in Figures 4.32 and 4.33, where similar accuracies in estimating the gradients are obtained. In spite of the similar performances, these two backpropagation methods have different implications in hardware implementations. This point is elaborated in Chapter 5.

4.3.5.3.3 MNIST Benchmark To demonstrate the efficacy of the proposed learning algorithm and to compare the algorithm with the state-of-the-art results, the MNIST task was employed in [75] to benchmark the new algorithm. All 60 000 training images were used for the learning and the trained networks were tested with the whole test set, which contained 10 000 images. The two neural networks presented in two previous sections were trained without any preprocessing step for a fair comparison.

The obtained classification accuracies are compared with the state-of-the-art results in Table 4.2. The three-layer neural network and the four-layer neural network trained with our learning algorithm achieve recognition accuracies of 97.2% and 97.8%, respectively. Such accuracies are higher than those achieved by ANNs with the same network configurations [83], which demonstrates the efficacy of our learning algorithm. In addition, compared to the state-of-the-art result obtained from an FCNN, 98.6% in [70], which was achieved through an off-line ANN-to-SNN conversion with a network that is nine times larger than ours, our result is only slightly inferior.

References

1 Maass, W. (1997). Networks of spiking neurons: the third generation of neural network models. *Neural Netw.* 10 (9): 1659–1671.
2 Abdalrhman, M. (2012). Spiking neural networks. *Int. J. Neural Syst.* 19 (1907): 1802–1809.
3 Hodgkin, A.L. and Huxley, A.F. (1952). A quantitative description of membrane current and its application to conduction and excitation in nerve. *J. Physiol.* 117 (4): 500–544.
4 Gerstein, G. and Mandlebrot, B. (1964). Random walk models for the spike activity of a single neuron. *J. Biophys.* 4 (1): 41–68.
5 Tuckwell, H.C. and Richter, W. (1978). Neuronal interspike time distributions and the estimation of neurophysiological and neuroanatomical parameters. *J. Theor. Biol.* 71 (2): 167–183.
6 FitzHugh, R. (1961). Impulses and physiological states in theoretical models of nerve membrane. *Biophys. J.* 1 (6): 445–466.
7 Gerstner, W. and van Hemmen, J. (1992). Associative memory in a network of 'spiking' neurons. *Netw. Comput. Neural Syst.* 3 (2): 139–164.
8 Gerstner, W., Kempter, R., van Hemmen, J.L., and Wagner, H. (1996). A neuronal learning role for sub-millisecond temporal coding. *Nature* 383 (6595): 76–78.
9 Kempter, R., Gerstner, W., and van Hemmen, J.L. (1999). Hebbian learning and spiking neurons. *Phys. Rev. E* 59 (4): 4498–4514.
10 Maass, W. and Markram, H. (2004). On the computational power of circuits of spiking neurons. *J. Comput. Syst. Sci. Int.* 69 (4): 593–616.
11 Maass, W. (1996). Lower bounds for the computational power of networks of spiking neurons. *Neural Comput.* 8 (1): 1–40.
12 Schliebs, S. and Kasabov, N. (2014). Computational modeling with spiking neural networks. In: *Springer Handbook of Bio-/Neuroinformatics*, 625–646. Springer.
13 Burkitt, A.N. (2006). A review of the integrate-and-fire neuron model: I. Homogeneous synaptic input. *Biol. Cybern.* 95 (1): 1–19.

14 Izhikevich, E.M. (2003). Simple model of spiking neurons. *IEEE Trans. Neural Netw.* 14 (6): 1569–1572.

15 Adrian, E.D. (1926). The impulses produced by sensory nerve endings. *J. Physiol.* 61 (1): 49–72.

16 Hu, J., Tang, H., Tan, K.C., and Li, H. (2016). How the brain formulates memory: a spatio-temporal model research frontier. *IEEE Comput. Intell. Mag.* 11 (2): 56–68.

17 Thorpe, S., Fize, D., and Marlot, C. (1996). Speed of processing in the human visual system. *Nature* 381 (6582): 520.

18 Heil, P. (1997). Auditory cortical onset responses revisited. I. First-spike timing. *J. Neurophysiol.* 77 (5): 2616–2641.

19 Meister, M. and Berry, M.J. (1999). The neural code of the retina. *Neuron* 22 (3): 435–450.

20 Florian, R.V. (2012). The chronotron: a neuron that learns to fire temporally precise spike patterns. *PLoS One* 7 (8): e40233.

21 Memmesheimer, R.-M., Rubin, R., Ölveczky, B.P., and Sompolinsky, H. (2014). Learning precisely timed spikes. *Neuron* 82 (4): 925–938.

22 Mohemmed, A., Schliebs, S., Matsuda, S., and Kasabov, N. (2012). Span: spike pattern association neuron for learning spatio-temporal spike patterns. *Int. J. Neural Syst.* 22 (04): 1250012.

23 Ponulak, F. (2008). Analysis of the ReSuMe learning process for spiking neural networks. *Int. J. Appl. Math. Comput. Sci.* 18 (2): 117–127.

24 Gütig, R. and Sompolinsky, H. (2006). The tempotron: a neuron that learns spike timing–based decisions. *Nat. Neurosci.* 9 (3): 420–428.

25 Ponulak, F. and Kasiński, A. (2010). Supervised learning in spiking neural networks with ReSuMe: sequence learning, classification, and spike shifting. *Neural Comput.* 22 (2): 467–510.

26 Ponulak, F., ReSuMe-new supervised learning method for Spiking Neural Networks, Inst. Control Inf. Eng. Pozn. Univ. Technol., vol. 42, 2005.

27 Markram, H., Gerstner, W., and Sjöström, P.J. (2012). Spike-timing-dependent plasticity: a comprehensive overview. *Front. Synaptic Neurosci.* 4.

28 Florian, R.V. (2008). Tempotron-like learning with ReSuMe. In: *International Conference on Artificial Neural Networks*, 368–375.

29 Caporale, N. and Dan, Y. (2008). Spike timing-dependent plasticity: A Hebbian learning rule. *Annu. Rev. Neurosci.* 31: 25–46.

30 Yu, Q., Tang, H., Tan, K.C., and Li, H. (2013). Rapid feedforward computation by temporal encoding and learning with spiking neurons. *IEEE Trans. Neural Netw. Learn. Syst.* 24 (10): 1539–1552.

31 Markram, H., Gerstner, W., and Sjöström, P.J. (2011). A history of spike-timing-dependent plasticity. *Front. Synaptic Neurosci.* 3: 4.

32 Hebb, D.O. (1949). *The Organization of Behavior: A Neuropsychological Theory*. New York: Wiley.

33 Shatz, C.J. (1992). The developing brain. *Sci. Am.* 267 (3): 60–67.

34 Bi, G. and Poo, M. (1998). Synaptic modifications in cultured hippocampal neurons: dependence on spike timing, synaptic strength, and postsynaptic cell type. *J. Neurosci.* 18 (24): 10464–10472.

35 Pfister, J.-P. and Gerstner, W. (2006). Triplets of spikes in a model of spike timing-dependent plasticity. *J. Neurosci.* 26 (38): 9673–9682.

36 Brader, J.M., Senn, W., and Fusi, S. (2007). Learning real-world stimuli in a neural network with spike-driven synaptic dynamics. *Neural Comput.* 19 (11): 2881–2912.

37 Sjöström, J. and Gerstner, W. (2010). Spike-timing dependent plasticity. In: *Spike-Timing Depend. Plast*, 35.

38 Diehl, P.U. and Cook, M. (2015). Unsupervised learning of digit recognition using spike-timing-dependent plasticity. *Front. Comput. Neurosci.* 9: 99.

39 Querlioz, D., Bichler, O., Dollfus, P., and Gamrat, C. (2013). Immunity to device variations in a spiking neural network with memristive nanodevices. *IEEE Trans. Nanotechnol.* 12 (3): 288–295.

40 Florian, R.V. (2007). Reinforcement learning through modulation of spike-timing-dependent synaptic plasticity. *Neural Comput.* 19 (6): 1468–1502.

41 Potjans, W., Diesmann, M., and Morrison, A. (2011). An imperfect dopaminergic error signal can drive temporal-difference learning. *PLoS Comput. Biol.* 7 (5): e1001133.

42 Potjans, W., Morrison, A., and Diesmann, M. (2009). A spiking neural network model of an actor-critic learning agent. *Neural Comput.* 21 (2): 301–339.

43 Frémaux, N., Sprekeler, H., and Gerstner, W. (2013). Reinforcement learning using a continuous time actor-critic framework with spiking neurons. *PLoS Comput. Biol.* 9 (4).

44 Hinton, G., "How to do backpropagation in a brain," in Invited talk at the NIPS'2007 Deep Learning Workshop, 2007.

45 Zheng, N. and Mazumder, P. (2017). Hardware-friendly actor-critic reinforcement learning through modulation of spike-timing-dependent plasticity. *IEEE Trans. Comput.* 66 (2): 299–311.

46 Spall, J.C. (1998). An overview of the simultaneous perturbation method for efficient optimization. *Johns Hopkins APL Tech. Dig.* 19 (4): 482–492.

47 Hinton, G.E. (2012). A practical guide to training restricted Boltzmann machines. In: *Neural Networks: Tricks of the Trade*, 599–619. Springer.

48 Broomhead, D. S., and D. Lowe, "Radial basis functions, multi-variable functional interpolation and adaptive networks," Royal Signals and Radar Establishment, Malvern, United Kingdom 1988.

49 Park, J. and Sandberg, I.W. (1991). Universal approximation using radial-basis-function networks. *Neural Comput.* 3 (2): 246–257.

50 Cassidy, A.S., Merolla, P., Arthur, J.V. et al. (2013). Cognitive computing building block: a versatile and efficient digital neuron model for neurosynaptic cores. In: *The 2013 International Joint Conference on Neural Networks (IJCNN)*, 1–10.

51 Sutton, R.S. and Barto, A.G. (1998). *Reinforcement Learning: An Introduction*. MIT Press, Cambridge.

52 Bengio, Y., Lamblin, P., Popovici, D., and Larochelle, H. (2007). Greedy layer-wise training of deep networks. In: *Advances in Neural Information Processing Systems* (eds. B. Schlkopf, J.C. Platt and T. Hoffman), 153–160. MIT Press.

53 LeCun, Y., Bengio, Y., and Hinton, G. (2015). Deep learning. *Nature* 521 (7553): 436–444.

54 Chollet, F. (2017). *Deep Learning with Python*. Manning Publications Co.

55 Bohte, S.M., Kok, J.N., and La Poutré, H. (2002). Error-backpropagation in temporally encoded networks of spiking neurons. *Neurocomputing* 48 (1): 17–37.

56 Schrauwen, B. and Van Campenhout, J. (2004). Extending spikeprop. In: *Proceedings. 2004 IEEE International Joint Conference on Neural Networks, 2004*, vol. 1, 471–475.

57 McKennoch, S., Liu, D., and Bushnell, L.G. (2006). Fast modifications of the Spike-Prop algorithm. In: *IJCNN'06. International Joint Conference on Neural Networks*, 3970–3977.

58 Haruhiko, T., Masaru, F., Hiroharu, K. et al. (2009). Obstacle to training SpikeProp networks: cause of surges in training process. In: *Proceedings of the 2009 International Joint Conference on Neural Networks*, 1225–1229.

59 Shrestha, S.B. and Song, Q. (2015). Adaptive learning rate of SpikeProp based on weight convergence analysis. *Neural Netw.* 63: 185–198.

60 Zhao, B., Ding, R., Chen, S. et al. (2015). Feedforward categorization on AER motion events using cortex-like features in a spiking neural network. *IEEE Trans. Neural Netw. Learn. Syst.* 26 (9): 1963–1978.

61 Kheradpisheh, S.R., Ganjtabesh, M., and Masquelier, T. (2016). Bio-inspired unsupervised learning of visual features leads to robust invariant object recognition. *Neurocomputing* 205: 382–392.

62 Liu, D. and Yue, S. (2017). Fast unsupervised learning for visual pattern recognition using spike timing dependent plasticity. *Neurocomputing* 249: 212–224.

63 Masquelier, T. and Thorpe, S.J. (2007). Unsupervised learning of visual features through spike timing dependent plasticity. *PLoS Comput. Biol.* 3 (2): 0247–0257.

64 Riesenhuber, M. and Poggio, T. (1999). Hierarchical models of object recognition in cortex. *Nat. Neurosci.* 2 (11): 1019–1025.

65 Serre, T., Wolf, L., Bileschi, S. et al. (2007). Robust object recognition with cortex-like mechanisms. *IEEE Trans. Pattern Anal. Mach. Intell.* 29 (3): 411–426.

66 Krizhevsky, A., Sutskever, I., and Hinton, G.E. (2012). Imagenet classification with deep convolutional neural networks. In: *Advances in Neural Information Processing Systems*, 1097–1105.

67 Neftci, E., Das, S., Pedroni, B. et al. (2014). Event-driven contrastive divergence for spiking neuromorphic systems. *Front. Neurosci.* 7: 272.

68 Neftci, E.O., Pedroni, B.U., Joshi, S. et al. (2016). Stochastic synapses enable efficient brain-inspired learning machines. *Front. Neurosci.* 10.

69 Cao, Y., Chen, Y., and Khosla, D. (2015). Spiking deep convolutional neural networks for energy-efficient object recognition. *Int. J. Comput. Vision* 113 (1): 54–66.

70 Diehl, P.U., Neil, D., Binas, J. et al. (2015). Fast-classifying, high-accuracy spiking deep networks through weight and threshold balancing. In: *2015 International Joint Conference on Neural Networks (IJCNN)*, 1–8.

71 O'Connor, P., Neil, D., Liu, S.-C. et al. (2013). Real-time classification and sensor fusion with a spiking deep belief network. *Front. Neurosci.* 7.

72 Hunsberger, E., and Eliasmith, C., "Training spiking deep networks for neuromorphic hardware," *arXiv Prepr. arXiv1611.05141*, 2016.

73 Rueckauer, B., Lungu, I.-A., Hu, Y. et al. (2017). Conversion of continuous-valued deep networks to efficient event-driven networks for image classification. *Front. Neurosci.* 11: 682.

74 Esser, S.K., Appuswamy, R., Merolla, P. et al. (2015). Backpropagation for energy-efficient neuromorphic computing. In: *Advances in Neural Information Processing Systems* (eds. C. Cortes, N.D. Lawrence, D.D. Lee, et al.), 1117–1125. Curran Associates, Inc.

75 Zheng, N. and Mazumder, P. (2018). Online supervised learning for hardware-based multilayer spiking neural networks through the modulation of weight-dependent spike-timing-dependent plasticity. *IEEE Trans. Neural Netw. Learn. Syst.* 29 (9): 4287–4302.

76 Lee, J.H., Delbruck, T., and Pfeiffer, M. (2016). Training deep spiking neural networks using backpropagation. *Front. Neurosci.* 10: 508.

77 Esser, S.K., Merolla, P.A., Arthur, J.V. et al. (2016). Convolutional networks for fast, energy-efficient neuromorphic computing. *Proc. Natl. Acad. Sci. U.S.A.*: 201604850.

78 Hubara, I., Courbariaux, M., Soudry, D. et al. (2016). Binarized neural networks. In: *Advances in Neural Information Processing Systems* (eds. D.D. Lee, M. Sugiyama, U.V. Luxburg, et al.), 4107–4115. Curran Associates, Inc.

79 Grossberg, S. (1987). Competitive learning: from interactive activation to adaptive resonance. *Cognit. Sci.* 11 (1): 23–63.

80 Neftci, E.O., Augustine, C., Paul, S., and Detorakis, G. (2017). Event-driven random back-propagation: enabling neuromorphic deep learning machines. *Front. Neurosci.* 11: 324.

81 Lillicrap, T.P., Cownden, D., Tweed, D.B., and Akerman, C.J. (2016). Random synaptic feedback weights support error backpropagation for deep learning. *Nat. Commun.* 7: 13276.

82 Merolla, P.A., Arthur, J.V., Alvarez-Icaza, R. et al. (2014). A million spiking-neuron integrated circuit with a scalable communication network and interface. *Science* 345 (6197): 668–673.

83 LeCun, Y., Bottou, L., Bengio, Y., and Haffner, P. (1998). Gradient-based learning applied to document recognition. *Proc. IEEE* 86 (11): 2278–2323.

84 Neelakantan, A., Vilnis, L., Le, Q. V., et al., "Adding gradient noise improves learning for very deep networks," *arXiv Prepr. arXiv1511.06807*, 2015.

85 Merolla, P., Arthur, J., Akopyan, F. et al. (2011). A digital neurosynaptic core using embedded crossbar memory with 45pJ per spike in 45nm. In: *2011 IEEE Custom Integrated Circuits Conference (CICC)*, 1–4.

86 Hussain, S., Liu, S.-C., and Basu, A. (2014). Improved margin multi-class classification using dendritic neurons with morphological learning. In: *2014 IEEE International Symposium on Circuits and Systems (ISCAS)*, 2640–2643.

5

Hardware Implementations of Spiking Neural Networks

Tell me and I'll forget. Show me and I may remember. Involve me and I learn".
— Benjamin Franklin

5.1 The Need for Specialized Hardware

The computational model for spiking neural networks (SNNs) is different from that of artificial neural networks (ANNs). In an ANN, most operations can be represented as matrix–vector multiplications, which can be performed conveniently with conventional computing platforms such as central processing units (CPUs) and graphics processing units (GPUs). In an SNN, however, computations are triggered by events, i.e. computations take place only when spikes occur. Such an event-driven computation has the potential to be energy-efficient as spikes in the network are often sparse. At the same time, however, it demands specialized hardware that can match this event-triggered computational model. Therefore, there has lately been a strong need for such customized hardware. Indeed, it was observed in [1] that in recent years the number of publications discussing the hardware implementation of SNNs had risen rapidly. In this section, advantages and features of SNN hardware are discussed, which serve as the motivation for the rest of this chapter.

5.1.1 Address-Event Representation

One advantage of SNNs is that many small networks can be easily interconnected to form a larger network through leveraging the address-event representation (AER) [2]. The operational speed of a modern CMOS circuit is on the order of a nanosecond, whereas the operational speed for many neuromorphic systems targeting real-time applications is on the order of a microsecond or even a millisecond. This gap in the operational speed makes it possible to exploit time-multiplexing to reduce the complexity of routing spikes. The idea is that through encoding the sparse spike events, the number of physical interconnects can be reduced significantly. The concept of AER is illustrated in Figure 5.1. On the transmitter side, the spike event is first encoded into an address. The encoded address is then routed to the target neuron via a bus. Through such an encoding scheme, the width of the bus can be reduced from N in a naïve implementation to $\log_2 N$, where N is the number of axons. The encoded address is

Learning in Energy-Efficient Neuromorphic Computing: Algorithm and Architecture Co-Design,
First Edition. Nan Zheng and Pinaki Mazumder.

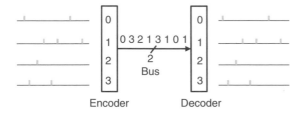

Encoder   Decoder

Figure 5.1 Illustration of the AER that is widely used in hardware SNNs. Sparse spike trains on the transmitter side are encoded by the address of the spiking neuron. On the receiver side, the addresses are decoded into the form of spikes. Such an encoding scheme is helpful in reducing the number of wires needed to transmit spikes.

then decoded at the receiver side and this encoding–decoding process could be totally transparent to the rest of the network. The use of an AER greatly reduces the wire routing and provides a much better scalability for the network. It is because of this good scalability that many large-scale SNN hardware platforms have been demonstrated over the past few years. A few representative examples are reviewed in this chapter.

5.1.2 Event-Driven Computation

One striking feature of an SNN hardware compared to its ANN counterpart is that computations in an SNN are often event-driven. The computational model for one layer of neurons in an event-based SNN is conceptually illustrated in Figure 5.2. In an ANN, activations from the previous layer are treated as a vector. The vector is sent to an array of processing elements (PEs) for matrix–vector multiplication, as discussed in Chapter 3. In contrast, each input spike in an SNN is abstracted as a computation request. More conveniently, by combining the aforementioned AER, the evaluation of an SNN could be entirely conducted in the address domain. Addresses for the spiking neurons are already

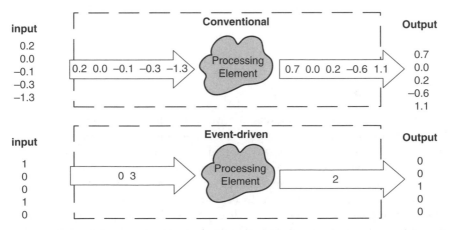

Figure 5.2 Comparison between the conventional frame-based computing and the event-driven computing. In the conventional scheme, matrix–vector multiplications are employed to conduct the computation. The throughput and energy consumption of the system is only weakly correlated to the sparsity of the input signal. In the event-driven scheme, sparser input signals require less time and energy to perform the computation.

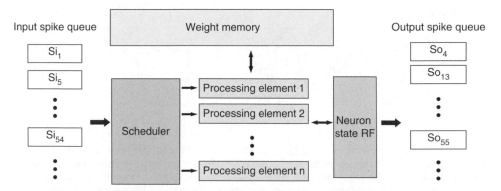

Figure 5.3 Block diagram illustrating the concept of an event-triggered computation. Each input spike is abstracted as a request and is enqueued into a buffer. The scheduler then distributes the computing requests to available processing units. The processing units update the neuron states and the resultant spikes are routed to the next layer.

extracted by the encoder used in AER. These addresses are then processed by the PE array to generate the addresses for the spiking neuron at the current layer. To achieve such a computational model, one example of the high-level hardware architecture is illustrated in Figure 5.3. All the computation requests are put into a queue according to their spike timings. A scheduler then assigns available computational resources and memory bandwidth to these requests. After a request is responded to and completed, the scheduler frees up previously assigned resources to the request. Each neuron in this layer is evaluated after all the requests pointed toward that neuron have been addressed. If a spike is generated by the neuron, it is routed to the next layer. Such a computational model significantly reduces the power consumption of the system, since only useful information is processed.

5.1.3 Inference with a Progressive Precision

The third advantage that an SNN provides is that it can infer with progressive precision [3]. The results presented in Section 4.3.5 are obtained from a way similar to how inference is done in conventional ANNs. In other words, we present the input to the SNN, let the SNN run for D_I ticks, and then read out the results. Even though such an operating method works well, SNNs actually provide some unique opportunities for a more rapid evaluation.

Figure 5.4 compares the difference between inferences in an ANN and an SNN. For a feedforward ANN, the inference result at the output layer is a vector. Therefore, the inference is made once the entire forward pass is completed. The output from an SNN, on the other hand, is a function of time. In other words, if we take a snapshot at any time while the SNN is running, we can make an inference based on the observation we have at that moment. Obviously, in an SNN, the earlier one makes the inference, the less time and energy is consumed for that inference. On the other hand, making an inference after letting the network run for a while has the advantage that more information is available, so a more precise and robust inference is expected. We call such a property the inference with a progressive precision.

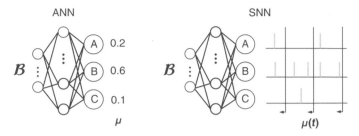

ANN   SNN

A	0.2
B	0.6
C	0.1

μ $\mu(t)$

Figure 5.4 Comparison between the output of an ANN and the output from an SNN. For the ANN, the output of the network is not a function of time and is fixed for a given input. Inference can be made after the entire forward pass is completed. For the SNN, on the other hand, since the output is a spatiotemporal pattern, one can make inferences at any time based on the current information available. The inference accuracy may vary in the process of evaluating the network.

To supply an example, suppose we have trained a neural network with the objective that when a digit i is presented to the network, the output neuron x_i^o fires with a firing density of μ_H, and all other output neurons spike with a firing density of μ_L, where $\mu_L < \mu_H$. When a rate coding is used, the spike train from a neuron is often similar to the output of a $\Sigma - \Delta$ modulator. In other words, the signal is buried in the high-frequency quantization noise. The purpose of counting spikes is to filter out the high-frequency noise so that the signal part can be read out. The longer the inference, the better the filter can remove any unbiased noise. This scheme is similar to the well-known progressive precision existing in stochastic computing [4].

One technique to accelerate the inference is to reduce the noise margin we have in the process of inference, as shown in Figure 5.5. Ideally, the output neuron with the correct label (in our example, neuron B) should have a firing rate around μ_H whereas all other neurons (in our example, neuron A and neuron C) should have a low firing rate μ_L. This is shown in Figure 5.5a. In reality, however, we do not have to wait until such an ideal pattern appears before making a correct decision. If we are willing to pay for the extra reduction in the noise margin, we can loose the decision boundary, as shown in Figure 5.5b. Under this circumstance, as long as the noise associated with the evaluation of the neural network is less than $\mu_H - \mu_L$, we can still obtain a reasonable inference.

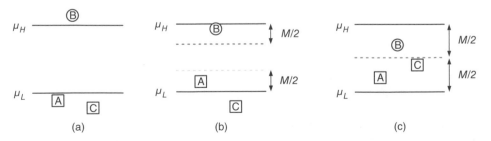

(a)   (b)   (c)

Figure 5.5 Comparison of three inference strategies: (a) with no reduced margin, (b) with a reduced margin, and (c) with a full reduced margin. The two solid lines with labels "μ_H" and "μ_L" are the target output of the neural network when the corresponding input category is present and absent, respectively. In the Inference phase, the criteria can be relaxed to accelerate the inference process. With a relaxed decision boundary, inferences can be made faster, yet the noise margin is reduced accordingly.

Such an extreme case corresponds to Figure 5.5c, where a zero noise margin is used. With such a technique, the inference latency of an SNN becomes data-dependent.

When we apply an image that is easy to be recognized, the corresponding output neuron should fire with a density of μ_H and all other neurons should spike with a density of μ_L. The signal strength is strong in this case. Therefore, we do not have to wait for a long time before removing the quantization noise. Consequently, the inference duration can be short. On the contrary, when the input image is hard to recognize, it is possible that more than one output neuron spikes with a high firing density because they are not sure which digit it is. Under this circumstance, the removal of the quantization noise has a significant impact on the final inference result. Therefore, we should extend the inference duration to allow the output to have enough time to settle. In a sense, an SNN can spread the computational precision over time. The ability of an SNN to terminate the inference early adaptively means it can choose the required precision to conduct the current inference. On the other hand, an ANN has to use its full precision every time it conducts the inference.

To demonstrate the feature of inference with a progressive precision, the three-layer neural network trained with the MNIST dataset in Section 4.3.5 is employed as an example. To provide a rapider inference, the firing density of each output neuron is evaluated at each tick. Whenever the firing density of one output neuron is larger than $\mu_H - M/2$ and all the other output neurons have firing densities less than $\mu_L + M/2$, the output neuron with the highest firing density is chosen and the inference process is terminated. If the condition on firing density is not met, the inference is continued until a maximum allowed inference duration is reached. In this process, M is a design parameter used to measure the reduced margin. A larger M is expected to speed up the inference, yet it might deteriorate the classification accuracy.

Three examples are compared in Figure 5.6. When the easy-to-recognize digit "7" is applied to the neural network, the output of the SNN converges quickly to μ_H and μ_L and the inference can be completed within a few ticks using the adaptive early termination condition. When a relative hard-to-recognize digit "7" is present, it takes the SNN longer to separate the outputs. Finally, when an even more difficult-to-recognize digit "9" is applied, the outputs from neuron 4 and neuron 9 are hard to distinguish and it takes almost 100 ticks to reach the final conclusion. Such an inference process is analogous to how human accomplish recognition. When the pattern presented is easy to recognize, the response is quick. When the pattern is complicated, it takes a longer time to come up with an answer. The classification accuracy and effective number of ticks needed for the 10 000 test images are shown in Figure 5.7 and Figure 5.8. As shown in Figure 5.8, the effective inference duration is reduced rapidly as the reduced margin M increases. In this figure, the effective inference duration is obtained by averaging the inference durations over the 10 000 test cases. In spite of the shortened inference duration, the inference accuracy is not affected until the reduced margin reaches 0.3.

Figure 5.9a illustrates a few examples of the "easy" digits, images that meet the early-termination criteria, whereas Figure 5.9b shows the "hard" digits, images that do not satisfy the requirement. A reduced margin of 0.2 is used for the simulations. One trend that can be observed from Figure 5.9 is that the "hard" digits are more difficult to recognize (by humans) compared to the "easy" digits. There are, in total, 8577 "easy" digits out of the 10 000 test images. For these easy digits, only 32.7 ticks, on average, are needed to accomplish the inference. In spite of such a short inference duration, only

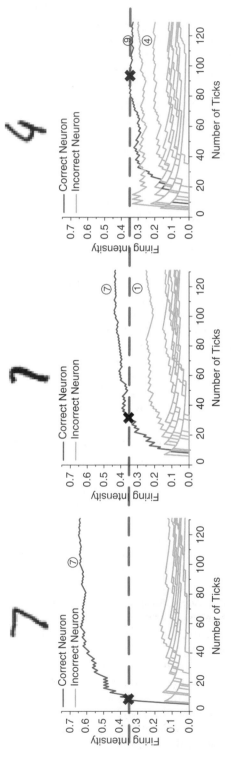

Figure 5.6 Comparison of the output of the SNN when different images are presented to the network. When a relatively simple image ("7") is presented to the network, the output from the correct neuron separates quickly from the outputs of other neurons. When more complicated patterns ("7" and "9") are presented, the outputs from the correct neuron and other neurons are not as far apart as desired. It takes a longer time to make a decision when the input is complex and confusing. In the figure, the correct answer is drawn in a darker color. The red dashed line is the decision boundary for the neural network to draw a conclusion.

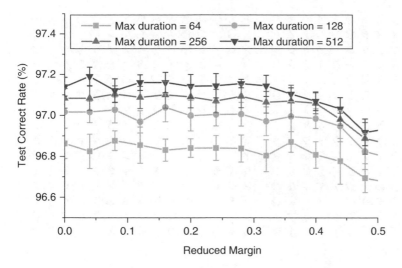

Figure 5.7 The recognition accuracy on the test set for different levels of reduced margin. Only a slight degradation in the correct rate is observed as the margin reduces [3]. Reproduced with permission of IEEE.

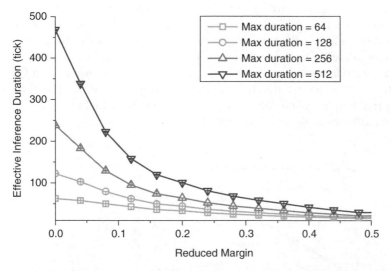

Figure 5.8 Comparison of the effective inference duration needed for classifications. The number of ticks that is needed to complete one inference decreases significantly as the margin reduces [3]. Reproduced with permission of IEEE.

25 digits are classified incorrectly, which translates to a 0.29% error rate on these easy digits. Clearly, the feature of inferring with a progressive precision can help dynamically adjust the inference duration in order to save energy and clock cycles when inputs with different levels of difficulty are present.

Another way to utilize the progressive precision is to exploit the competition among output neurons. A readout scheme, which picks the neuron that first emits K spikes as the winning neuron, is explored. The effective inference duration and the inference

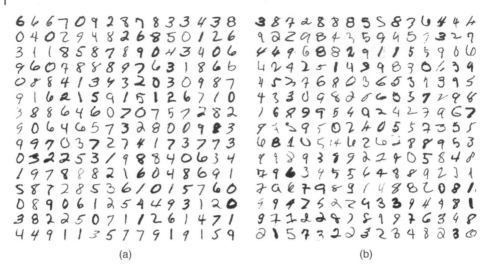

(a)   (b)

Figure 5.9 Examples of the digits that (a) meet the early-termination criteria and (b) do not meet the early-termination criteria. A reduced margin of 0.2 is assumed.

accuracy are compared in Figure 5.10 for different K. As expected, as K decreases, the effective inference duration is shortened with the cost of a reduced classification accuracy. Regardless of the reduced accuracy, one can still achieve an 89% recognition rate with an effective inference duration of only 5.6 ticks. Such a rapid evaluation is valuable in many applications where a fast response is desired yet the requirement for accuracy is not strict. Furthermore, the feature of inferring with a progressive precision provides one extra freedom in optimizing the system performance. The implication of this feature in SNN hardware is discussed in Section 5.2.4.

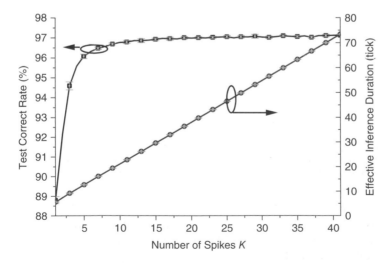

Figure 5.10 The recognition accuracy on the test set with the first-to-spike-K-spikes readout. The digit corresponding to the output neuron that first emits K spikes is read out. The black curve and the blue curve show the recognition correct rate and the average inference duration with different values of K [3]. Reproduced with permission of IEEE.

5.1.4 Hardware Considerations for Implementing the Weight-Dependent STDP Learning Rule

The hardware-friendly learning algorithm presented in Section 4.3.5 is implemented in hardware in this chapter. There are two possible ways of implementing the algorithm, and the main difference lies in how the memory is organized and used in the system [3]. Examples of these two types of architectures are illustrated in Figures 5.11 and 5.12. We call the first type of architecture a centralized memory architecture. This architecture resembles the architectures used in conventional processors where the memory and processing units are separate. The synaptic weights are stored in a centralized memory and can be accessed through buses. We call the second architecture, shown in Figure 5.12, a distributed memory architecture since the memory elements in this architecture are distributed along with the processing units. The memory element itself can often function as a processing unit. Therefore, the in-memory processing is often carried out in this architecture to achieve a high energy efficiency.

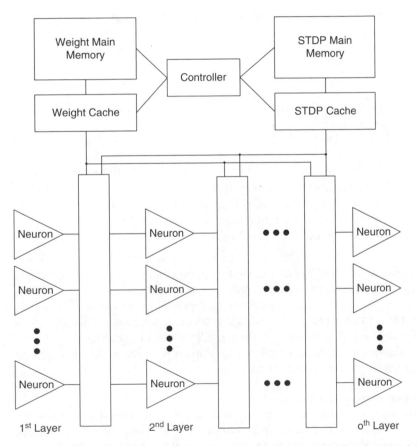

Figure 5.11 Illustration of an example of the centralized memory architecture. Weights and STDP information are stored in a centralized memory and can be accessed through buses. This architecture is similar to a conventional von Neumann architecture in the sense that processing elements and memory are separate [3]. Reproduced with permission of IEEE.

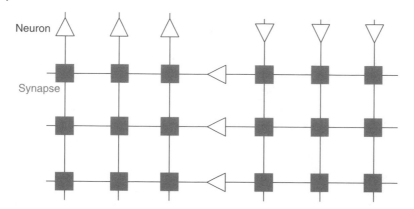

Figure 5.12 An example of the distributed memory architecture-based neuromorphic system. A crossbar structure is used for demonstration. The triangles represent neurons and the squares represent synapses. The storage units are distributed along with the processing elements.

For the learning algorithm presented in Section 4.3.5, the synaptic weight can be updated in different ways. Two examples are

$$\Delta W_{ij}^l = -\alpha \cdot e_j^{l+1} \cdot \frac{\overline{stdp_{ij}^l}}{w_{ij}^l (1 - \overline{x_i^l})} \tag{5.1}$$

$$\Delta W_{ij}^l = \sum_n \left(\frac{-\alpha \cdot e_j^{l+1} \cdot stdp_{ij}^l[n]}{w_{ij}^l (1 - \overline{x_i^l})(D_L - T)} \right) \tag{5.2}$$

where ΔW_{ij}^l is the total weight change in one learning iteration and $e_j^{l+1} = \sum_{k=1}^{N_o} e_k^o \cdot (\partial \mu_k^o / \partial \mu_j^{l+1})$ is the back-propagated error at neuron x_j^{l+1}.

We call the style of updating weight according to (5.1) a cumulative update style, whereas the update shown in (5.2) is an incremental update. For the cumulative update style, the spike timing information is first accumulated to form the averaged spike timing information $\overline{stdp_{ij}^l}$, and then the weight update is conducted in a batch mode. Such a way of updating weights is preferred in a centralized memory architecture where the access cost for the large memory array is high. In contrast, the weight update in an incremental update occurs whenever there is a spike-timing-dependent plasticity (STDP) event. Consequently, it is not necessary to store the spike timing in a separate memory as this information is accumulated in the weight memory. This way of updating the synaptic weight may be suitable for a distributed memory architecture where reading and writing the memory are relatively low-cost operations.

5.1.4.1 Centralized Memory Architecture

The centralized memory architecture, as shown in Figure 5.11, is widely used in the literature, especially for CMOS-based implementations. Even though physical neurons, i.e. neurons with dedicated computational resources, are illustrated in the figure, virtual neurons can also be used to reduce the circuit area.

In a centralized memory architecture, a special memory array is necessary to store the spike timing information. The number of bits (NOB) needed to represent the spike

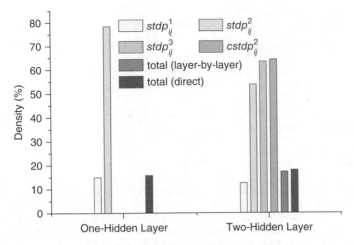

Figure 5.13 Percentage of the active (non-zero) STDP information $stdp_{ij}^l$ and cross-layer STDP information $cstdp_{ij}^l$ for synapses in different layers. The results are normalized to the number of synaptic weights in the corresponding layer [3]. Reproduced with permission of IEEE.

timing information is at most $\lceil \log_2(2D_L \cdot WIN_{STDP} + 1) \rceil$. In reality, the NOBs needed is much less than this upper bound because of the sparseness of spikes. Through numerical simulations, it is observed that a word length of 5 is enough to represent the spike timing information when $WIN_{STDP} = 1$ and $D = 128$ are used.

Another point worth mentioning is that even though in a straightforward design, each synapse can have its dedicated STDP field. This way of storing the spike timing information is not economic, considering that only recently active synapses has non-zero spike timing information. Indeed, sparsity is an important feature of many neural networks. Figure 5.13 illustrates the level of spike density existing in a three-layer neural network and a four-layer neural network when the images in the Modified National Institute of Standards and Technology (MNIST) dataset are presented to the network. It is observed in this figure that only around 15% of the synapses in the first layer are active. A synapse is called active if the corresponding presynaptic neuron ever fires. Such a low activity in synaptic weights can be leveraged to reduce the footprint of the STDP memory significantly through a cache structure. This technique is discussed in more detail in Section 5.2.4. A typical timing diagram of the centralized memory architecture is shown in Figure 5.14. During learning, both the synaptic weight memory and the STDP memory are active. The neuron block is only running for the feedforward phase. During inference, the STDP memory is no longer used so it can be power-gated to save energy. The centralized memory architecture is implemented in a synchronous digital circuit in Section 5.2.4.

5.1.4.2 Distributed Memory Architecture

One example of the distributed memory architecture-based system is shown in Figure 5.12. It is based on a crossbar structure. In the figure, triangles represent neurons, whereas squares represent the synapse connecting two neurons. Due to the nature of this architecture, analog synapses are often employed. A memristor device is a popular choice as an artificial synapse in recent years. When a memristor

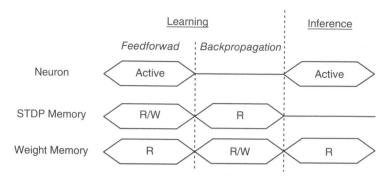

Figure 5.14 An example of the timing diagram of the proposed learning algorithm employed in a centralized memory system. Neurons are only active during a forward pass. The STDP memory is updated in the forward iteration in the learning phase and is used for a weight update in the backward iteration. The weight memory is needed in both the inference and learning phases but is only updated in the backward pass when learning is performed [3]. Reproduced with permission of IEEE.

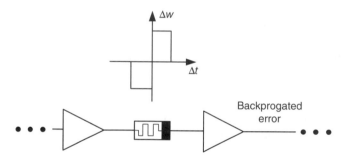

Figure 5.15 Illustration of a memristor-based synapse sandwiched by two neurons. An STDP protocol can be employed for learning. The STDP protocol can be implemented by either the neuron circuitry or the device itself [3]. Reproduced with permission of IEEE.

is used, each cross point in Figure 5.12 is a memristor-based synapse. The matrix vector multiplication associated with the neural network is conducted with the help of physical laws, as demonstrated in Chapter 3. The update of the synaptic weights in a distributed memory architecture can occur when the neural network is operating as illustrated in Figure 5.15. The synaptic weight is updated based on the timings of pre- and postsynaptic spikes. The STDP protocol can be implemented either through the memristor itself or through the neuron circuits. This is elaborated in Section 5.3.4.

As discussed in Section 4.3.5, both the conventional layer-by-layer backpropagation and the direct backpropagation methods can be used for the proposed learning scheme. They might achieve similar performance in learning, yet they have different implications in hardware implementation. In terms of the conventional backpropagation, $N_l N_{l+1}$ multiply-accumulate (MAC) operations are required for the lth layer in the network. The number of MAC operations needed is changed to $N_l N_o$ for the direct backpropagation. As the idea of many deep neural networks is to refine the information in the raw input layer by layer, most neural networks have far less output neurons compared to their hidden layer neurons. Therefore, many fewer MAC operations are involved in the direct backpropagation scheme, considering $N_o \ll N_l$. Nevertheless, in order to apply the direct backpropagation, more storage spaces are needed to hold the cross-layer spike information. The newly added memory size is on the order of $N_o \cdot \sum_{i=1}^{o-1} N_i$. In other words, we trade off memory spaces for computations in the direct backpropagation method. The answer to which method is more proper depends on the actual implementation style and technology.

One concern of building SNNs with analog memory is that the division-by-weight operation is hard to implement efficiently in hardware. Acquiring the synaptic weight value accurately in an analog synapse is already difficult enough, not to mention the division operation. One solution to this problem is to use an approximate division. In other words, we can aggressively quantize the synaptic weight in the denominator before the division has taken place. To illustrate this idea, simulations are conducted with the weight in the denominator being quantized according to

$$
w_{ij}^l = \begin{cases}
\ \cdots \\
w_{q1}, w_{q1} \leq w_{ij}^l < w_{q2} \\
w_{q0}, 0 \leq w_{ij}^l < w_{q1} \\
-w_{q0}, -w_{q1} \leq w_{ij}^l < 0 \\
-w_{q1}, -w_{q2} \leq w_{ij}^l < -w_{q1} \\
\ \cdots
\end{cases}
\tag{5.3}
$$

where w_{q0}, w_{q1}, and so on are manually chosen constants.

With such a quantization scheme, the learning performance achieved with different levels of quantization is shown in Figure 5.16. As shown in the figure, even though the best results are achieved when no quantization is used, aggressively quantizing the weight denominator does not deteriorate the performance significantly. The results

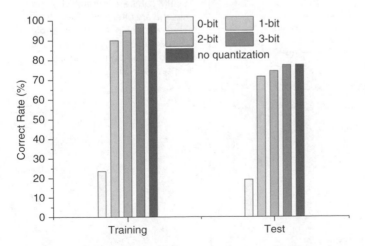

Figure 5.16 Comparison of the classification accuracy when different levels of quantization in the weight denominator are used. As the quantization becomes more aggressive, the recognition correct rate after learning drops, yet impressive results can still be achieved even with a one-bit precision [3]. Reproduced with permission of IEEE.

obtained from even one-bit accuracy in the denominator is still acceptable. How this quantization strategy can lead to an economic implementation of a memristor crossbar-based neural network is discussed in Section 5.3.5.

5.2 Digital SNNs

5.2.1 Large-Scale SNN ASICs

SNNs implemented in the form of digital ASICs can benefit more from the CMOS technology. Recently, many large-scale neuromorphic engines based on digital SNNs have been developed to help understand how the brain works and also to provide an efficient way of computing. In this section, a few examples of large-scale SNNs in digital form are reviewed.

5.2.1.1 SpiNNaker

SpiNNaker is a massively parallel multicore system designed to model and simulate large-scale SNNs in real time. It was mainly developed by Furber et al. at the University of Manchester. The computing machine can contain up to 1 036 800 ARM cores that are distributed in 57K nodes [5]. A high-level diagram illustrating key components in a SpiNNaker node is shown in Figure 5.17 [6]. Each node in SpiNNaker is a system-in-package that consists of a chip multiprocessor (CMP) and a 128-MB off-die synchronous dynamic random-access memory (SDRAM). Low-power ARM cores are used in order to obtain a good energy efficiency. Furthermore, through leveraging the event-triggered nature of an SNN, the SpiNNaker system is able to simulate large-scale

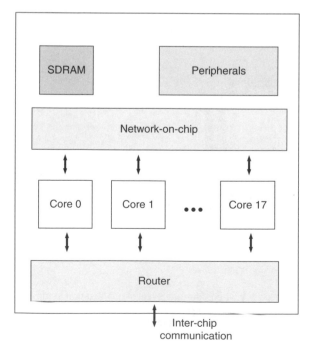

Figure 5.17 High-level overview of key components in a SpiNNaker node. 18 ARM cores are integrated on one chip and are interconnected by a network-on-chip. A 128-MB off-die SDRAM is used to store information needed in computing. Source: adapted from [6].

SNNs efficiently. The power budget for each node in the system is 1 W. With other components in the system, the board is expected to consume 75 W of power. Scaling this system to a million-neuron network increases the power consumption to around 90 KW.

Each CMP in the SpiNNaker is built with 18 low-power ARM cores. The chip, which is fabricated in a 130-nm CMOS process, contains more than 100 million transistors. The chip is designed with the objective to minimize the power consumption. Such an objective is derived from the philosophy that the designers of SpiNNaker have: in a massively parallel computing system, the cost incurred by powering and cooling the system over its lifetime can be much more than the cost of the machine itself [7]. With such a design strategy, low-power ARM processors as well as mobile SDRAM are employed [6]. Furthermore, the implementation utilizes architecture-level and logic-level clock gating and power-aware synthesis throughout the design flow. Asynchronous communication and event-driven computation are also exploited. Each processing core is put into a low-power sleep mode while being idle in order to save power. The CMP chip consumes 1 W of power when operating at a clock frequency of 180 MHz, whereas it consumes 360 mW when idle. In addition to the ARM cores, routers are needed to route packets representing spikes to different destinations, such as another on-chip processor or another SpiNNaker node. There are four different types of packets that need to be handled by the router: multicast packets, point-to-point packets, nearest-neighbor packets, and fixed-route packets [8]. The router has a 72-bit databus, which can provide a maximum bandwidth of 14.4 Gb/s when running at a clock frequency of 200 MHz.

Each processor core is directly connected to a 32-kb instruction memory and a 64-kb data memory. Program code and data that might be used frequently are stored in the tightly coupled instruction and data memory. The access latency for the off-die SDRAM is usually long because this is done through the system network-on-chip (NoC). As the neurons are periodically updated, the neuron states are kept in the memory locally at the process cores. The larger and less frequently used synaptic weights are stored in the larger memory, which can be accessed through a direct memory access.

As a powerful platform for modeling and simulating SNNs, the SpiNNaker system has been deployed in many applications. For example, four SpiNNaker chips have been interfaced with a dynamic video sensor silicon retina to demonstrate a feature-based attentional selection system in [9]. A SpiNNaker system was employed as an event-driven backend processing platform in order to process the visionary input. Another example of utilizing a SpiNNaker system is that a robot was built with a 768-core SpiNNaker board in [10], and real-time tasks performed by the mobile robot have been demonstrated.

Even though the SpiNNaker system belongs to the category of ASICs, it is still largely based on the concept of general-purpose processors. The flexibility of the hardware comes with a cost of a relatively high power consumption. In recent years, to account for many emerging low-power applications, more and more ASICs have been built to sacrifice certain flexibility in the hardware in order to gain more energy efficiency. The TrueNorth chip presented in the next section is such an example.

5.2.1.2 TrueNorth

5.2.1.2.1 Motivations The computational speed of computers used for scientific computations has passed the limit of the human brain a long time back, yet the human brain turns out to be incredibly energy efficient in certain tasks. To simulate a brain that consists of 100 trillion synapses and consumes 20 W of power, 96 supercomputers will be needed, which will require 12 GW of power [11]. Such a huge gap in energy efficiency motivates researchers to look into new architectures that differ from the conventional von Neumann based one. TrueNorth is an SNN accelerator consisting of one million spiking neurons [12], which was produced by IBM under the Defense Advanced Research Projects Agency (DARPA) Systems of Neuromorphic Adaptive Plastic Scalable Electronics (SyNAPSE) project [11]. TrueNorth is more application-specific compared to SpiNNaker, discussed in Section 5.2.1.1. For SpiNNaker, one has more freedom in programming the hardware as general-purpose processors are implemented. For TrueNorth, however, more limitations are posed as many functions are hardwired in order to achieve a high energy efficiency. The objective of TrueNorth is to find a balance between energy efficiency and flexibility. Therefore, in spite of the high energy efficiency it can achieve, it is still reasonably reconfigurable and programmable.

5.2.1.2.2 **The Chips** In order to achieve a large-scale spiking system, 4096 cores are interconnected, where each core implements 256 neurons and 64K synapses, as shown in Figure 5.18 [13]. It features a peak performance of 58 giga-synaptic operations per second (GSOPS) and a maximum energy efficiency of 400 GSOPS per watt. In order to reduce the power consumption, an event-triggered architecture with asynchronous circuits is used in TrueNorth. Computation is conducted only when it is needed. The power-hungry clock network is therefore eliminated in the neuron core, which can easily dominate the power consumption of modern digital ASICs.

Though at each clock-less neuron operations are asynchronous, evaluations of SNNs are synchronous across all cores with a frequency of 1 KHz. Such a constraint is there mainly for the hardware–software one-to-one mapping. Even though the internal evaluations of the neuron states are asynchronous to save energy, states of the SNN at the

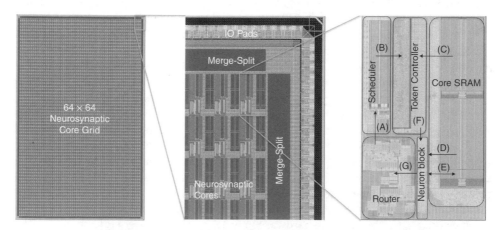

Figure 5.18 Overview of the TrueNorth chip. Each chip (leftmost) contains 64 × 64 neurosynaptic cores (rightmost) [13]. The scheduler stores input spikes in a queue for further processing. The token controller schedules the computations of neurosynaptic cores. The router of a core communicates with the routers of neighboring cores. The core SRAM stores the synaptic connectivity, neuron parameters, neuron states, etc. The neuron block implements various neuron dynamics. The detailed implementation of the neuron block is illustrated in Figure 5.19. Reproduced with permission of IEEE.

end of each tick, which is the minimum timing resolution in TrueNorth, are the same as the states simulated in software. This hardware–software coherence is the key for good programmability and the ease of prototyping and debugging.

In contrast to many other neural network accelerators where a large DRAM is often used to hold the synaptic weights, the TrueNorth chip employs on-chip SRAM arrays. Figure 5.18 shows the layout and floorplan of the TrueNorth accelerator. It consists of a 64×64 array of neurosynaptic cores. Each core has its own SRAM array. In other words, SRAMs are distributed across the chip with the cores. Such a near-memory processing helps reduce the data movements, improving the energy efficiency of the TrueNorth system.

5.2.1.2.3 *Neuron Models* The neuron model employed in the TrueNorth chip is based on the leaky integrate-and-fire (LIF) neuron model because this neuron model can be implemented in CMOS efficiently while preserving most features of a spiking neuron, as discussed in Chapter 4. TrueNorth supports various neural codes, such as rate coding, population coding and time-to-spike coding, and so on. Like other neuromorphic hardware, TrueNorth also employs a fixed-point number representation. Such a choice yields a neuron model implementation with 1272 logic gates.

Inspired by the philosophy behind the reduced instruction set computer (RISC), the strategy used in TrueNorth is that each neuron only implements simple models. However, by combining several simple neurons, complex computations and behaviors can be synthesized when needed. The circuit diagram of the neuron block is shown in Figure 5.19. There are mainly five basic operations in the neuron mode: integration, leak, threshold, spike, and reset. For each operation, there are different modes that can

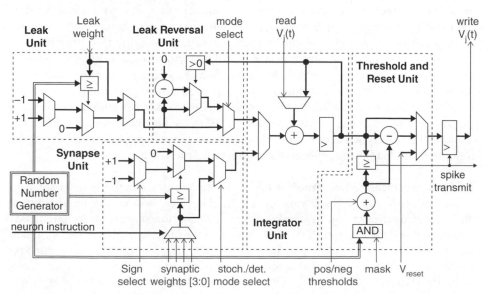

Figure 5.19 Schematic of the neuron implemented in TrueNorth [13]. Five major units are shown in the figure. The synapse unit implements both the deterministic and stochastic synaptic input. The leak and leak reversal unit supply leakage to the neuron. The integrator unit accumulates the membrane potential over time. The threshold and reset unit perform the thresholding and reset operation. Reproduced with permission of IEEE.

be configured by the user, enriching the behavior of the neuron model. More detailed information on the models that TrueNorth are capable of simulating can be found in [14]. By using only one neuron or by combining multiple neurons, many complex behaviors can be synthesized. Twenty common biological feasible spike behaviors can be implemented with the neurons in TrueNorth. Among these 20 behaviors, 11 of them use only one neuron, whereas 7 of them use two neurons, and the last two complex behaviors can be generated with three neurons.

One strategy employed in TrueNorth is that there are only four types of axons available. In other words, the synaptic weight has only four possible values. This is illustrated in the synapse unit shown in Figure 5.19 where a multiplexor is employed to pick one out of four synaptic weights depending on the type of axon. Thanks to such an aggressive quantization scheme, the synaptic weights can fit into on-chip SRAM arrays without resorting to power-hungry external DRAM chips. This strategy is similar to that of a binary-weight/binarized network discussed in Chapters 2 and 3.

5.2.1.2.4 *Programming and Simulation*

5.2.1.2.4 Programming and Simulation TrueNorth is a piece of powerful hardware that can be reconfigured to conduct many different computations. To take advantage of the flexibility and reconfigurability of the hardware, programming is needed, and this can be achieved through an object-oriented Corelet Language [15]. A corelet is an abstraction of a network. It encapsulates all the connections and operations in the network. Only the input and output of the network can be observed externally. The Corelet Language is composed of four main classes: the Neuron class, the Core class, the Connector class, and the Corelet class. With the help of this Corelet Language, programmers can conveniently realize desired networks on TrueNorth.

In addition to the programming language, a simulator called Compass was also developed in order to simulate the behavior of TrueNorth on a conventional processor [16]. Compass can be employed to simulate large networks comprising tens of millions of TrueNorth cores. It has been demonstrated in [16] that with a 16-rack IBM Blue Gene/Q supercomputer, Compass was able to simulate 256M TrueNorth cores containing 65B neurons and 16T synapses. The number of synapses simulated was comparable to the number of synapses in a monkey cortex, despite the fact that the simulation was 388 times slower than what was needed in real time. With the help of Compass, many algorithms can be prototyped and tested on conventional processors, which accelerates the development cycle for new applications running on TrueNorth.

5.2.1.2.5 Applications Since TrueNorth was released, much effort has been spent to develop both hardware and software for the TrueNorth ecosystem. The main target of TrueNorth is to conduct cognitive tasks with a reasonably low power consumption. IBM has launched several TrueNorth systems, including the neurosynaptic system 1 million neuron evaluation (NS1e) platform, the NS1e-16 platform, and the neurosynaptic system with a 16 million neuron evaluation (NS16e) platform [17]. The NS1e consists of a TrueNorth chip and a Xilinx SoC. With the help of the ARM cores in the Xilinx SoC, the NS1e platform can operate in a standalone fashion. The NS1e-16 platform consists of 16 NS1e boards, which increases the number of neurons and the number of synapses of the NS1e-16 platform to 16 million and 4 billion, respectively. For the NS16e board, a 4×4 array of TrueNorth chips are connected with a chip-to-chip asynchronous communication interface [18].

Various applications have been demonstrated with the TrueNorth system to achieve real-time low-power cognitive computing, including speaker recognition, composer recognition, digit recognition, collision avoidance, and so on [19]. In [12] and [13], a real-world application of multiobject detection and classification in a fixed-camera setting was demonstrated. The chip consumed 63 mW of power on a 30-frame-per-second three-color video with each frame being 400 pixels by 240 pixels. In [17], three examples have been demonstrated on the TrueNorth platforms, including recognition of a handwritten character written on a tablet, text extraction and recognition, and defect detection in additive manufacturing.

Tsai et al. demonstrated an always-on speech recognition application with the TrueNorth system in [20] and [21]. A feature extractor, called low-power audio transform with a TrueNorth ecosystem (LATTE) was developed. In this work, LATTE leverages the low-power capability of the TrueNorth to conduct feature extraction on an audio signal. With the help of the low-power computation provided by the NS1e platform, the speech recognition system was able to operate continuously for up to 100 hours with a button cell battery.

The learning algorithm presented in Section 4.3.4 was demonstrated on the TrueNorth chip as well. A record-breaking classification accuracy of 99.42% was reported while dissipating 108 μJ per image [22]. More studies on implementing convolutional neural networks (CNNs) on TrueNorth were presented in [23]. Eight popular benchmarks widely employed in deep learning were examined. Since the TrueNorth chip was designed to deal with spikes, the real numbers in the dataset were first converted into binary numbers. The obtained benchmark results were compared with the state-of-the-art results, and similar performances were achieved with the TrueNorth platform. This result is quite impressive, considering the limited precision of the weights in a TrueNorth chip. One important conclusion drawn by Esser et al. in [23] was that the difference between a spike-based and a non-spike-based computing was not fundamental, and through introducing the correct training method, spike-based computing could also yield impressive results.

5.2.1.3 Loihi

Loihi is a neuromorphic multicore processor developed by Intel's Microarchitecture Research Lab [24]. Loihi consists of 128 neuromorphic cores where each core implements 1024 primitive spiking neural units. Considering the variety of demands in connecting neurons in different networks, Loihi strives to provide a flexible solution which supports (i) sparse network compression, (ii) core-to-core multicast, (iii) variable synaptic formats, and (iv) population-based hierarchical connectivity. Davies et al. believes that Loihi is the first fully integrated SNN chip that can support any of these features [24]. One feature that Loihi strives to achieve is the on-chip learning. It was claimed in [25] that Loihi was the first of its kind to equip on-chip learning capability through a microcode-based learning rule engine within each neuron core. On every learning epoch, the synaptic weights are updated according to certain programmable learning rules in sum-of-products form. With such a scheme, basic pairwise STDP and other more advanced learning rules can be implemented.

On the microarchitecture side, all logic in Loihi is implemented in an asynchronous bundled data design style [24], which allows spikes to be generated, transported, and consumed in an event-triggered fashion. This choice of asynchronous implementation has a large impact on power consumption of the system. With asynchronous logic, activity gating is automatically applied, which eliminates the excessive power consumed by the continuously-running clock network. Even though the clock power can also be reduced through a very carefully designed hierarchical clock gating scheme in a synchronous design, an event-driven asynchronous circuit is more convenient and finer-grained. In addition, an asynchronous design effectively eliminates the need for a timing margin as everything is in a self-timed fashion. The elimination of overdesign in timing is often considered as one of the most attractive features of an asynchronous circuit in terms of reducing power consumption [26].

Loihi is fabricated in a 14-nm process. The chip with a die size of $60\,mm^2$ contains 128 neuromorphic cores and three x86 cores. In total, 16 MB of synaptic memory is included in Loihi, which provides 2.1 million unique synaptic variables per mm^2 when the densest 1-bit synapse format is used. Such a number was reported to be over 3 × higher than that of TrueNorth, which is already known as a very dense SNN chip. The neuron density of Lodhi was reported to be 2 × smaller than that of TrueNorth. This reduction in the neuron density may be attributed to the expanded feature set in Lodhi [24].

To demonstrate the superiority of Loihi in solving large-scale parallelizable problems, an L1 minimization problem was benchmarked. Compared to a baseline CPU running a numerical solver, Loihi was able to achieve 5760 × boost in the energy-delay product (EDP) when the size of the problem was large. To allow more algorithms to be prototyped with Loihi, a Python-based API was also developed. It allows researchers and developers to quickly implement and map SNNs onto the hardware for both learning and inference. With this toolchain being introduced, it is expected Loihi will find more valuable applications in the near future.

5.2.2 Small/Moderate-Scale Digital SNNs

In addition to many large-scale general-purpose SNN hardware that target various problems, as discussed in Section 5.2.1, there are also many efforts devoted to developing relatively small-scale digital SNN hardware either on a Field-Programmable Gate Array (FPGA) or on an ASIC. Some of the work mainly develops the building blocks of an SNN, whereas others focus more on the system-level learning capability. In this section, a few examples are briefly reviewed based on these two themes.

5.2.2.1 Bottom-Up Approach

The basic building components of SNNs generally have more complicated dynamics compared to those of ANNs. In an ANN, neurons are memoryless summing nodes whose output is abstracted as activation levels and can be conveniently stored in memory. Synapses are multiplicative edges that can be naturally implemented as multipliers. Therefore, the evaluation of an ANN is usually a series of MAC operations with a few look-up tables based on non-linear operation. Most research, therefore, focuses on the dataflow and the arrangement of computations, as discussed in Chapter 2. In an SNN, on the other hand, neurons have more complicated dynamics even for a simple LIF neuron, not to mention other biologically more feasible neurons such as the Izhikevich model and the Hodgkin-Huxley model. In addition, the synaptic plasticity, which plays an important role in an SNN, also needs special handling. Therefore, compared to hardware-based ANNs, there are more interests in implementing basic building blocks, such as neurons and synapses, for hardware SNNs. Once these fundamental components are ready, a network of spiking neurons can then be constructed.

The most popular neuron model in hardware SNNs is probably the LIF model due to its simplicity [27–30]. Even though the most bare-bones LIF model has a limited capability in producing certain biological feasible responses, there exist many variants of LIF models that have more complicated dynamics. Several LIF-related neuron models were implemented and compared in [29]. As one would expect, there is a clear tradeoff between the level of complexity of a neuron model and its associated computational efficiency. The simplest linear LIF neuron model only requires fewer than one-tenth of the operations of the more complicated LIF model with decaying synaptic conductances.

Even though many digital SNNs rely on a simple LIF neuron model that outputs binary events, there are also much effort spent on developing efficient and flexible spiking neurons that can be used for a more accurate simulation of biological SNNs [31–34]. For example, 12 biologically common features grouped into five categories that exist in various neuron models were identified by Lee et al. [33]. With the common features being spotted out, datapaths that realize these features were designed and were integrated to form Flexon, a flexible and efficient digital neuron that can be used in large-scale SNN simulations.

An interesting work with the objective of breaking Liebig's law was presented in [28]. Wang and Schaik noticed that virtually all existing neuromorphic hardware systems were limited by the components in the shortest supply in the systems. In order to tackle this problem, a strategy of using generic components that can be configured as a neuron, a synapse, or an axon as needed is used. This strategy is based on the observation that neurons, synapses, and axons all require similar resources such as local memory, which can be implemented with SRAM arrays. A virtual prototype was implemented in a 28 nm technology. Despite the generic nature of the components in that design, neuron and synapse density as well as energy efficiency that were comparable to what could be achieved by dedicated components were reported [28].

5.2.2.2 Top-Down Approach

In addition to basic building blocks, many hardware architectures optimized for various learning algorithms or tasks have been reported in recent years. To name a few, there are a neuromorphic processor performing an unsupervised online spike-clustering task in the context of deep-brain sensing and stimulation systems [35], a neuromorphic chip with low-precision synaptic weights for conducting auto-association through STDP learning [36], a low-power neural network ASIC that conducts sparse learning for feature extraction [37], and so on [38–42].

As stated in previous sections, one of the most important features of an SNN is its sparsity. Many hardware architectures in the literature attempt to improve the system performance and energy efficiency through leveraging this sparsity. For example, in Minitaur, an event-driven FPGA accelerator for SNNs [43], the algorithm for updating the membrane voltage is event-triggered. The algorithm stores unprocessed spikes in a queue based on the input spike timings. The processing unit then evaluates spikes in the queue and updates the membrane voltage for the destination neurons. When a new spike is generated in the process of updating the membrane voltage, the newly generated spike is sent to the spike timing queue. Such an architecture is similar to the one shown in Figure 5.3. It takes advantage of the event-trigger nature of an SNN and is highly scalable and flexible. The performance of the accelerator was measured as 18.73 million postsynaptic currents per second while consuming 1.5 W of power.

Another example of leveraging the sparsity in an SNN is a digital ASIC aiming at sparse learning [37, 39], which was developed by Kim et al. at the University of Michigan. The ASIC implements a specific learning algorithm called SAILnet, which was proposed in [44]. The target of sparse learning is to learn a sparse representation for a given input. Neurons in the network are locally connected through grids. Since the sparse learning is conducted on chip, the activity of the neural network is expected to be low. To leverage the sparse spike trains, the size of the grid structure is chosen to be small enough such that the collision rate for the spikes in the same cluster is low. With such a low collision rate, the arbitration scheme is no longer needed. In other words, the collided spikes can be discarded right away without introducing too much error. Such an arbitration-free bus was shown to have little impact on the performance of the chip. Memory partitioning was also explored in [37] to save power during inference. The synaptic weight memory in the design is partitioned into core memory and auxiliary memory. The core memory is used during both the learning and inference phases, whereas the auxiliary memory is only needed during learning. This arrangement is motivated by the finding that the precision needed for learning and inference differ significantly.

The chip presented in [37] relies on unsupervised learning and is not capable of conducting object recognition directly. In [40], classifiers are integrated together with the sparse feature extraction inference module to form an object recognition processor. Nevertheless, due to the fully connected nature of the network used in [37] and [40], the supported size of the input image patch is limited. To accommodate larger problems, a convolutional sparse coding accelerator was demonstrated in [38] by the same group. Through sharing the kernel weights, the accelerator is able to support a kernel size up to 15×15 and an image size up to 32×32.

5.2.3 Hardware-Friendly Reinforcement Learning in SNNs

In this section, a hardware architecture proposed in [45] for the algorithm outlined in Section 4.2.4 is presented. The high-level description of the architecture is illustrated in Figure 5.20. The on-chip memory is partitioned into two parts according to how frequently the memory is visited. Memory A and memory B can be visited in every tick and every unit interval, respectively. The definitions of tick and unit interval are illustrated in Figure 5.21. A tick is the minimal temporal resolution when a neuron is allowed to spike, following the convention in TrueNorth [12], whereas a unit interval is the time unit for decimating. Memory A stores the data that need to be visited frequently, such as synaptic weights and the raw spike timing information. Memory B stores data that are accessed at a much lower frequency, such as the decimated ST information.

Figure 5.21 shows the timing diagram of the system. Within a tick, all input neurons are scanned through. Two clock cycles are needed for evaluating each input neuron. During the first clock cycle, the synaptic weights are read out from the memory if the input neuron that is under evaluation spikes in the current or last tick. During the second clock cycle, the ST field is updated. In the current design, a virtual neuron is used. In other words, all the arithmetic circuitry, such as adders and comparators needed in the neuron model, are shared among neurons. Nevertheless, the inherent parallelism in

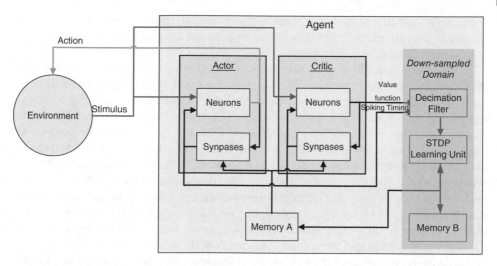

Figure 5.20 A hardware architecture for the STDP learning rule-based actor–critic reinforcement learning. Actor and critic are two SNNs with the same input. Memory A and memory B are partitioned according to the frequency that they are accessed. Memory A stores synaptic weights and tick-level spike timing information, and it may be read or written at every tick. Memory B stores decimated spike timing information, and its chance of being accessed is much lower than that of memory A [45]. Reproduced with permission of IEEE.

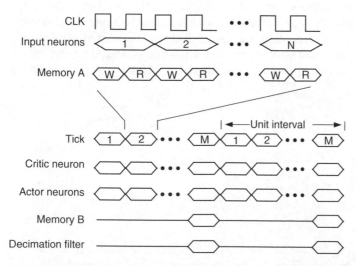

Figure 5.21 An example of a timing diagram for the proposed hardware architecture. Each input neuron takes two clock cycles to be evaluated: one cycle for reading weight and another cycle for writing spike timing information. States of all neurons are updated at every tick. The blocks in the down-sampled domain operate at a much lower frequency [45]. Reproduced with permission of IEEE.

the neural network can be readily exploited by spending more computational resources to speed up the computation. For each unit interval, the output from the critic neuron is filtered and decimated. The temporal-difference (TD) error is then calculated and the synaptic weights are updated accordingly. The ST information needed in this process is read from memory B. In the figure, the filtration and access of memory B is shown to occur at the last tick of each unit interval. Nevertheless, these operations can actually be spread over one unit interval in a time-interleaved fashion in order to reduce the critical path delay as well as to save hardware resources.

In the algorithm outlined in Section 4.2.4, a division operation is needed. The division operation for fixed-point numbers is a relatively expensive operation when implementing in hardware. The operation can take several clock cycles to complete, or otherwise a pipelined divider is needed, which takes a lot of area. To circumvent this difficulty, an approximate division is adopted in our design. With the approximate division, we set the divisor to the closest power of 2. With such an approximation, the division can be performed by first rounding the divisor to a power of 2 and then shifting the dividend according to the rounded divisor. To study how this approximate division affects the performance of the learning, numerical simulations are carried out and the obtained results are shown in Figure 5.22. In the figure, we compare the root-mean-square error (RMSE) in the one-dimensional learning problem under three sets of configurations. In the first configuration and the second configuration, the exact division and the approximate division are used, whereas in the third configuration, no division is used at all. One can observe that the difference between the results obtained with the exact and approximate division is hardly noticeable, which demonstrates the effectiveness of the approximate division. The results obtained without division seemed to work fairly in the diagram. To study how the denominator shown in Eq. (4.25) affects the learning performance, we compare the RMSEs obtained with and without division in Figure 5.23. In the figure, three different learning rates are also compared. The trend we notice is that learning with a weight denominator is less sensitive to the learning rate. Furthermore, the converging process is also smoother and faster in this case. In contrast, learning without the division is slower and sensitive to the learning rates. Despite being suboptimal, learning without the weight denominator might be useful in cases where analog

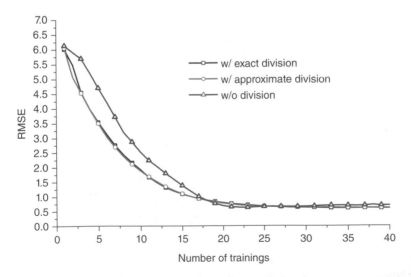

Figure 5.22 Comparison of the RMSE of the critic network obtained when exact division, approximate division, and no division are used. The approximate division does not yield noticeable performance degradation compared to the exact division. The results obtained with no division is slightly worse [45]. Reproduced with permission of IEEE.

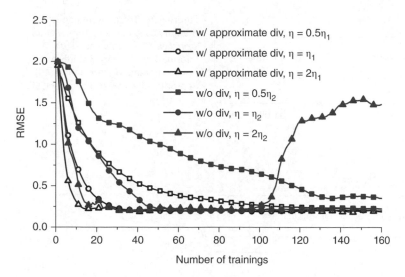

Figure 5.23 Comparison of the learning performance achieved with multiplicative STDP learning rules (w/ dividing by weights) and non-weight-dependent STDP learning rules (w/o dividing by weights). Three different learning rates are employed [45]. Reproduced with permission of IEEE.

synapses are employed. In that case, it might be difficult to know the exact value of the synaptic weight, which makes the division unrealistic. This situation is considered in more detail in Section 5.3.5.

Most operations incorporated in our system are pertaining to the reading from (Load operation by the processor) and the writing into (Store operation by the processor) memory, as illustrated in Figure 5.21. The memory access is likely to be the most energy-consuming operation in the system, especially when a large memory array is used. Fortunately, the number of memory accesses is significantly reduced through an event-driven operation. The memory is accessed only when there is a spike. In this design, only the memory access is event-triggered, yet each neuron is still evaluated at each tick even when there is no spike, as shown in Figure 5.21. In the next section, we demonstrate how an event-triggered neuron evaluation can be exploited to improve the throughput as well as the energy efficiency of the system. Figure 5.24 illustrates the content of each word in memory A and memory B. Memory A stores the synaptic weight and the ST information before decimation, whereas memory B stores the downsampled ST information.

The memory requirement of the system can be estimated as

$$N_{weight} = \prod_k N_k(1 + N_{action})w \tag{5.4}$$

where N_k is the number of states associated with dimension k and N_{action} is the number of possible actions.

Clearly, the memory requirement is affected by the well-known curse of dimensionality [46]. To avoid the speed and energy consumption penalty brought by the large memory array, the memory hierarchy employed in modern processors can be leveraged. In addition, the memory can be partitioned and placed near the processing unit to improve the performance and energy efficiency of the system, similar to the approach

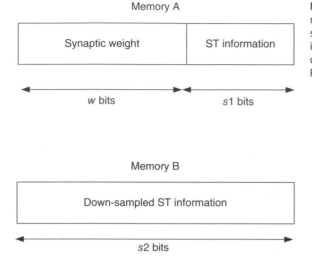

Memory A

Synaptic weight	ST information

w bits   s1 bits

Figure 5.24 The content of each memory used in the system. Memory A stores w-bit weights and s1-bit ST information. Memory B stores s2-bit down-sampled ST information [45]. Reproduced with permission of IEEE.

Memory B

Down-sampled ST information

s2 bits

taken in TrueNorth [12]. Furthermore, the size of each dimension of the problem can be compressed through a self-adapting kernel similar to what has been used in radial basis function networks [47]. To provide guidelines on determining the NOBs needed to represent the synaptic weights, a parametric simulation is conducted on the synaptic weights for the one-dimensional learning task. The obtained results are shown in Figure 5.25. One can conclude that the precision needed to represent the synaptic weight varies with the type of task. The learning process demands a wider bit width to represent the weight. Such a conclusion is consistent with what has been observed in ANN

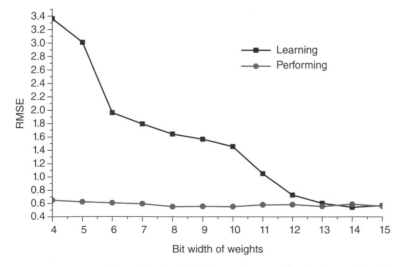

Figure 5.25 Comparison of the RMSE of the critic network versus the number of bits used to represent the synaptic weights in the network. Only a few bits are necessary to perform a task after being well trained, whereas many more bits are needed for successful learning. This is mainly because small changes in weights need to be accumulated during learning, which requires a high resolution in weights [45]. Reproduced with permission of IEEE.

hardware, as discussed in Chapter 2. Neural networks can typically conduct inference that is effective even when only a few bits are used to represent the weight, yet more bits are needed for the learning. Such a difference in the requirement of bit width can be readily exploited to partition the memory into different banks [37], as discussed in Section 5.2.2.

The word lengths needed for the spike timing information, $s1$ and $s2$, vary with the choice of the decimation filter and the down-sampling ratio. In this design, a cascaded integrator-comb (CIC) filter is employed for $stdp_{ij}$. Such a choice is helpful in reducing the memory requirement. When a first-order CIC filter is used, $\log_2 M$ NOBs are needed to represent the ST information in memory A, where M is the down-sampling ratio; $s2$ in this case is equal to $(N_{tap}/M + 1)\log_2 M$, where N_{tap} is the number of taps used in the finite impulse response (FIR) filter for the state-value function. Memory B is used as a data buffer for the spike timing information. The delay associated with the buffer is helpful in matching the group delay caused by filtering the output of the critic network. The worst-case memory requirement of the ST field is similar to that shown in Eq. (5.4). The only difference is that w in the equation is replaced by $(s1 + s2)$. The memory requirement can be greatly reduced if the sparsity in an SNN can be exploited. More specifically, there are only a few states in a large state space that are active for a particular period of time in our problem. In other words, only a portion of $stdp_{ij}(t)$ is non-zero. Therefore, a cache structure can be created to hold the spike timing information. Whenever a spike arrives, the existing spike timing information is updated if the entry is already in the cache, otherwise a new entry will be created. This idea is elaborated in more detail in the next section.

5.2.4 Hardware-Friendly Supervised Learning in Multilayer SNNs

In this section, an efficient hardware architecture is presented for the learning algorithm introduced in Section 4.3.5. The material in this section is based on our previous work [48] with more supplementary results and more in-depth discussion included. In order to make the hardware more efficient, many adaptations to the original algorithm are made. Several design techniques are also introduced to leverage the inherent pattern in an SNN and the proposed learning algorithm to boost the energy efficiency.

5.2.4.1 Hardware Architecture

5.2.4.1.1 Algorithm Adaptations A few adaptations and simplifications are first introduced to improve the performance and energy efficiency of the system in this section. In order to study the newly proposed ideas, the MNIST benchmark is employed for evaluating various aspects of the proposed algorithm adaptations and techniques. The original 28×28 images were down-sampled to 16×16 images in order to reduce the size of the network we evaluated [49]. Such an arrangement was helpful in accelerating the simulation process and allowed us to complete the evaluation within a reasonable amount of time. A few digits from the down-sampled dataset are compared with the original images in the standard MNIST dataset in Figure 5.26. It may be noted that the main features of the images are preserved regardless of the down-sampling. The neural network that we study in this section is a three-layer network with 256 input neurons, 50 hidden neurons, and 10 out neurons that correspond to 10 digits.

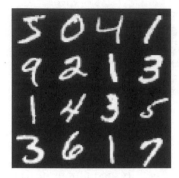

Original MNIST
images (28 by 28)

Down-sampled MNIST
images (16 by 16)

Figure 5.26 Comparison of the original MNIST images and the down-sampled MNIST images. The down-sampled images preserve most of the features in the hand-written digits, regardless of the reduced size of the images [49]. Reproduced with permission of IEEE.

Storage units in a hardware neural network can easily dominate both the area and power consumption of the system. Therefore, it is always desirable to minimize the amount of memory that is needed to store the synaptic weights. To study the trade-off existing in choosing the bit width for the synaptic weight, a parametric study is conducted. Figure 5.27 compares the test accuracy achieved by SNNs with different numbers of bits used to represent the synaptic weights. In this set of tests, the maximum weight of the network is kept as 1023 and the resolution of the weight is swept. It can be observed from the figure that a bit width of 20 is needed to avoid significantly deteriorating the performance of the system. This result is consistent with the empirical value that is widely used in the machine learning community: the weight update should be maintained to be around 10^{-3} of the weight itself [50]. For a task such as handwritten

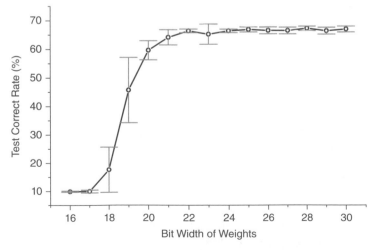

Figure 5.27 Effect of shortening bit width of synaptic weights on classification results. A bit width of 20 is enough to perform the learning [48]. Reproduced with permission of IEEE.

digits recognition, a weight dynamic range of 10^3 is enough. Therefore, after taking into account the weight update resolution, a weight range of 1 to 10^6 was selected. This translates to a word length of 20. We picked 24 as the bit width of the weight memory in our design to allow some margins.

In the original learning algorithm, division operations as shown in Eqs. (4.51) and (4.52) in Chapter 4 are needed. The term $(1 - \overline{x_i^l})$ can be approximated with 1 without introducing significant error, especially when considering that spikes in SNNs are often sparse. For other division operations, the approximate division presented in Section 5.2.3 can be used. More specifically, the divisor in a division operation can be approximated as

$$w_{ij}^{l\prime} = \mathrm{sgn}(w_{ij}^l) \cdot 2^{\left\lfloor \log_2 |w_{ij}^l| \right\rceil} \tag{5.5}$$

where $\mathrm{sgn}(\cdot)$ denotes the sign function. In Eq. (5.5), w_{ij}^l is used as an example. The same idea can be employed for $\overline{x_i^l}$ as well. The approximate division operation can be implemented as a low-cost round-and-shift operation. This can greatly reduce the hardware complexity as well as the power consumption of the system.

As discussed in Section 4.3.5, random refractoriness can be helpful in boosting the learning performance. With the proposed refractory mechanism, two consecutive spikes rarely occur, so the term $(1 - x_i^l[n - T - 1])$ in Eq. (4.40) can be omitted to reduce the implementation complexity. In a straightforward implementation, random number generators are needed to achieve such a randomness in the refractory period. It is possible to leverage the pseudorandomness in the system to achieve a similar refractory effect without actually resorting to true/pseudo random number generators. It is noted that the randomness is only needed when a neuron is firing in order to help decide whether that neuron should stay silent due to the refractoriness. Therefore, we can use the last few digits in the membrane voltage as the source of the randomness.

The test accuracies achieved with neural networks that adopt above-mentioned simplifications are compared in Figure 5.28. It can be observed from the figure that none of the proposed adaptations to the baseline algorithm introduces a noticeable performance degradation.

5.2.4.1.2 *Layer Unit* The diagram for a layer of the neural network is given in Figure 5.29. The architecture used here is a realization of the abstract idea illustrated in Figures 5.2 and 5.3. The input to this layer is the address representing spikes from the preceding layer. Corresponding synaptic weights are read out from the weight memory upon a new spike event and the membrane potential for each neuron in the layer is updated. After all spikes from the preceding layer for the current tick are committed, the states of the neurons in this layer are evaluated. Spikes are generated according to the employed neuron dynamics. A priority encoder is then employed to encode output spikes into corresponding addresses.

The proposed neuron circuit is also shown in Figure 5.29. Since adders are normally the most bulky and power-hungry components in a spiking neuron, the target here is to minimize the number of adders. In the proposed neuron circuit, only one adder and three comparators are needed. Among these comparators, two of them are trivial sign detectors.

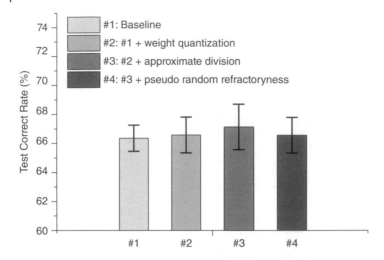

Figure 5.28 Comparison of the recognition rate achieved with the proposed simplifications on the original algorithm. None of the proposed simplifications on the original learning algorithm degrades the learning performance noticeably [48]. Reproduced with permission of IEEE.

5.2.4.1.3 Leveraging Sparsity In the proposed learning algorithm, memories are required to store the ST information in order to estimate the gradients. At first glance, every synaptic weight requires a dedicated entry to store the associated ST information. ST information typically requires a much shorter bit width compared to what is needed by synaptic weights. Nevertheless, sparsity existing in the network can be exploited to reduce the amount of memory needed. Sparsity is an important feature of many neural networks. It is not only observed in a real neural network but is also controlled in many ANNs [51]. Indeed, many real-world signals are sparse in nature, such as image, audio, and video. The inherent sparsity in these signals can result in sparsely activated neurons when these signals are presented to neural networks. The density of the spike trains in the network are shown in Figure 5.30, where density is defined as the percentage of neurons that actually output a spike during the whole inference duration. The data are generated from 10 000 images in the MNIST test set. It is shown in the figure that the average density for the spike trains is as low as 0.2. In other words, only 20% of neurons, on average, are active for one test image. The worst-case density of 0.4 is obtained across all 10 000 test images. In other words, at most 40% of neurons are activated for any image in the test set. In the simulations, there is no requirement on the sparsity of the hidden-layer units. In the literature, sparsity regularization is sometimes utilized to regularize hidden-layer neurons in order to meet a certain sparsity target. To fully exploit this sparsity, we propose a cache structure to hold ST information, considering that only recently active synapses have non-zero ST fields. This idea is illustrated in Figure 5.31. The operational principle of this cache structure is similar to the cache used in modern processors. When a new entry comes, the cache first searches if there is any entry sitting in the cache that has the same address. If there is one, the old entry is read out and updated. If such an entry does not exist, a new entry is created.

5.2.4.1.4 Background ST Update In the proposed learning scheme, the ST information needs to be collected and updated at run-time. There are different strategies in choosing

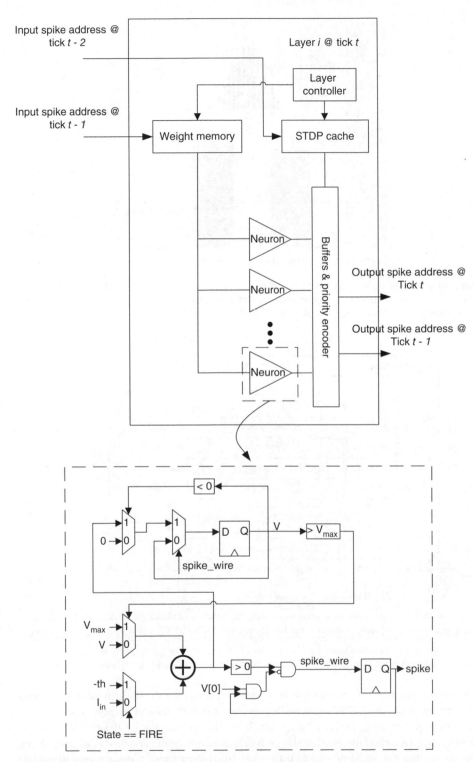

Figure 5.29 Schematic of one layer of neural network. The input to the network is the addresses of spiking neurons from the preceding layer. The neuron circuit is optimized to reuse the adder in the neuron as much as possible in order to save the area and energy consumption.

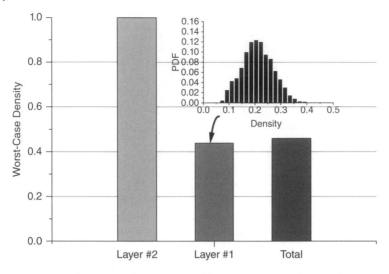

Figure 5.30 Illustration of the sparsity of the employed neural network over 10 000 test images. The density is defined as the portion of active neurons. The input layer has the highest sparsity due to the sparse nature of the MNIST image [48]. Reproduced with permission of IEEE.

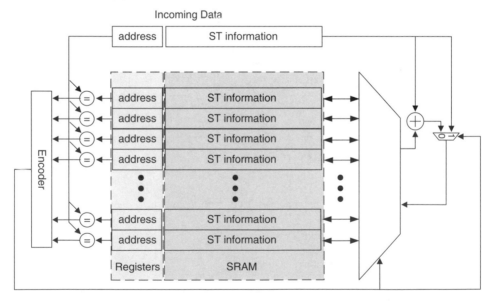

Figure 5.31 Diagram of the cache structure employed to store active ST information. ST information associated with recently active synapses is stored in a fully associative cache. Every synapse has a unique identifier which is recorded in the address field [48]. Reproduced with permission of IEEE.

the updating frequency. One extreme is to store all the spike timings for all the neurons in buffers and flush the STDP cache with new data once at the end of the iteration. Such a method requires a big buffer to store all the spikes outputted by the neurons in the network. Another extreme is to flush the STDP cache whenever there is a spike. Such a scheme avoids the need of a large buffer yet requires a more frequent update to the

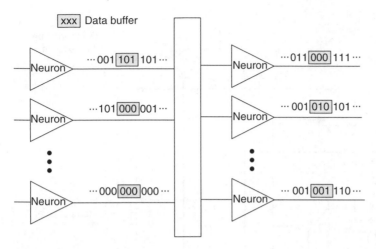

Figure 5.32 Illustration of the concept of utilizing local buffers to speed up the process of updating ST information. Local buffers are allocated at the output of each neuron to hold the spike history temporarily.

STDP cache. Synapse memory and STDP cache are visited at similar frequencies with this scheme, yet only the read operation is needed for the synapse memory, whereas both read and write are needed for the STDP cache. Therefore, this way of updating the ST information inevitably slows down the system. To find a balance between these two extremes, we proposed a background ST update scheme that can hide the update process of the ST information in the background while reducing the size of the buffer that is needed to hold the spike timings.

The basic idea and the circuit schematic are illustrated in Figures 5.32 and 5.33, respectively. Buffers that are local to neurons are employed to hold the spikes from the neurons. These spike timings are processed and updated to the STDP cache at a proper time. In order to determine what is a proper time to flush the STDP cache, a finite-state machine (FSM) with a state diagram shown in Figure 5.34 is used to control the update. The idea is to try to conduct an ST update only when a membrane voltage update is currently going on. By doing this, we can hide the latency of updating the ST field in the background. There are certain cases where an ST update has to be carried out. One example is one when the capacity of the buffer is reached. The FSM raises an "urgent" flag in this case and the ST update is forced to be conducted. Obviously, such an exception can be avoided by choosing a buffer size that is sufficiently large. Figure 5.35 compares the number of cycles per tick when different buffer depths are used. The dashed line in the figure is the reference where no ST update is conducted at all. When the buffer is shallow, the duration of a tick is prolonged as the system has to wait for the ST update to finish. Once the buffer size reaches 5, the ST update has little impact on prolonging the tick duration.

5.2.4.2 CMOS Implementation Results
The proposed hardware architecture is implemented in a 65-nm technology. All the adaptations and techniques proposed in previous sections are also included. The floor plan of the chip is shown in Figure 5.36. The chip occupies an area of 1.8 mm^2 including

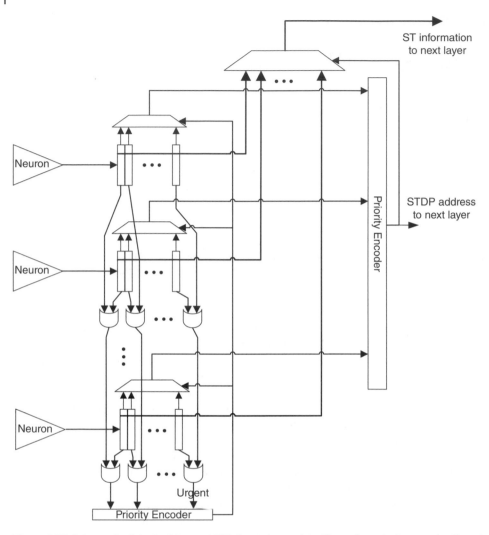

Figure 5.33 Schematic of the background ST information update. Through a priority encoder, there is a pointer that always points to the first element in the buffer. When the buffer is full, an "urgent" flag is set, which forces the ST cache to update in order to avoid loss of data [48]. Reproduced with permission of IEEE.

pads. Synaptic weight memories, which are implemented by SRAM arrays, take most of the area. It is also noted that the STDP cache is much smaller compared to the weight memory. The estimated power consumption breakdown is illustrated in Table 5.1. The numbers are estimated from a post-layout netlist annotated with the circuit-switching activities obtained from gate-level simulations. It can be seen from the table that the majority of the power is dissipated in the second layer of the network. This is because this layer contains most of the synapses and hence most of the synaptic operations in the network. Furthermore, inference consumes less power compared to the learning since the ST information and the synaptic weights do not need to be updated in the inference mode.

Figure 5.34 State diagram of the
background ST updating scheduler. The
scheduler attempts to perform ST field
update when there is currently neuron
evaluation going on. The latency of the
ST update can be hidden in the
background with such a scheme.

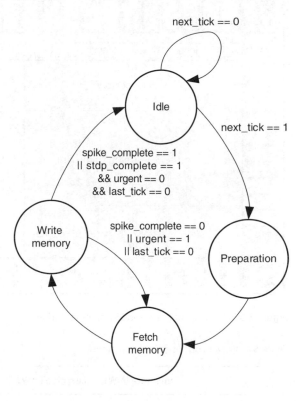

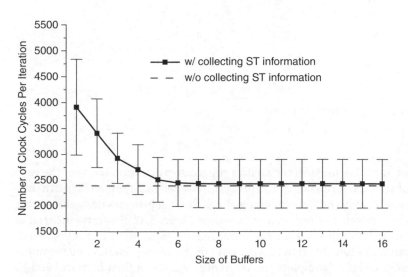

Figure 5.35 Comparison of the number of clock cycles per forward pass for different buffer depths.
When the size of the buffer reaches 5, the process of updating ST information does not affect the
system throughput noticeably [48]. Reproduced with permission of IEEE.

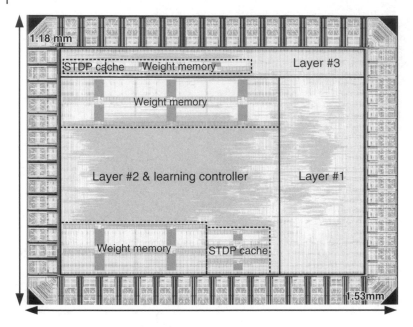

Figure 5.36 Chip layout of the CMOS implementation.

Table 5.1 Power consumption breakdown.

	Inference (mW)	Learning (mW)
Total	91.30	104.12
Layer 1	21.15	20.97
Layer 2	43.64	49.11
Layer 3	7.47	7.34
Learning controller	4.74	12.13
Others	14.3	13.28

Source: data are from [48]. Reproduced with permission of
IEEE.

As stated in Section 5.1.3, the inference with a progressive precision is one of the very useful features of an SNN. It can adjust the precision needed in different tasks dynamically in order to save energy and inference time. In addition, it provides designers with a knob to optimize the system performance at run-time. Figure 5.37 illustrates the results obtained with the first-to-spike-K-spikes readout scheme introduced in Section 5.1.3. For different K values, we can obtain the corresponding inference accuracy and the number of clock cycles needed to complete the inference, which can then be translated to the inference latency and energy. As shown in the figure, one can configure the SNN hardware at run-time on a per image basis for different K values in order to trade off between energy/latency and inference accuracy. Such a convenient knob can be helpful in optimizing the system performance.

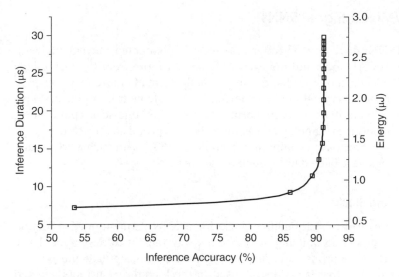

Figure 5.37 Time and energy needed per inference as a function of the inference accuracy when the first-to-spike-K-spikes readout scheme is used. By varying the value of K, the inference accuracy can be readily traded off for a faster inference speed and a lower energy consumption [48]. Reproduced with permission of IEEE.

Table 5.2 Summary and comparison with prior arts.

	This work	**[37]**	**[52]**	**[36]**
Network configuration	256-50-10	256–256	484–256	256–256
Bit width of synapses	24 bit	8 bit and 13 bit	1 bit	4 bit
On-chip memory size	358.3 kb	1.31 Mb	256 kb	256 kb
Technology	65 nm	65 nm	45 nm	45 nm
Core area	1.1 mm²	3.1 mm²	4.2 mm²	4.2 mm²
Supply voltage	1.2 V	1 V	0.85 V	—
Clock frequency	166.7 MHz	235 MHz @ learning, 310 MHz @ inference	1 kHz	1 MHz
Power consumption	104.12 mW @ learning 91.3 mW @ inference	228.1 mW @ learning 218 mW @ inference	45 pJ/spike	—

Source: data are from [48]. Reproduced with permission of IEEE.

Table 5.2 summarizes the performance of the CMOS implementation and compares it to the state-of-the-art SNN implementation in [36], [37], and [52]. One striking feature of the design presented in this section is that it is a multilayer neural network that is capable of conducting on-chip supervised learning.

5.3 Analog/Mixed-Signal SNNs

While the digital SNNs discussed in Section 5.2 can be implemented conveniently with the help of modern CAD tools and they generally benefit more from the technology scaling, analog SNNs, on the other hand, can normally be made more compact [53]. This density advantage is very helpful in achieving large-scale neuromorphic systems, as millions or even billions of neurons are normally needed for large-scale applications. In addition, many emerging technologies, e.g. memristor, is promising to achieve lower power consumption when analog computing is employed [53]. With such a hope, various analog SNN chips have been developed in recent years.

5.3.1 Basic Building Blocks

Neurons and synapses are the basic building blocks for an SNN. As mentioned in Section 5.2.2.1, one important step in building hardware SNN systems is to develop its basic building blocks, such as neurons and synapses. Figure 5.38 conceptually illustrates a typical analog LIF neuron that is commonly used in the literature. Currents injected from presynaptic neurons are summed up and integrated on a capacitor that is used to store the membrane voltage of the neuron. A comparator is usually used to determine whether the neuron should spike. Depending on the implementation, this comparator can be a clocked regenerative comparator or a non-clocked operational amplifier. The charge stored on the capacitor can be discharged through two paths in addition to the inputs. One path is the leakage path, which models the leakage current. Both state-dependent and state-independent leakage can be modeled, depending on whether the current source is controlled by the membrane voltage. Another path for the charge to sink is the current source used to model the reset operation and the refractoriness. Once the membrane voltage of the neuron exceeds the threshold voltage, a voltage pulse is emitted, and is used as a trigger to reset the membrane voltage. It is noted that in a non-clocked design, either hysteresis in the comparator [54] or enough delay on the resetting path is needed to reset the membrane voltage properly. A refractory period can be achieved by controlling the resetting current source with a memory element such as a capacitor, which can temporarily hold the information that the neuron just spiked [55, 56]. Even though the circuit shown in Figure 5.38 only implements a basic LIF neuron model, many biological plausible features can be implemented by adding more

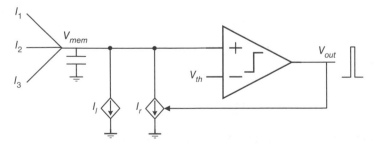

Figure 5.38 Diagram of a typical analog LIF neuron. Input currents are summed and integrated on a capacitor to form the membrane voltage. The comparator outputs a spike when the membrane voltage exceeds the threshold. A resetting current source is used to reset the membrane voltage and to implement a refractory period.

components into this circuit. For example, one more memory element was introduced in [56] to achieve the effect of spike-frequency adaptation. More complicated neuron models such as the Hodgkin-Huxley model have also been demonstrated on silicon using various circuit design techniques [57–59]. These complicated neuron models are usually useful in applications where the detailed modeling of neuron activities is needed.

Many CMOS-based analog SNNs either store the synaptic weights dynamically on capacitors [54, 56, 60] or statically in SRAM cells [61]. Sometimes non-volatile memory (NVM) can also be employed for storing weight parameters [62]. A capacitor-based synapse has the advantage of being compact, yet it suffers from leakage. Therefore, this type of synapse is more commonly seen in a system where certain synaptic plasticity is implemented so that the synaptic weight stored on the capacitor can be updated and refreshed once in a while. Updating weights for these synapses in an analog system is also easier as analog signals can be directly applied on the capacitors. The weights are often modified through charging or discharging the corresponding capacitors [54, 56, 60, 63, 64]. SRAM-based synapses, on the other hand, tend to take more area, yet it can hold the value as long as the system is being powered. Another downside of using digital memory is that one inevitably needs to quantize the weight, which might deteriorate the system performance.

Synapses that are capable of changing weights according to an STDP protocol have received the most attention [54, 56, 60, 62–68], as STDP learning is hypothesized to be the underlying learning mechanism of biological SNNs. Figure 5.39a shows an example of an analog synapse. A typical STDP protocol is implemented. The synaptic weight is conveniently stored on a capacitor. The voltage on the capacitor represents the value of the stored synaptic weight. This information is converted into a synaptic current through a transconductance amplifier. Depending on whether the conductance of this amplifier is a function of the membrane voltage of the postsynaptic neuron, a conductance-based or a current-based synapse can be realized. An STDP protocol can be implemented with the help of two leaky integrators, which are used to hold the information of recently occurred spikes [54, 56, 64, 65]. The leaky integrator shown in Figure 5.39a consists of a transconductance amplifier, a capacitor, and a resistor. When a presynaptic/postsynaptic spike arrives, an exponentially decaying voltage trace is formed at the output of the integrator. The voltage trace is then converted into current and sampled by the corresponding postsynaptic/presynaptic spike, as shown in Figure 5.39b. Depending on the relative timing of the pre- and postsynaptic spikes, the voltage stored on the weight capacitor changes accordingly.

Besides long-term plasticity, short-term plasticity was also explored by many researchers. For example, a dynamic charge transfer synapse circuit with a depressing response was reported in [69]. The depressing behavior was shown to be modulated by a control voltage. The synaptic charge recovery time was controlled to produce a depressing effect over a presynaptic spiking frequency range of more than four orders of magnitude. Such a wide tuning range was comparable to those that have been observed in biological synapses.

5.3.2 Large-Scale Analog/Mixed-Signal CMOS SNNs

5.3.2.1 CAVIAR
CAVIAR is the outcome of the project "Convolution AER Vision Architecture for Real-Time" [70]. It is a system consisting of several analog ASICs. The main goal of

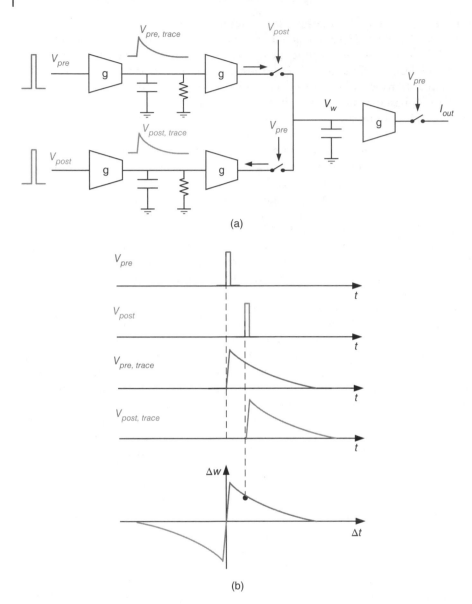

(a)

(b)

Figure 5.39 An example of an analog synapse implementing a typical STDP protocol. (a) Schematic of the synapse. The synaptic weight is stored in the form of charge on a capacitor. A transconductance amplifier converts this voltage into synaptic current upon a presynaptic spike. Two leaky integrators are employed to implement an STDP protocol. Decaying voltage traces are formed at the output of the integrator and are sampled by the presynaptic and postsynaptic spikes. The amount of net charge going to the capacitor is then proportional to the weight change according to the STDP protocol. (b) Illustration of the voltage waveform at each node and the resultant STDP rule.

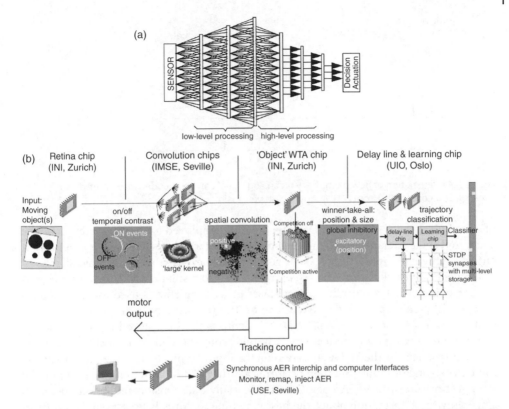

Figure 5.40 Overview of the CAVIAR system. (a) Abstract view of the system architecture. (b) Detailed view of the system architecture, which consists of silicon retina, convolution chips, WTA chips, and learning chips [70]. Reproduced with permission of IEEE.

the project is to build hardware infrastructure with a bio-inspired architecture and to utilize this infrastructure to implement brain-like computing. As shown in Figure 5.40, the CAVIAR system mainly consists of four building blocks: a temporal contrast retina, a programmable convolutional processing chip, a winner-take-all (WTA) object-detection chip, and learning chips.

The temporal contrast retina is built to generate spikes for moving objects in the field. This can be achieved through a 128 × 128 pixel CMOS vision sensor [71, 72]. The vision sensor is event-based and data-driven. It can achieve a wide dynamic range of more than 120 dB, a low latency of 15 μs, and a small power consumption of 23 mW. Such a wide dynamic range enables the sensor to be used with uncontrolled natural lighting. Each pixel in the vision sensor responds to the change of intensity in the input asynchronously. The output from the vision sensor is a spike address-event, which represents both the address of the pixel and the information of whether the input intensity has significant changes. In the pixel circuit, a photoreceptor circuit, which serves as a transimpedance amplifier, converts the photocurrent into voltage. A following differencing circuit is then utilized to calculate a voltage that is proportional to the difference of pixel intensity in the log domain. After the difference voltage is generated, it is compared with two predefined threshold voltages to generate the "ON" and "OFF" signals. With these two signals, the actual images can be reconstructed approximately, as illustrated in Figure 5.41. This way

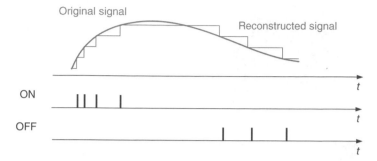

Figure 5.41 Illustration of how a time-varying analog signal can be encoded into event-driven bit streams. Source: adapted from [71]. Adapted with permission of IEEE.

of encoding signal is similar to a Delta modulation employed for data transmission. The generated spike streams are event-triggered. In other words, when there is a significant change in the input intensity, a spike is generated, otherwise the sensor stays silent.

The generated spikes from the silicon retina lead to a splat at the convolutional processing chips. A neuromorphic cortical-layer processing chip is used for this purpose [73, 74]. The chip computes convolutions on the two-dimensional input data represented in the AER form. An AER 2-D WTA chip is then employed to determine the location and the type of the feature outputted from the convolution chips [75]. The WTA chip performs the WTA operation on the input from a feature map first to determine the strongest input, and then performs another WTA operation to determine the strongest feature. The WTA operation is essentially a dimension-reduction operation since only the information about the best matching features is preserved. Learning in CAVIAR is conducted by AER learning chips [70]. The first chip expands the time into a spatial dimension by using delay lines, whereas another chip implements a spike-based timing-dependent learning rule [76].

5.3.2.2 BrainScaleS

The BrainScaleS wafer-scale system is the outcome of the European research project "Fast Analog Computing with Emergent Transient States (FACETS)" and its following project "Brain-inspired multiscale computation in neuromorphic hybrid systems (BrainScaleS)." One of the most noticeable features of the BrainScaleS system is its wafer-scale integration. Up to 196 608 neurons can be emulated on a single wafer [77]. The main intuition behind this wafer-scale integration originates from the gigantic communication bandwidth needed by the system. A large acceleration factor of 10^3 to 10^5 is targeted in the FACETS project. By doing so, the simulations of large-scale neural networks modeling neural systems with billions of synapses can be finished within a few million seconds. Such an aggressive acceleration rate demands a huge communication bandwidth as high as 10^{11} neural events per second [78]. In order to avoid the cumbersome and expensive package and printed-circuit board cost associated with the huge number of I/O pads that are needed, a wafer-scale integration is used to directly connect chips on the wafer.

The main building block of the BrainScaleS system is the high input count analog neural network (HICANN), which contains mixed-signal neurons and synapse circuits. Synaptic weights used in the BrainScaleS system are stored in 4-bit SRAM cells. Such a resolution is chosen based on the trade-off between precision and the required hardware resources. It was shown that a 4-bit resolution was sufficient for certain benchmark tasks [79]. The neuron model implemented in HICANNs is the so-called adaptive exponential integrate-and-fire neuron model, which is able to generate various spiking patterns [80, 81]. The full-wafer system, which consists of 384 HICANNs, can implement approximately 45 million synapses and 200 000 neurons [82].

Due to the large-scale integration, about 2 pF wire capacitance is seen by a 10 mm wire needed to traverse an HICANN die. Such long wires for communication would have induced a 1.7 kW power consumption for a system containing 450 HICANNs if a conventional full-swing digital signal had been used for transmitting the neural events. To lower the power consumption, low-swing differential signals are employed everywhere outside the HICANNs. This reduced the power consumption by 300× compared to the naïve full-swing case [78]. In addition to dynamic power reduction, leakage power has also been carefully reduced. With these techniques, the average power consumption of the system is expected to stay below 1 kW.

5.3.2.3 Neurogrid

Neurogrid is a neuromorphic system developed by researchers at Stanford University [83]. It is mainly used for simulating large-scale neural models in real time. Neurogrid is a complete system consisting of software for interactive visualization and hardware that performs the actual computations.

Neurons in Neurogrid are implemented as analog circuits, as analog neurons are much more compact compared to their digital counterparts. The size of the neurons matters to Neurogrid since the ultimate goal for Neurogrid is to simulate large-scale neural networks. Through utilizing transistors operating in the subthreshold region, complex neuron models can be implemented with only a few transistors. Biasing transistors in the subthreshold region normally leads to a low operating speed, which is something that many circuit designers try to avoid. This, on the other hand, may actually be advantageous for Neurogrid as it targets real-time simulations of large neural systems. To match the time constant of a silicon neuron with that of a biological neuron, small current flowing through transistors that are biased in the subthreshold region can be harnessed to avoid the use of large capacitors [63, 84]. This is in contrast to the approach taken by the BrainScaleS system where the superthreshold region is exploited in order to achieve a large acceleration factor.

One drawback of implementing analog neurons is the mismatch between the desired neuron model and the actual model due to process variation. This non-ideality is overcome through calibration in Neurogrid. Four circuit responses, namely dynamic current, steady-state current, steady-state spike rate, and spike-rate discontinuity, can be measured and employed for calibration purposes.

The event-based nature of SNNs leads to inefficient simulations in a non-event-based simulation environment. Special transmitters, receivers, and routers are used in Neurogrid to communicate spike information [85–87]. All these circuits are event-triggered asynchronous circuits, which helps to save power when the network is inactive. It was shown in [83] that Neurogrid consumed only a few watts while simulating a million neurons in real time, which was five orders of magnitude more energy efficient than that of a personal computer.

5.3.3 Other Analog/Mixed-Signal CMOS SNN ASICs

In addition to large-scale analog SNNs that mainly target simulating and modeling biological neural networks, many customized analog ASICs have also been demonstrated over the past few years. Even though the scalability might not be an issue for mid- to small-scale hardware SNNs, many of these accelerators still use an AER-based spike routing and an event-based computing scheme in order to leverage the sparsity in an SNN. For example, an event-triggered architecture is presented in [61]. The main motivation of this work is to combine fast digital circuits with slow analog components in order to obtain a high overall performance. An asynchronous SRAM array is employed to store synaptic weights for an event-triggered access. To bridge the digital synapse SRAM and analog neural cores, current-mode event-driven digital-to-analog converters are used to produce the synaptic current.

For SNN accelerators of smaller sizes, even though they might not be able to simulate large-scale neural systems like an entire brain, they are useful in performing many smaller-scale learning tasks such as pattern recognition [88, 89], odor classification [90], assisting cardiac delay prediction [91], and so on. A neuromorphic chip was presented by Mitra et al. in [88] for real-time classification tasks. The chip consists of 16 integrate-and-fire neurons and 2048 synapses. Learning of the system is conducted locally with a presynaptic weight-update module and a postsynaptic weight control module. The synaptic weights are incremented or decremented at each presynaptic spike event depending on the state of the neuron's membrane voltage. The modification of the synaptic weight is realized through charging or discharging the charge stored on a capacitor that is used to derive the weight, similar to the scheme discussed in Section 5.3.1. The implemented chip was used for learning classifying images of numbers consisting of binary pixels as well as graded patterns.

Another example is the reconfigurable online learning spiking (ROLLS) neuromorphic processor, a mixed-signal SNN hardware capable of doing online learning [89]. The system contains 256 neurons and 128K analog synapses. 64K synapses are used for modeling long-term plasticity, whereas the other 64K synapses are for modeling short-term plasticity. The ROLLS neuromorphic processor implements the spike-driven synaptic plasticity rule [92]. The learning algorithm relies on the spike timings, the membrane voltage of the postsynaptic neuron, as well as the recent spiking activity. All the operations of the ROLLS processor are conducted through asynchronous address-event streams. With the chip that was built in a 180-nm process, an example of emulating the biophysics of cortical neurons and another example of a conducting pattern recognition task in conjunction with a spiking vision sensor were demonstrated in [89].

5.3.4 SNNs Based on Emerging Nanotechnologies

In Chapter 3, the possibility of employing an emerging nanoscale NVM as analog synapses for ANNs has been explored. Similarly, many recent studies have focused on employing these devices in SNNs. Compared to their ANN counterpart, using analog synapses in an SNN is more advantageous when the non-linearity in the device is taken into consideration. This is because many SNNs essentially utilize a one-bit spike to carry the information. For such a one-bit signal, the analog synapse always behaves linearly. This is similar to the design concept in a $\Sigma - \Delta$ modulator where a 1-bit digital-to-analog converter (DAC) is often preferred as it is always linear. To stay close to the theme of this book, this section mainly discusses the algorithm and architecture aspects of SNNs with nanoscale NVM devices. Interested readers are referred to [93] for a review on nanoscale synapses based on different device physics.

One of the main advantages of using nanoscale NVM devices in implementing neuromorphic systems is that a high integration density can be achieved. Compared to SRAMs that are widely used in a digital implementation, nanoscale devices often come with a much smaller size when serving as synapses [53]. The advantage in device size is important, especially considering the ultimate goal of building a brain-like machine where trillions of synapses are involved. Building SNNs with NVM devices share something in common with building NVM-based ANNs, which is discussed in Chapter 3. Therefore, only the aspects that are closely related to SNNs are discussed in this section.

5.3.4.1 Energy-Efficient Solutions

In addition to the higher integration density provided by many emerging nanotechnologies, one important advantage of these NVM-based synapses is the associated high energy efficiency. One of the reasons for this good energy efficiency comes from the fact that in-memory computing can often be effectively employed when NVM-based synapses are used, as discussed in Chapter 3. Through in-memory processing, the computations are conducted directly in the memory. Therefore, the energy-consuming memory readout can be eliminated, which helps lower the power consumption of the system.

In Section 5.2.2, accelerating sparse coding through a digital ASIC is discussed. The possibility of accelerating this task with an even higher energy efficiency was explored by Sheridan et al. in [94] with the help of WO_x-based memristors. A locally competitive algorithm is used to implement sparse learning. The input image is fed to the rows of the memristor array with a pulse-width modulation encoding scheme. By doing so, the charges that are injected into the memristor array are proportional to the intensities of the input image pixels. The currents from the inputs then flow through memristors and are collected by the output neurons. Through such an arrangement, a matrix–vector multiplication scheme similar to the one shown in Chapter 3 can be achieved. Compared to a CMOS digital baseline that was implemented to achieve a similar function, the memristor crossbar was reported to be 16× more energy efficient while achieving a similar performance in terms of image reconstruction error.

Many spin-based neurons and synapses have also been proposed in the literature as building blocks for future low-power neuromorphic systems [95–104]. For example, in [104], Sengupta et al. leveraged the spin-orbit driven domain wall motion to build spintronic synapses. A magnetic tunnel junction (MTJ) is formed by a free layer (ferromagnet) and a pinned layer (a tunneling oxide barrier). By moving the position of the domain wall, the resistance of the MTJ can be modulated between two extreme conductance values. The conductance can be programmed through the current flowing through the heavy metal underneath the MTJ. A spiking neuron can be implemented with a similar structure. The input current, which is modulated by the synaptic weight, is integrated by the neuron through the movement of the domain wall. An MTJ is put at the end of the ferromagnet to detect whether the domain wall is moved to the edge of the device. Once the domain wall is moved to the other side of the ferromagnet, the reference MTJ is switched to the parallel state, which leads to a spike at the output.

Many different spiking network structures [105, 106] have been realized with these spin-based nanoscale devices [98, 100, 101, 104], and similar recognition accuracies have been achieved compared to the software baseline. In addition to being good at conserving the classification correct rate, extreme low-power consumptions were often reported. The low-power advantage of spin-based neurons and synapses is mainly a result of a low operating voltage, a low reading current and a fast programming time needed to operate the device. For example, it was shown in [101] through simulation that the current flowing through the ferromagnetic–heavy metal bilayer resistance consumes merely 5.7 fJ of energy, which is almost two orders of magnitude lower than that consumed by a CMOS neuron implemented in a 45 nm technology.

5.3.4.2 Synaptic Plasticity

As an important feature of an SNN, synaptic plasticity in nanoscale NVM-based synapses has been a feature pursued by many researchers in recent years. For many memristive devices presented in the literature, programming of the conductance is achieved through applying a voltage across the device. The voltage magnitude needs to reach a certain threshold in order to change the resistance of the device. For example, in the popular VTEAM model [107], the internal state of a memristor device, s, is modeled as

$$\frac{ds(t)}{dt} = f[s(t), v(t)] \tag{5.6}$$

$$s(t) = G[s(t), v(t)]v(t) \tag{5.7}$$

In Eqs. (5.6, 5.7), $v(t)$ and $i(t)$ are the voltage and current applied on the device and $G(s, v)$ is conductance of the memristor device. The dependence of G on s reflects the programmability of a memristor device. Through changing the internal state of the device, its conductance can be changed accordingly. The dependence of G on v is to model the non-linear characteristic of a memristor device. The function $f(s, v)$ is used to model how the internal state of a memristor device alters according to the applied voltage. In the VTEAM model, this function has the form of

$$\frac{ds}{dt} = \begin{cases} k_{off}\left(\frac{v(t)}{v_{off}} - 1\right)^{\alpha_{off}} f_{off}(s), & 0 < v_{off} < v \\ 0, & v_{on} < v < v_{off} \\ k_{on}\left(\frac{v(t)}{v_{on}} - 1\right)^{\alpha_{on}} f_{on}(s), & v < v_{on} < 0 \end{cases} \tag{5.8}$$

where k_{on}, k_{off}, α_{on}, α_{off}, v_{on}, and v_{off} are model parameters and $f_{on}(s)$ and $f_{off}(s)$ are window functions used to represent the dependence of the derivative of the state variable on the state variable itself. They are often used to bound the state variable within a reasonable range.

As can be seen from Eq. (5.8), the derivative of the state variable is 0 when the magnitude of the applied voltage is smaller than the threshold voltage $|v_{on}|$ or v_{off}. Consequently, the state variable and thereby the conductance of the memristor device stay unchanged. This property has also been observed in many experiments [107] where the programming voltage needs to reach a certain level in order to change the resistance of the device noticeably. Such a feature provides a convenient way to utilize the

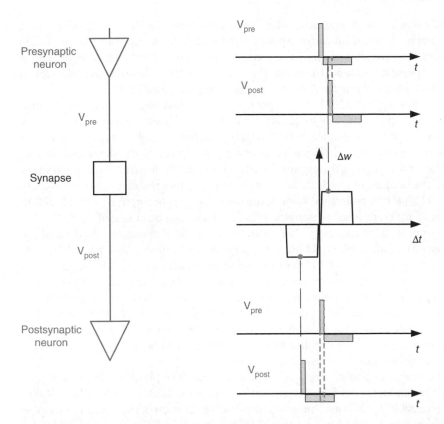

Figure 5.42 How the outputs from a presynaptic and a postsynaptic neuron can help achieve synaptic plasticity. When the spikes from the two neurons are properly aligned, the voltage across the memristor-based synapse is larger than the threshold voltage, and consequently the synaptic weight is changed.

memristor. Whenever a "read" operation is needed, such as using the synaptic weight to update the membrane voltage, a small voltage is applied to avoid disturbing the stored value, whereas a large voltage is applied when one wants to alter the resistance of the device. Such an operational model is used widely in the literature. Consequently, many researchers have leveraged this non-linear characteristic in programming the conductance to achieve synaptic plasticity [108–114]. To help understand this, Figure 5.42 shows a simplified example. In the figure, a synapse is sandwiched in between two neurons. In this example, instead of a simple pulse, a slightly more complicated shape is used to represent a spike. When the two spikes are far apart, there is no overlap between the presynaptic and the postsynaptic spikes, and the voltage across the synapse is lower than the threshold voltage that can trigger a change. When the two spikes are close to each other, on the other hand, the overlap between these two spikes can yield a voltage difference across the synapse that is larger than the threshold voltage. Consequently, the synaptic weight can be altered accordingly.

One of the earliest works of experimentally demonstrating the STDP on memristor-based synapses was presented by the researchers at the University of Michigan [111]. The time division multiplexing technique was employed to organize communications between neurons. Three time slots were divided to serve different purposes, such as communicating spikes, achieving long-term potentiation (LTP) and long-term depression (LTD). The overlap between a presynaptic pulse and a postsynaptic pulse is leveraged to change the synaptic weight. The exponential shape in a typical STDP protocol is achieved through adjusting the pulse widths.

It was also shown [108, 113, 114] that by properly engineering the shape of the pre- and postsynaptic spikes, the weight update in a memristor-based synapse approximately followed the shape of a biological STDP, which provides a feasible way to implement STDP-based learning. Indeed, through deliberately manipulating the shape and overlap between the action potentials outputted from a presynaptic neuron and a postsynaptic neuron, the amount of charges that flow through the memristor can be modulated, which then changes the resistance value of the memristor accordingly. As a result, STDP protocols with different weight-changing characteristics can be obtained.

Similar concepts have also been demonstrated in artificial synapses based on phase change memory (PCM). PCM is another emerging NVM device that has been studied extensively in recent years for neuromorphic applications [115]. One of the main appealing features of PCM that makes it well suited to being an artificial synapse is its capability of being programmed to intermediate resistance between the amorphous (high-resistivity) and crystalline (low-resistivity) states [110, 116]. In [110] and [112], STDPs were implemented for PCM-based synapses. Similarly, overlap between presynaptic spikes and postsynaptic spikes were leveraged to form a voltage difference that was large enough to alter the resistance of the cell.

Stochastic programming can also be employed to achieve plasticity. Instead of programming the conductance of the device into intermediate values, only "ON" and "OFF" states are accessible. In this case, one can either use a collection of these binary synapses to form synapses with a multilevel value [117] or directly use the binary synapses for computing [102, 118]. For example, in [102], MTJs were employed as stochastic synapses in an SNN. A simplified STDP rule was adopted to conduct unsupervised learning on a crossbar array consisting of MTJ synapses. Successful learning was observed and again the learning was robust against device variation due to the feedback mechanism in the learning.

In addition to implementing the STDP learning protocol through circuit-level techniques, the possibility of directly engineering the devices has also been investigated [119–121]. Second-order memristor effects have been studied and demonstrated through experiments [119, 120]. It was shown that memristors were dynamic devices that could implement various dynamic synaptic behaviors. Another example of building dynamic synapses, called memristor-based dynamic synapses, was illustrated in [122]. Two memristors were utilized to form one synapse where one memristor was for storing the weight whereas another one served as a selector. With the newly proposed dynamic synapses, learning could be readily carried out for both STDP and ReSuMe.

5.3.5 Case Study: Memristor Crossbar Based Learning in SNNs

5.3.5.1 Motivations

One of the biggest problems that most analog/mixed-signal neural networks encounter is the process variation. As discussed in Chapter 4, a popular way to provide an SNN with intelligence is to convert a well-trained ANN into an SNN. Even though such a

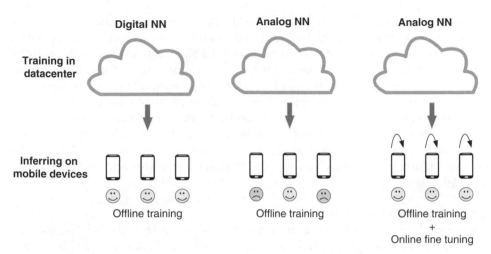

Figure 5.43 Different training and utilization models for analog/digital neural networks. For digital hardware, the neural network model can be trained in a datacenter and downloaded to the edge devices thanks to the one-to-one mapping in the digital system. Analog hardware, however, suffers from device variations, which need to be compensated through on-chip fine-tuning.

method can exploit the well-established theories and techniques in the field of ANNs, it often fails to achieve the optimal performance, especially for memristor-based SNNs. The reason behind this is that memristors suffer a lot from device variations. Therefore, the mismatch between the assumed device parameters at the training time and the actual parameter at the inference time could lead to a significant performance degradation. Therefore, a chip-in-the-loop configuration where the actual hardware containing memristors are tuned through feedback is highly desirable. Figure 5.43 illustrates this conceptually. When a digital neural network is used, the network parameters obtained through training in the datacenter can be downloaded directly to the mobile devices for the purpose of inference. This, unfortunately, might not be directly applicable to analog neural networks due to severe process variations. The downloaded weights might generate wrong inference results, depending on how the actual devices are different from the devices used for training. A method to counteract this mismatch effect is to utilize online learning in order to fine-tune the weights based on the actual hardware. Such an online learning can be achieved through the learning algorithm presented in the previous chapter.

In this section, we study how the learning algorithm presented in Section 4.3.5 can be applied to a memristor-based neural network. Algorithms and hardware architectures presented in [49] are discussed in this section to serve as a case study for implementing analog/mixed-signal SNNs with a crossbar structure. The dynamics of the memristor device and the memristor network are generally hard to capture accurately, which poses difficulties for many learning algorithms that require exact information of the device and the network. The proposed algorithm is well suited in this case, as it does not rely on much information of the exact device parameters. Most gradient information can be derived from spike-timing information, which can be readily observed externally. Even though the algorithm presented in Section 4.3.5 is already hardware-friendly, a few adaptations are needed, considering some limitations of memristor-based SNNs.

For example, in the original algorithm, a divide-by-weight operation is needed to carry out the learning. Even though such a division operation has been simplified to an approximate division, discussed in Sections 5.2.3 and 5.2.4, it is still problematic in a memristor-based neural network as the weight information is stored as an analog value in memrisitive devices. Another example is that a memristor device, by its nature, can only represent one type of synapse, either excitatory or inhibitory. In the original algorithm, however, there is no such limitation. Furthermore, nanoscale memristors based on resistive oxide and metallic nanowire are known to suffer from significant spatial and temporal variations compared to silicon-based devices [123, 124]. How to conduct reliable computing with these unreliable devices is also a problem that needs to be addressed.

5.3.5.2 Algorithm Adaptations

One typical configuration of a memristor crossbar-based SNN is illustrated in Figure 5.44. Memristors are placed at the cross point of the horizontal line and the vertical line. Neurons in the network are represented by the triangles in the figure. A voltage pulse is emitted by a neuron when the neuron fires, and the current flows through the memristor at the cross point. The charge injected to the postsynaptic neuron is weighted by the conductance of the memristor with the help of Ohm's law. The charge packets created by presynaptic neurons are accumulated at the input of the corresponding postsynaptic neuron. For each postsynaptic neuron, it fires when its membrane voltage exceeds a threshold. Clearly, the synaptic weights of the network are proportional to the conductances of the memristors used. Therefore, one can always scale the parameters of the network such that w_{ij}^l and G_{ij}^l are equal in value. For the convenience of discussion, we use these two quantities interchangeably in the rest of this section.

The encoding and decoding of information from an SNN are similar to what was done in Section 4.3.5. To briefly recap, real values from the dataset are injected as incremental

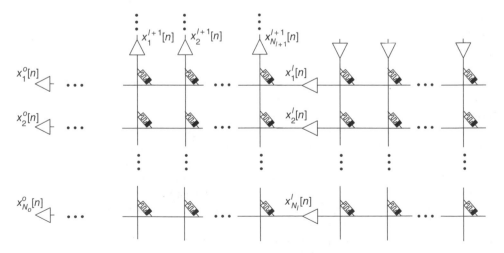

Figure 5.44 Illustration of a multilayer memristor crossbar-based spiking neural network. The memristor devices located at the cross-point weigh and convert the applied voltage pulses into currents, which then accumulate at the input of the neurons [49]. Reproduced with permission of IEEE.

membrane voltages into the input-layer neurons. Ten output neurons that correspond to the ten digits are used in the network. The goal of the learning is to train the SNN such that the correct output neuron that corresponds to the correct digit fires with a high firing density whereas all other output neurons stay relatively quiet. Such a goal can be achieved by calculating the error at the output layer and propagating the error back to each synapse in the network. To evaluate the trained network, the images in the test set are fed to the SNN and the output neuron with the highest firing density is picked as the winning neuron, and its corresponding digit is read out as the inference result.

One difficulty of mapping the original learning algorithm directly on to a memristor crossbar fabric is the required division operation in the original algorithm. The term $(1 - \overline{x_i^l})$ in the denominator can be neglected without introducing much error. This is true because the spike trains in an SNN are often sparse, which makes the value of $(1 - \overline{x_i^l})$ close to 1. The divide-by-weight operation, on the other hand, is more problematic. Such a division operation can be conveniently carried out in a digital implementation as the synaptic weights are stored in digital form. In a memristor network, the weights are stored in memristor devices as analog values. Therefore, how to conduct division without readout and quantization is not obvious. A workaround on this is briefly discussed in Section 5.1.4.2. We only need to include the sign information of the synaptic weight in the computation. Mathematically, $stdp_{ij}^l / \mathrm{sgn}(w_{ij}^l)$ is used for estimating the gradient in learning, where $\mathrm{sgn}(\cdot)$ is the sign function. To demonstrate this, learning is performed on the MNIST benchmark task. In order to accelerate the evaluation of the algorithm adaptations and hardware architecture, a down-sampled MNIST dataset similar to the one presented in Section 5.2.4 is employed unless otherwise stated. For all the results presented in this section, the first 500 images from the down-sampled MNIST training set are used for learning, whereas all the 10 000 images in the down-sampled MNIST test set are used for testing. For each configuration, 10 runs of learning are conducted to obtain a more confident evaluation. Each run of learning includes many iterations where an iteration is a process of going through all 500 images. The error bars shown in the results correspond to the 95% confidence interval.

Figure 5.45 compares the test correct rate achieved with three configurations. The first set of results is obtained from the baseline setting where the original algorithm is used. The second configuration assumes that the term $(1 - \overline{x_i^l})$ can be neglected safely. In the third configuration, the weight in the denominator is replaced by its sign information. From Figure 5.45, we can observe that regardless of the simplification made in the division operation, comparable learning results can be achieved. The results obtained from the division-free operation is slightly worse, yet it provides the advantage that no division is needed at all. Under this circumstance, an STDP-like rule is used for excitatory synapses and an anti-STDP-like rule is employed for inhibitory synapses. Such an arrangement significantly reduces the computational effort needed in the learning.

Another difficulty in using the original learning algorithm is that both positive and negative weights are assumed in the algorithm. By its nature, a memristor device can only represent a positive weight. By reversing the polarity of the injected voltage pulse, a negative weight can be achieved. It is possible to use two memristors to form a synapse that can span both the positive and negative ranges. However, since the sign information of the synaptic weight is still needed in the learning algorithm, a method of

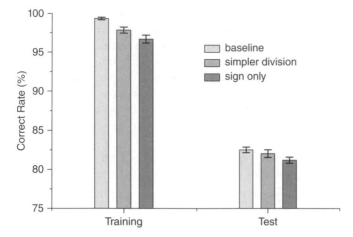

Figure 5.45 Comparison of the learning performance achieved with and without exact division. The approximate divisions can significantly reduce the design complexity while only degrading the learning performance slightly [49]. Reproduced with permission of IEEE.

detecting the polarity of the weight is required, which is likely to be cumbersome and power-consuming.

One solution to this problem is to fix the polarities of the weights in the network, as shown in Figure 5.46a. Each memristor in the array behaves either as an excitatory synapse or an inhibitory synapse, depending on which row it is located at. For rows where a positive pulse is applied, the presynaptic neuron injects charge into the post-synaptic neuron through the memristor device. Thereby the memristors on this row behave as excitatory synapses. Similarly, for memristors that are located at rows where a negative pulse is applied, they behave as inhibitory synapses. The polarities of the

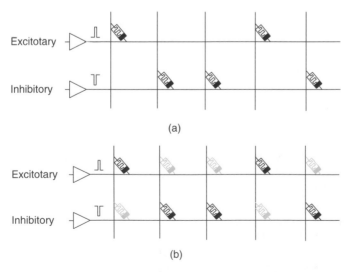

Figure 5.46 Illustration of the fixed-polarity memristor crossbar-based neural network. (a) Polarities of synapses weights are hardcoded. (b) Polarities of synaptic weights are run-time programmable [49]. Reproduced with permission of IEEE.

synapses in the network are determined before learning takes place. The polarity can be chosen purely randomly or with some strategy. For example, one can first use the original learning algorithm to simulate the training of a crossbar-based neural network and then determine the polarity of the actual network based on the simulated training results. For simplicity, we use a random-picking strategy in this section. In other words, the polarities of the synaptic weights are decided randomly at the beginning of the learning and are fixed throughout the learning process.

The hard-coded configuration shown in Figure 5.46a is conceptually simple yet not flexible. A more generic approach is illustrated in Figure 5.46b. For excitatory synapses, the memristor on the inhibitory row is set to G_{off}, the low conductance state of the memristor. The advantage of this configuration is that it stays compatible with a conventional densely connected crossbar and it can be programmed arbitrarily. One thing worth mentioning is that the "off" memristors need to be refreshed once in a while during the learning in order for them to stay at the "off" state.

Figure 5.47 compares the results obtained from the flexible-sign configuration where the polarities of the synaptic weights can be altered arbitrarily and the results obtained from the fixed-sign configuration where the polarities are fixed. One conclusion we can draw from the figure is that the performance achieved by the fixed-sign configuration is comparable yet slightly worse than the original algorithm. This is expected as fixing the sign of the weights essentially reduces the entropy of the network. Therefore, the results obtained from larger networks are also compared for fairness. As can be seen from the figure, as the size of the network increases, the performance improves. The performance loss due to restricting the polarities of the weights is compensated by adding more synapses into the network.

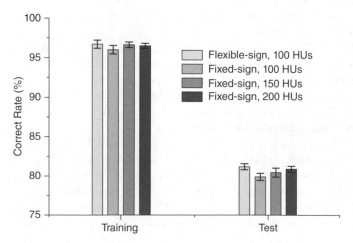

Figure 5.47 Comparison of the classification correct rate for several network configurations. The flexible-sign configuration is the case where synaptic weights can be programmed arbitrarily, whereas the fixed-sign configuration corresponds to configurations shown in Figure 5.46. The performance of the fixed-polarity network is slightly worse than the flexible-sign network with the same number of hidden units. This is mainly because of the loss of capacity in the network by restricting the polarity of each synapse. This loss can be compensated by increasing the number of hidden units [49]. Reproduced with permission of IEEE.

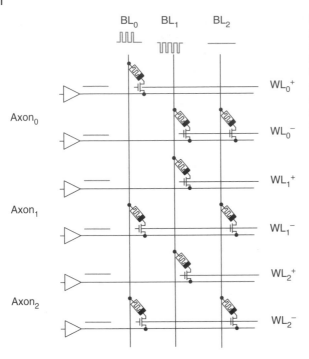

Figure 5.48 An example of programming the memristor crossbar during learning. The word-lines are used to select memristors that need to be programmed, whereas bit-lines are used to control how much change needs to be updated to the selected memristor devices [49]. Reproduced with permission of IEEE.

As discussed in Section 5.1.4, there are two methods of updating synaptic weights: the cumulative style and the incremental style. These two updating styles should lead to similar results regardless of how the weight update is conducted. This is because the weight change in one learning iteration is usually small. Indeed, a good rule of thumb in deep learning is to control the learning rate such that the weight changes in one iteration are less than one thousandth of the weights [50].

Examples of the possible realizations of these two updating styles are illustrated in Figures 5.48 and 5.49. In the figures, a one-transistor-one-memristor configuration is assumed. Each axon has two word-lines (WLs), which correspond to the excitatory synapses and inhibitory synapses, respectively. When an axon is activated, the WL is asserted. The current then flows to the bit-lines (BLs) and eventually to postsynaptic neurons through the memristors located at the crosspoint.

In order to program the memristors, a pulse with a modulated width can be used. Alternatively, a pulse train with a modulated number of pulses, which is used for demonstration in Figure 5.48, can also be employed. The cumulative and incremental updating methods result in different memory-writing patterns, as shown in Figure 5.49. Updating weights in a cumulative style resembles the process of writing a memristor-based memory array. The synaptic weights stored in the memristor devices are altered row by row.

One thing worth noting is that since the memristors in the excitatory row and the inhibitory row are complementary, these two rows can be updated simultaneously, thereby improving the system throughput. For the incremental update, on the other hand, the excitatory and the inhibitory rows cannot be updated at the same time because the polarity of the update depends on the polarity of the synaptic weight. Nevertheless, weights associated with different axons that have the same polarity can

Figure 5.49 Comparison of the updating schedules for the cumulative and the incremental update style. For the cumulative updating style, synapses connected to each axon need to be updated separately. For the incremental updating style, synapses with the same polarity can be updated simultaneously [49]. Reproduced with permission of IEEE.

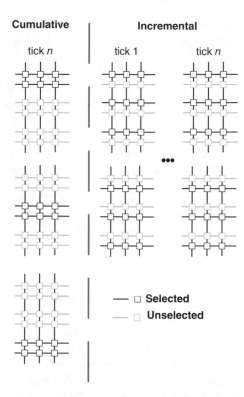

be updated together as they share the same postsynaptic activities. This is illustrated in Figure 5.49. Even though an incremental update can be parallelized, it needs to be performed at every tick.

The results for these two different updating methods are compared in Figure 5.50. Both the programmable and the hardcoded configurations are simulated. Clearly,

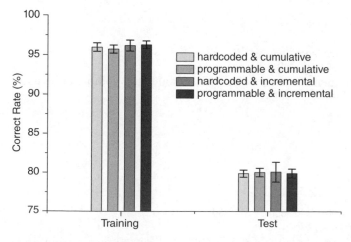

Figure 5.50 Comparison of the learning performance obtained with the hardcoded/programmable fixed-polarity network configurations when the cumulative/incremental weight update style is employed [49]. Reproduced with permission of IEEE.

all four cases yield similar results. From the perspective of a hardware designer, an incremental update style has the advantage that it does not need any extra storage space to hold the spike timing information as this information is directly accumulated in the memristor. On the other hand, the cumulative update style is more efficient since it only needs to make one update at the end of each learning iteration. This significantly reduces the number of write operations to the memristor array. Furthermore, in Section 5.3.5.3.2, we show another advantage of the cumulative update: it is much more robust against variations and noises in the devices. It is worth mentioning that the way of updating weights presented in Figures 5.48 and 5.49 is just one example. There may exist other ways of conducting the weight update, depending on the actual requirement.

As mentioned in Chapter 3, one difficulty of using NVM devices as synapses is the non-idealities in programming the weights. In general, the conductance change of a memristor device is a function of the conductance itself. Mathematically, it can be expressed as

$$\Delta G_{ij}^l = g(\Delta s_{ij}^l, G_{ij}^l) \tag{5.9}$$

where s_{ij}^l is a linearly or approximately linearly controllable state variable.

In order to model more realistic memristor devices, we resort to the VTEAM model in [107], which is briefly discussed in Section 5.3.4.2. It may be noted that s_{ij}^l shown in Eq. (5.9) can be thought of as the state variable in the VTEAM model. One can achieve linear control over s_{ij}^l through applying pulse-width modulated pulses or pulse trains. Figure 5.51 compares several conductance-changing characteristics. They correspond to different $g(\cdot)$ in Eq. (5.9). Only cases where incrementing and decrementing weights are symmetrical are illustrated. In other words, $g(\cdot)$ is an even function of Δs_{ij}^l. The more general case is investigated in Section 5.3.5.3.2. In the figure, a G_{off} of $10^{-6}S$ and G_{on} of $10^{-3}S$ are assumed, which represent the maximum and minimum conductances the employed memristor can achieve. In Figure 5.51, the linear-R, which is defined as linear with respect to resistance, and the exponential models are popular models used in

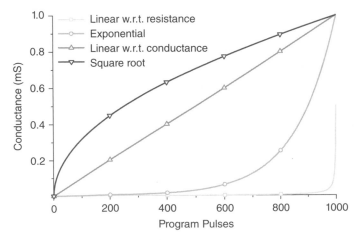

Figure 5.51 Illustration of different conductance-changing characteristics. The update of the conductance is a function of the current conductance value [49]. Reproduced with permission of IEEE.

VTEAM [107]. Mathematically, they can be described by

$$G_{ij}^l = \left(R_{ON} + \frac{R_{OFF} - R_{ON}}{s_{off} - s_{on}} (s_{ij}^l - s_{on}) \right)^{-1} \tag{5.10}$$

$$G_{ij}^l = \frac{exp\left(-\frac{\lambda(s_{ij}^l - s_{on})}{s_{off} - s_{on}} \right)}{R_{ON}} \tag{5.11}$$

The linear-G model, which is defined as linear with respect to conductance, is the trivial case where conductance can be programmed linearly through different pulse widths. As both the exponential model and the linear-R model follow a convex dependency on the conductance, we also create an artificial model that has a concave dependency on the weight, called a square-root model, which can be described as

$$G_{ij}^l = G_{OFF} + (G_{ON} - G_{OFF}) \sqrt{\frac{s_{ij}^l - s_{on}}{s_{off} - s_{on}}} \tag{5.12}$$

The results obtained after learning with these different memristor models are compared in Figure 5.52. It is shown that all the models yield acceptable results, even for the ill-behaved linear-R model. This result demonstrates the efficacy of the proposed learning algorithm.

In addition to the linearity in controlling the weight, the tuning granularity might not be arbitrarily fine for real devices either. In principle, the conductance of a memristor device is analog in nature, and it should be able to be tuned continuously through applying specific voltages. Nevertheless, such programming, in reality, might be limited by the programming circuit.

For example, if we choose to use a pulse-number modulation scheme to program the weight, we are apparently limited by the amount of change in conductance that one voltage pulse can induce. As mentioned earlier, in training a neural network, a good

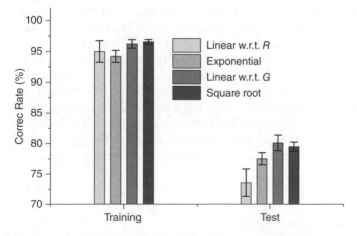

Figure 5.52 Comparison of the learning performances with different conductance-changing characteristics shown in Figure 5.51 [49]. Reasonable classification rates can be achieved with all the models, even though the recognition rate obtained from a more linear model is higher. Reproduced with permission of IEEE.

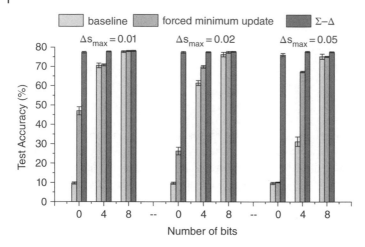

Figure 5.53 Comparison of the test accuracies achieved with different programming granularities. The $\Sigma - \Delta$ modulation scheme can help avoid the performance penalty when a coarse programming granularity is used [49]. Reproduced with permission of IEEE.

rule of thumb is to control the weight update to be three orders of magnitude smaller than the weight itself. This might not be a problem for neural networks training in the datacenter where floating-point numbers are used to represent information but it is a problem for a memristor device as such a fine granularity is difficult to achieve.

To study how the programming granularity affects the learning performance, we simulated learning with different maximum and minimum allowed changes in the state variable, namely Δs_{max} and Δs_{min}. These two quantities are related through another quantity called the NOB. For example, an NOB of 3 indicates that the absolute change in conductance can only be a multiple of $2^{-3} \cdot \Delta s_{max}$.

Different Δs_{max} and NOB quantities are swept, and the obtained results are compared in Figure 5.53. It can be observed that the performance is degraded when the granularity in programming the weight is coarse. The degradation is mainly because small weight changes are ignored due to limited programming granularity in the learning.

One straightforward solution is to force an update even when the change is small. The obtained results are also compared in Figure 5.53 and it can be seen that this method is somewhat effective in improving the learning performance. To tolerate even coarser programming granularities, a more advanced approach is to use $\Sigma - \Delta$ modulation. It is a popular technique that is widely used in the digital signal processing community to allow using low-precision data to represent high-precision data. The procedure of conducting $\Sigma - \Delta$ modulation in our case can be represented mathematically as

$$\Delta s'[n] = \left\lfloor \frac{\Delta s[n] + \Delta s_{int}[n]}{\Delta s_{min}} \right\rfloor \Delta s_{min} \tag{5.13}$$

$$\Delta s_{int}[n+1] = \Delta s_{int}[n] + \Delta s[n] - \Delta s'[n] \tag{5.14}$$

where $\Delta s_{int}[n]$ is the output of the integrator used to store the residue that is generated in $\Sigma - \Delta$ modulation and $\Delta s'[n]$ is the actual amount of change that needs to be applied to the state variable.

By introducing a $\Sigma - \Delta$ modulation in the weight update, the requirement on the programming granularity can be remarkably relaxed. This is shown in Figure 5.53. Thanks to the integrator that helps accumulate the residue value, which is too small to be updated, it is even possible to use only one programming level to achieve good learning results.

5.3.5.3 Non-idealities

As stated previously, one of the major concerns of implementing neural networks with analog synapses is that the variations in the analog components might deteriorate the performance. A well-trained network might malfunction after being mapped to real hardware. On-chip learning can effectively mitigate this problem by compensating the variations through feedback. Nevertheless, some non-idealities in real devices might still cause problems in learning. The non-idealities can be categorized into two types: neuron non-idealities and synapse non-idealities. They are studied separately in the following sections.

5.3.5.3.1 Neuron Non-idealities One possible realization of the neuron circuits that can be employed in our proposed architecture is shown in Figure 5.54. This realization is similar to the neuron shown in Figure 5.38. Currents weighted by the memristor synapses are summed up and accumulated by an analog integrator. Since the learning algorithm assumes a current-based synapse, a virtual ground is forced at the summing node. After the charge is accumulated, a clocked comparator is used to determine whether the membrane voltage exceeds the firing threshold. A spike is emitted if this condition is met. Once a neuron fires, a charge packet is delivered to the input of the integrator through a digital-to-analog converter (DAC). Such a feedback DAC implements the subtraction of the threshold voltage specified in the modified LIF model, as shown in Eq. (4.57) in Chapter 4.

Two parameters are subject to variations in this neuron circuit: the threshold and the leakage. The threshold voltage is mainly determined by the weight of the feedback DAC and the leakage mainly consists of three terms. (i) L_1 is the leakage from any unselected

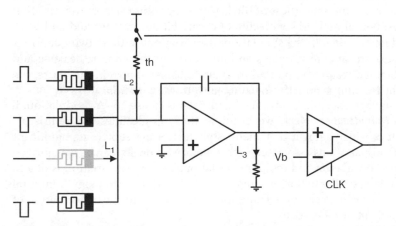

Figure 5.54 An example of the analog neuron that can be employed in a memristor crossbar-based neural network. An analog integrator is used to integrate the current weighted by the memristor-based synapse. A spike is outputted by the clock comparator when the membrane voltage exceeds the threshold [49]. Reproduced with permission of IEEE.

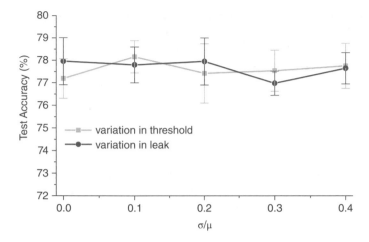

Figure 5.55 Comparison of the test accuracy obtained with different levels of leakage and threshold variations. Learning is robust against variations in these two parameters [49]. Reproduced with permission of IEEE.

or half-selected memristor. It can model the sneak-path leakage in a transistor-free implementation or the subthreshold leakage in implementations where transistors are needed. (ii) L_2 represents any leakage at the input of the integrator, e.g. the leakage from the feedback DAC. (iii) L_3 models the leakage at the output of the analog integrator. Possible sources for L_3 are leakage that is intentionally placed at the output the integrator, kickback noise from the clocked comparator, and the low-frequency error due to the finite gain of the operational amplifier [125].

Since the actual values of the threshold and leakage are not used in the learning algorithm, it is anticipated that the learning should be somewhat robust against the variations in these two parameters. Indeed, the weight-dependent STDP learning algorithm treats neurons as black boxes and all relevant information is derived from the spike timings. To demonstrate the robustness of the learning algorithm against the variation in neuron circuits, neural networks with different variabilities in threshold and leakage are employed for training. In the simulation, the variations in these two quantities are treated as Gaussian random variables and are added to the nominal leakage and threshold. The obtained results are plotted in Figure 5.55. As shown in the figure, the performance of the learning is not affected noticeably by the variations.

5.3.5.3.2 Synapse Non-idealities Next, we examine the non-idealities in memristor-based synapses. It is widely known that memristor devices are subject to significant variations. In general, variations associated with a memristor device can be categorized into two types: spatial variation and temporal variation. The spatial variation is due to the fabrication process, similar to that existing in modern CMOS technology. Temporal variation, on the other hand, can come from, for example, the stochastic formation of the conducting filament in a memristor.

To study how the temporal variation affects learning, Figure 5.56 compares the results obtained after learning with two types of noises. The first type of noise is a white noise with a Gaussian distribution. The actual conductance change is scaled by a normally distributed random variable with a mean of 1 and a standard deviation of σ. The second

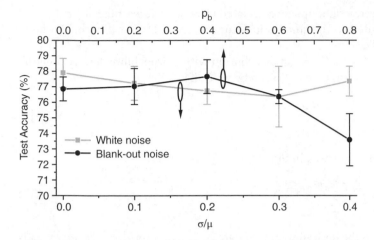

Figure 5.56 Comparison of the test accuracy obtained with different levels of noise in changing the conductances of memristors. Two types of noise are considered: a normally distributed white noise with a standard deviation of σ and a blank-out noise with a blank-out probability of p_b [49]. Reproduced with permission of IEEE.

type of noise is a blank-out noise with a blank-out rate of p_b that is directly masked on the weight change. In other words, the weight update is set to 0 with a probability of p_b. As can be observed from the figure, due to the stochastic nature of the learning algorithm, it is robust against these two types of temporal noises. Such robustness is highly mandated in a memristor implementation where temporal variations are substantial.

Ideally, when applying voltage pulses with the same amplitude and duration to two memristors on the chip with the same conductance, the changes in the conductances for these two devices should be the same. This, unfortunately, is not true due to spatial variations. Consequently, each memristor has its own $g(\cdot)$ (see Eq. (5.9)) that governs its conductance-changing characteristics.

To study the effect of the spatial variations, we scale the conductance change for each memristor synapse by a Gaussian variable with a mean of 1 and a standard deviation of σ. Since we are only interested in the spatial variation, the random scaling factors are fixed throughout one learning period in the simulation. In other words, the scaling factor is random across different devices but is not a function of time.

The obtained results are compared in Figure 5.57. We separate the case where the variations in incrementing and decrementing are symmetrical from the more general case. The symmetrical case is denoted as "Sym" in the figure where the more general asymmetrical case is labeled with "Asym." It can be seen from the figure that learning is quite robust against the symmetrical variation, yet significant performance degradation is observed when the variations are asymmetrical. Such a degradation is more remarkable when the incremental update style is used. Note that such a phenomenon is not unique to the learning algorithm used here. It is actually common to all stochastic gradient descent (SGD) learning. The reason for such performance degradation is the bias term generated by the imbalance in the conductance update. Mathematically, let us assume that the weight update can be written as

$$\Delta w = \Delta w_0 + n_w \tag{5.15}$$

where Δw_0 is the desired weight update calculated based on the learning algorithm and n_w is the noise attributed to the stochastic nature of an SGD algorithm or the noisy estimation of gradients.

We assume that n_w obeys a Gaussian distribution with a probability density function of $\phi(n_w/\sigma_n)$. We further assume that $g(\cdot)$ can be written as

$$g(\Delta w, w) = \begin{cases} g_p \Delta w, \Delta w > 0 \\ g_n \Delta w, \Delta w < 0 \end{cases} \tag{5.16}$$

Then the expectation of the weight update can be derived as

$$E(\Delta w) = \Delta w_0 + \frac{(g_p - g_n)\sigma_n}{\sqrt{2\pi}} \tag{5.17}$$

Clearly, in addition to the desired weight update Δw_0, a bias term $(g_p - g_n)\sigma_n/\sqrt{2\pi}$ is created due to the imbalance in incrementing and decrementing the weight. An SGD learning is quite robust against any unbiased noise because this noise can be effectively filtered out as the learning continues. Unfortunately, when the noise is biased, a bias term is generated. The cost function never reaches zero in this case as a non-zero error is needed to generate a corresponding weight update that can cancel out the bias term in the steady state.

Apparently, in order to improve the performance of the learning, we have to minimize this bias term. The most straightforward way is to make the weight change as symmetrically as possible to reduce $(g_p - g_n)$. This can be done through engineering the device. Alternatively, we can reduce the term σ_n through algorithm- and architecture-level techniques. For example, we can use a cumulative update style to effectively reduce σ_n. The two updating styles are compared in Figure 5.57. The cumulative update style is helpful in improving the performance because it first averages the gradients before applying them to the memristor devices. Such a step is effective in reducing the noise associated with the gradient estimation.

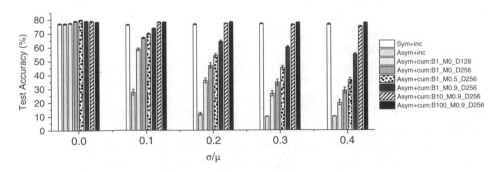

Figure 5.57 Comparison of the test accuracies achieved with different non-idealities and learning hyperparameters. In the figure, "Sym" and "Asym" indicate that the variations in incrementing and decrementing the conductance are symmetrical and asymmetrical, respectively. Styles "inc" and "cum" represent the incremental and the cumulative update styles, respectively. The numbers after letter "B," "M," and "D" represent the size of the mini-batch, the coefficient for the momentum, and the learning duration, respectively [49]. Reproduced with permission of IEEE.

When the variations become larger, more advanced techniques are required to restore the performance. Since the proposed learning algorithm is based on a stochastic approximation, a longer learning duration leads to a more reliable gradient estimation, which can improve the performance. Another helpful technique is the momentum [50], which is discussed in Section 2.5.3.1.4. Applying a momentum to the weight update is similar to applying a filter. A larger momentum coefficient can result in a lower cut-off frequency, which is helpful in getting rid of the high-frequency noise.

Updating weight with a momentum is essentially an average operation in the time domain. Similarly, we can use a mini-batch to average the weight update across multiple inputs. These two techniques are orthogonal to each other. The learning results obtained with the above-mentioned techniques are compared in Figure 5.57. One can conclude from the figure that a longer learning duration, a larger momentum coefficient, and a larger batch size are effective in boosting the learning performance in the presence of asymmetrical variations.

Besides the asymmetrical spatial variations, a systematic asymmetrical conductance-changing characteristic has been observed in many fabricated devices as well. For example, in the model used in [109], the increase and decrease in the conductance can be represented as

$$\Delta G_p = a_p \exp\left(-b_p \frac{G - G_{off}}{G_{on} - G_{off}}\right) \tag{5.18}$$

$$\Delta G_n = a_n \exp\left(-b_n \frac{G_{on} - G}{G_{on} - G_{off}}\right) \tag{5.19}$$

Such a model is illustrated in Figure 5.58. Clearly, the slope for incrementing and decrementing the conductance are generally not equal. This asymmetry causes a similar problem for the learning. The techniques proposed previously can be employed

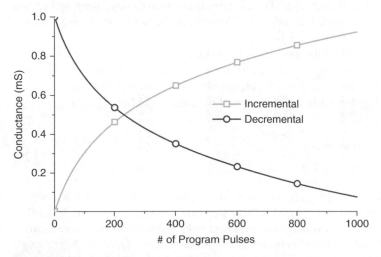

Figure 5.58 Illustration of the weight-changing characteristics of the memristor model employed in [109]. The increment and decrement in the conductance, in general, is not symmetrical for this memristor model [49]. Reproduced with permission of IEEE.

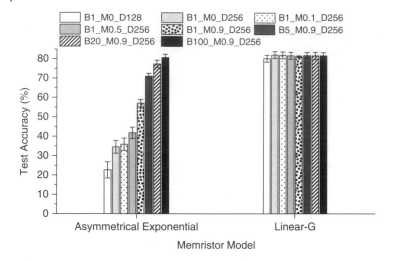

Figure 5.59 Comparison of the test accuracies obtained with different hyperparameters. The numbers after letters "B," "M," and "D" represent the size of the mini-batch, the coefficient for the momentum, and the learning duration, respectively [49]. Reproduced with permission of IEEE.

to counteract this non-ideality as well. The obtained learning results are compared in Figure 5.59.

In the figure, the results obtained from a symmetrical linear-G model is also plotted for the purpose of comparison. It can be observed from the figure that even though the proposed techniques do not yield much improvement in the symmetrical case, they are very effective in boosting the performance for the asymmetrical case.

In order to further improve the performance when the changes in weights are asymmetrical, one more technique can be used. The fundamental reason that an asymmetrical conductance change is detrimental to learning is because the bias term serves as a regularization term that attempts to adjust the weights of the network to a point where incrementing and decrementing weights are balanced.

Recognizing this, one possible way to improve the performance is to overwrite the regularization term by creating a stronger but more meaningful one. This can be done through, for example, encouraging small weights, which is similar to the regularization widely used in training deep ANNs, as discussed in Chapter 2. The regularization can be written as

$$\Delta w' = \begin{cases} \Delta w & \Delta w \cdot w > 0 \\ \lambda \Delta w & \Delta w \cdot w < 0 \end{cases} \tag{5.20}$$

where $\Delta w'$ is the actual weight update applied to the neural network and λ is a hyperparameter used to promote small weights. The introduction of λ is also very helpful in reducing the power consumption of the neural network. A significant power consumption in a memristor crossbar-based neural network is due to the current flowing through the memristors. This kind of power consumption scales linearly with $\sum |w_{ij}|$. Therefore, by encouraging small conductances, the currents that flow through the memristors can be reduced.

The test accuracy and the sum of the absolute values of all the weights in the neural network are compared in Figures 5.60 and 5.61. It can be observed from the figures that the test accuracy is improved for the asymmetrical exponential case when a large λ is used. In addition, it is clearly shown that a large λ can reduce the sum of the absolute values of the conductances of the network, which is beneficial for the energy efficiency of the system.

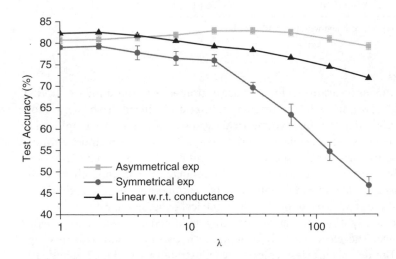

Figure 5.60 The obtained test accuracy as λ increases. The learning performance is not sensitive to the value of λ. A larger λ is helpful for the asymmetrical case by encouraging small weights [49]. Reproduced with permission of IEEE.

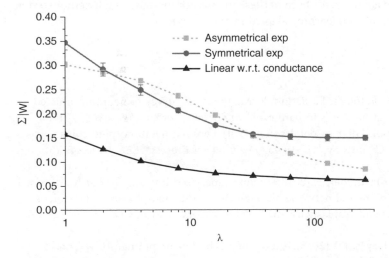

Figure 5.61 The obtained sum of the absolute synaptic weights in the neural network when different λ are used. A larger λ can help achieve better energy efficiency by encouraging small weights [49]. Reproduced with permission of IEEE.

Table 5.3 Benchmark performance achieved with different memristor models.

Memristor model	Network configuration	Recognition accuracy (%)
Symmetrical exponential	784-300-10	97.03
	784-400-10	97.10
Linear-G	784-300-10	96.51
	784-400-10	96.76

Source: data are from [49]. Reproduced with permission of IEEE.

5.3.5.4 Benchmarks

The learning algorithm and various techniques are examined with the standard MNIST benchmark. The 28×28 images in the original dataset are utilized. Two memristors models, the linear-G model and the asymmetrical exponential model, are employed in the simulations to demonstrate the efficacy of the design. The minimum and maximum conductance for the memristors are set to be 1 μS and 1 mS, respectively. To reduce the detrimental effect posed by the asymmetrical-conductance-changing characteristics, a mini-batch size of 100 and a momentum coefficient of 0.9 are used. In addition, a regularization constant of 32 is used for the asymmetrical exponential model.

As discussed previously, the cumulative updating style generally leads to a better learning performance. Therefore, we only benchmark network trained with this type of updating style. The test accuracies achieved are illustrated in Table 5.3, where 100 iterations of learning are carried out. In each iteration, all the 60 000 images in the training set are fed to the neural network for learning. The test accuracies reported in Table 5.3 are obtained by averaging the test accuracies recorded between the 91st iteration and the 100th iteration. It can be concluded from the table that good performance can be achieved for both of these memristor models, which demonstrates the efficacy of the employed learning algorithm and techniques.

References

1 Schuman, C. D., Potok, T. E., Patton, R. M., et al., "A survey of neuromorphic computing and neural networks in hardware," *arXiv Prepr. arXiv1705.06963*, 2017.

2 Boahen, K. (2000). Point-to-point connectivity between neuromorphic chips using address events. *Circuits Syst. II Analog Digit. Signal Process. IEEE Trans.* 47 (5): 416–434.

3 Zheng, N. and Mazumder, P. (2018). Online supervised learning for hardware-based multilayer spiking neural networks through the modulation of weight-dependent spike-timing-dependent plasticity. *IEEE Trans. Neural Networks Learn. Syst.* 29 (9): 4287–4302.

4 Alaghi, A. and Hayes, J.P. (2013). Survey of stochastic computing. *ACM Trans. Embed. Comput. Syst.* 12 (2s): 92.

5 Furber, S.B., Lester, D.R., Plana, L.A. et al. (2013). Overview of the spinnaker system architecture. *IEEE Trans. Comput.* 62 (12): 2454–2467.

6 Furber, S.B., Galluppi, F., Temple, S., and Plana, L.A. (2014). The spinnaker project. *Proc. IEEE* 102 (5): 652–665.

7 Painkras, E., Plana, L.A., Garside, J. et al. (2013). SpiNNaker: A 1-W 18-core system-on-chip for massively-parallel neural network simulation. *IEEE J. Solid-State Circuits* 48 (8): 1943–1953.

8 Wu, J., Furber, S., and Garside, J. (2009). A programmable adaptive router for a GALS parallel system. In: *Asynchronous Circuits and Systems, 2009. ASYNC'09. 15th IEEE Symposium on*, 23–31.

9 Galluppi, F., Brohan, K., Davidson, S. et al. (2012). A real-time, event-driven neuromorphic system for goal-directed attentional selection. In: *International Conference on Neural Information Processing*, 226–233.

10 Denk, C., Llobet-Blandino, F., Galluppi, F. et al. (2013). Real-time interface board for closed-loop robotic tasks on the spinnaker neural computing system. In: *International Conference on Artificial Neural Networks*, 467–474.

11 Modha, D. S., "IBM Research: Brain-inspired chip." [Online]. Available at: http://www.research.ibm.com/articles/brain-chip.shtml [Accessed: 8 April 2018].

12 Merolla, P.A., Arthur, J.V., Alvarez-Icaza, R. et al. (2014). A million spiking-neuron integrated circuit with a scalable communication network and interface. *Science* 345 (6197): 668–673.

13 Akopyan, F., Sawada, J., Cassidy, A. et al. (2015). Truenorth: design and tool flow of a 65 mW 1 million neuron programmable neurosynaptic chip. *IEEE Trans. Comput. Aided Des. Integr. Circuits Syst* 34 (10): 1537–1557.

14 Cassidy, A.S., Merolla, P., Arthur, J.V. et al. (2013). Cognitive computing building block: a versatile and efficient digital neuron model for neurosynaptic cores. In: *Neural Networks (IJCNN), The 2013 International Joint Conference on*, 1–10.

15 Amir, A., Datta, P., Risk, W.P. et al. (2013). Cognitive computing programming paradigm: a corelet language for composing networks of neurosynaptic cores. In: *Neural Networks (IJCNN), The 2013 International Joint Conference on*, 1–10.

16 Preissl, R., Wong, T.M., Datta, P. et al. (2012). Compass: a scalable simulator for an architecture for cognitive computing. In: *Proceedings of the International Conference on High Performance Computing, Networking, Storage and Analysis*, 54.

17 Sawada, J., Akopyan, F., Cassidy, A.S. et al. (2016). Truenorth ecosystem for brain-inspired computing: scalable systems, software, and applications. In: *High Performance Computing, Networking, Storage and Analysis, SC16: International Conference for*, 130–141.

18 Cheng, H.-P., Wen, W., Wu, C. et al. (2017). Understanding the design of IBM neurosynaptic system and its tradeoffs: a user perspective. In: *2017 Design, Automation & Test in Europe Conference & Exhibition (DATE)*, 139–144.

19 Esser, S.K., Andreopoulos, A., Appuswamy, R. et al. (2013). Cognitive computing systems: algorithms and applications for networks of neurosynaptic cores. In: *Neural Networks (IJCNN), The 2013 International Joint Conference on*, 1–10.

20 Tsai, W.-Y., Barch, D.R., Cassidy, A.S. et al. (2016). LATTE: low-power audio transform with truenorth ecosystem. In: *Neural Networks (IJCNN), 2016 International Joint Conference on*, 4270–4277.

21 Tsai, W.-Y., Barch, D.R., Cassidy, A.S. et al. (2017). Always-on speech recognition using truenorth, a reconfigurable, neurosynaptic processor. *IEEE Trans. Comput.* 66 (6): 996–1007.

22 Esser, S.K., Appuswamy, R., Merolla, P. et al. (2015). Backpropagation for energy-efficient neuromorphic computing. In: *Advances in Neural Information Processing Systems*, 1117–1125.

23 Esser, S.K., Merolla, P.A., Arthur, J.V. et al. (2016). Convolutional networks for fast, energy-efficient neuromorphic computing. In: *Proceedings of the National Academy of Science USA*, 201604850. National Academy of Science.

24 Davies, M., Srinivasa, N., Lin, T.H. et al. (2018). Loihi: a neuromorphic manycore processor with on-chip learning. *IEEE Micro* 38 (1): 82–99.

25 Lin, C., Wild, A., Chinya, G.N. et al. (2018). Programming spiking neural networks on Intel's Loihi. *Computer* 51 (3): 52–61.

26 Rabaey, J. (2009). *Low Power Design Essentials*. Springer Science & Business Media.

27 Cassidy, A.S., Georgiou, J., and Andreou, A.G. (2013). Design of silicon brains in the nano-CMOS era: spiking neurons, learning synapses and neural architecture optimization. *Neural Networks* 45: 4–26.

28 Wang, R. and van Schaik, A. (2018). Breaking Liebig's law: an advanced multipurpose neuromorphic engine. *Front. Neurosci.* 12.

29 Smith, J.E. (2014). Efficient digital neurons for large scale cortical architectures. In: *Proceeding of the 41st Annual International Symposium on Computer Architecture*, 229–240.

30 Cassidy, A., Andreou, A.G., and Georgiou, J. (2011). Design of a one million neuron single FPGA neuromorphic system for real-time multimodal scene analysis. In: *Information Sciences and Systems (CISS), 2011 45th Annual Conference on*, 1–6.

31 Nouri, M., Karimi, G.R., Ahmadi, A., and Abbott, D. (2015). Digital multiplierless implementation of the biological FitzHugh–Nagumo model. *Neurocomputing* 165: 468–476.

32 Luo, J., Nikolic, K., Evans, B.D. et al. (2017). Optogenetics in silicon: a neural processor for predicting optically active neural networks. *IEEE Trans. Biomed. Circuits Syst.* 11 (1): 15–27.

33 Lee, D., Lee, G., Kwon, D. et al. (2018). Flexon: a flexible digital neuron for efficient spiking neural network simulations. In: *Proceedings of the 45th Annual International Symposium on Computer Architecture*, 275–288.

34 Soleimani, H., Ahmadi, A., and Bavandpour, M. (2012). Biologically inspired spiking neurons: piecewise linear models and digital implementation. *IEEE Trans. Circuits Syst. Regul. Pap.* 59 (12): 2991–3004.

35 Zhang, B., Jiang, Z., Wang, Q. et al. (2015). A neuromorphic neural spike clustering processor for deep-brain sensing and stimulation systems. In: *Low Power Electronics and Design (ISLPED), 2015 IEEE/ACM International Symposium on*, 91–97.

36 Seo, J., Brezzo, B., Liu, Y. et al. (2011). A 45nm CMOS neuromorphic chip with a scalable architecture for learning in networks of spiking neurons. In: *Custom Integrated Circuits Conference (CICC), 2011 IEEE*, 1–4.

37 Knag, P., Kim, J.K., Chen, T., and Zhang, Z. (2015). A sparse coding neural network ASIC with on-chip learning for feature extraction and encoding. *IEEE J. Solid-State Circuits* 50 (4): 1070–1079.

38 Liu, C., Cho, S., and Zhang, Z. (2018). A 2.56-mm^2 718GOPS configurable spiking convolutional sparse coding accelerator in 40-nm CMOS. *IEEE J. Solid-State Circuits* 53 (10): 2818–2827.

39 Kim, J.K., Knag, P., Chen, T., and Zhang, Z. (2014). A 6.67mW sparse coding ASIC enabling on-chip learning and inference. In: *2014 Symposium on VLSI Circuits Digest of Technical Papers*, 1–2.

40 Kim, J.K., Knag, P., Chen, T., and Zhang, Z. (2015). A 640M pixel/s 3.65mW sparse event-driven neuromorphic object recognition processor with on-chip learning. In: *2015 Symposium on VLSI Circuits (VLSI Circuits)*, C50–C51.

41 Pearson, M.J., Pipe, A.G., Mitchinson, B. et al. (2007). Implementing spiking neural networks for real-time signal-processing and control applications: a model-validated FPGA approach. *IEEE Trans. Neural Networks* 18 (5): 1472–1487.

42 Wang, R., Thakur, C.S., Cohen, G. et al. (2017). Neuromorphic hardware architecture using the neural engineering framework for pattern recognition. *IEEE Trans. Biomed. Circuits Syst.* 11 (3): 574–584.

43 Neil, D. and Liu, S.-C. (2014). Minitaur, an event-driven FPGA-based spiking network accelerator. *IEEE Trans. Very Large Scale Integr. Syst.* 22 (12): 2621–2628.

44 Zylberberg, J., Murphy, J.T., and DeWeese, M.R. (2011). A sparse coding model with synaptically local plasticity and spiking neurons can account for the diverse shapes of V1 simple cell receptive fields. *PLoS Comput. Biol.* 7 (10): e1002250.

45 Zheng, N. and Mazumder, P. (2017). Hardware-friendly actor-critic reinforcement learning through modulation of spike-timing-dependent plasticity. *IEEE Trans. Comput.* 66 (2): 299–311.

46 Powell, W.B. (2011). *Approximate dynamic programming solving the curses of dimensionality*, Hoboken, NJ: Wiley.

47 Lian, J., Lee, Y., Sudhoff, S.D., and Zak, S.H. (2008). Self-organizing radial basis function network for real-time approximation of continuous-time dynamical systems. *IEEE Trans. Neural Networks* 19 (3): 460–474.

48 Zheng, N. and Mazumder, P. (2018). A low-power hardware architecture for on-line supervised learning in multi-layer spiking neural networks. In: *2018 IEEE International Symposium on Circuits and Systems (ISCAS)*, 1–5.

49 Zheng, N. and Mazumder, P. (2018). Learning in memristor crossbar-based spiking neural networks through modulation of weight dependent spike-timing-dependent plasticity. *IEEE Trans. Nanotechnol.* 17 (3): 520–532.

50 Hinton, G.E. (2012). A practical guide to training restricted Boltzmann machines. In: *Neural Networks: Tricks of the Trade*, 599–619. Springer.

51 Glorot, X., Bordes, A., and Bengio, Y. (2011). Deep sparse rectifier neural networks. In: *Proceedings of the Fourteenth International Conference on Artificial Intelligence and Statistics*, 315–323.

52 Merolla, P., Arthur, J., Akopyan, F. et al. (2011). A digital neurosynaptic core using embedded crossbar memory with 45pJ per spike in 45nm. In: *2011 IEEE custom integrated circuits conference (CICC)*, 1–4.

53 Rajendran, B., Liu, Y., Seo, J.S. et al. (2013). Specifications of nanoscale devices and circuits for neuromorphic computational systems. *IEEE Trans. Electron Devices* 60 (1): 246–253.

54 Cruz-Albrecht, J.M., Yung, M.W., and Srinivasa, N. (2012). Energy-efficient neuron, synapse and STDP integrated circuits. *IEEE Trans. Biomed. Circuits Syst.* 6 (3): 246–256.

55 Livi, P. and Indiveri, G. (2009). A current-mode conductance-based silicon neuron for address-event neuromorphic systems. In: *2009 IEEE International Symposium on Circuits and Systems*, 2898–2901.

56 Indiveri, G., Chicca, E., and Douglas, R. (2006). A VLSI array of low-power spiking neurons and bistable synapses with spike-timing dependent plasticity. *IEEE Trans. Neural Networks* 17 (1): 211–221.

57 Indiveri, G., Linares-Barranco, B., Hamilton, T.J. et al. (2011). Neuromorphic silicon neuron circuits. *Front. Neurosci.* 5: 73.

58 Yu, T. and Cauwenberghs, G. (2010). Analog VLSI biophysical neurons and synapses with programmable membrane channel kinetics. *IEEE Trans. Biomed. Circuits Syst.* 4 (3): 139–148.

59 Simoni, M.F., Cymbalyuk, G.S., Sorensen, M.E. et al. (2004). A multiconductance silicon neuron with biologically matched dynamics. *IEEE Trans. Biomed. Eng.* 51 (2): 342–354.

60 Azghadi, M.R., Al-Sarawi, S., Abbott, D., and Iannella, N. (2013). A neuromorphic VLSI design for spike timing and rate based synaptic plasticity. *Neural Networks* 45: 70–82.

61 Moradi, S. and Indiveri, G. (2014). An event-based neural network architecture with an asynchronous programmable synaptic memory. *IEEE Trans. Biomed. Circuits Syst.* 8 (1): 98–107.

62 Ramakrishnan, S., Hasler, P.E., and Gordon, C. (2011). Floating gate synapses with spike-time-dependent plasticity. *IEEE Trans. Biomed. Circuits Syst.* 5 (3): 244–252.

63 Chicca, E., Stefanini, F., Bartolozzi, C., and Indiveri, G. (2014). Neuromorphic electronic circuits for building autonomous cognitive systems. *Proc. IEEE* 102 (9): 1367–1388.

64 Azghadi, M.R., Iannella, N., Al-Sarawi, S.F. et al. (2014). Spike-based synaptic plasticity in silicon: design, implementation, application, and challenges. *Proc. IEEE* 102 (5): 717–737.

65 Ebong, I.E. and Mazumder, P. (2012). CMOS and memristor-based neural network design for position detection. *Proc. IEEE* 100 (6): 2050–2060.

66 Indiveri, G. (2003). Neuromorphic bisable VLSI synapses with spike-timing-dependent plasticity. *Adv. Neural Inform. Process. Sys.*: 1115–1122.

67 Bofill-i-Petit, A. and Murray, A.F. (2004). Synchrony detection and amplification by silicon neurons with STDP synapses. *IEEE Trans. Neural Networks* 15 (5): 1296–1304.

68 Bamford, S.A., Murray, A.F., and Willshaw, D.J. (2012). Spike-timing-dependent plasticity with weight dependence evoked from physical constraints. *IEEE Trans. Biomed. Circuits Syst.* 6 (4): 385–398.

69 Dowrick, T., Hall, S., and McDaid, L.J. (2012). Silicon-based dynamic synapse with depressing response. *IEEE Trans. Neural Networks Learn. Syst.* 23 (10): 1513–1525.

70 Serrano-Gotarredona, R., Oster, M., Lichtsteiner, P. et al. (2009). CAVIAR: A 45k neuron, 5M synapse, 12G connects/s AER hardware sensory-processing-learning-actuating system for high-speed visual object recognition and tracking. *IEEE Trans. Neural Networks* 20 (9): 1417–1438.

71 Lichtsteiner, P., Posch, C., and Delbruck, T. (2008). A 128 × 128 120 dB 15 micro sec latency asynchronous temporal contrast vision sensor. *IEEE J. Solid-State Circuits* 43 (2): 566–576.

72 Lichtsteiner, P., Posch, C., and Delbruck, T. (2006). A 128 X 128 120 dB 30 mW asynchronous vision sensor that responds to relative intensity change. In: *Solid-State Circuits Conference, 2006. ISSCC 2006. Digest of Technical Papers*, 2060–2069. IEEE International.

73 Serrano-Gotarredona, R., Serrano-Gotarredona, T., Acosta-Jimenez, A., and Linares-Barranco, B. (2006). A neuromorphic cortical-layer microchip for spike-based event processing vision systems. *IEEE Trans. Circuits Syst. Regul. Pap.* 53 (12): 2548–2566.

74 Serrano-Gotarredona, R., Serrano-Gotarredona, T., Acosta-Jiménez, A. et al. (2008). On real-time AER 2-D convolutions hardware for neuromorphic spike-based cortical processing. *IEEE Trans. Neural Networks* 19 (7): 1196–1219.

75 Oster, M., Wang, Y., Douglas, R., and Liu, S.-C. (2008). Quantification of a spike-based winner-take-all VLSI network. *IEEE Trans. Circuits Syst. Regul. Pap.* 55 (10): 3160–3169.

76 Hafliger, P. (2007). Adaptive WTA with an analog VLSI neuromorphic learning chip. *IEEE Trans. Neural Networks* 18 (2): 551–572.

77 Millner, S., Grübl, A., Meier, K. et al. (2010). A VLSI implementation of the adaptive exponential integrate-and-fire neuron model. *Adv. Neural Inform. Process. Sys.*: 1642–1650.

78 Schemmel, J., Fieres, J., and Meier, K. (2008). Wafer-scale integration of analog neural networks. In: *Neural Networks, 2008. IJCNN 2008(IEEE World Congress on Computational Intelligence). IEEE International Joint Conference on*, 431–438.

79 Pfeil, T., Potjans, T.C., Schrader, S. et al. (2012). Is a 4-bit synaptic weight resolution enough? – constraints on enabling spike-timing dependent plasticity in neuromorphic hardware. *Front. Neurosci.* 6: 90.

80 Schemmel, J., Briiderle, D., Griibl, A. et al. (2010). A wafer-scale neuromorphic hardware system for large-scale neural modeling. In: *Proceedings of 2010 IEEE International Symposium on Circuits and Systems*, 1947–1950.

81 Naud, R., Marcille, N., Clopath, C., and Gerstner, W. (2008). Firing patterns in the adaptive exponential integrate-and-fire model. *Biol. Cybern.* 99 (4): 335.

82 Brüderle, D., Petrovici, M.A., Vogginger, B. et al. (2011). A comprehensive workflow for general-purpose neural modeling with highly configurable neuromorphic hardware systems. *Biol. Cybern.* 104 (4): 263–296.

83 Benjamin, B.V., Gao, P., McQuinn, E. et al. (2014). Neurogrid: a mixed-analog-digital multichip system for large-scale neural simulations. *Proc. IEEE* 102 (5): 699–716.

84 Indiveri, G. and Liu, S. (2015). Memory and information processing in neuromorphic systems. *Proc. IEEE* 103 (8): 1379–1397.

85 Boahen, K.A. (2004). A burst-mode word-serial address-event Link-I: transmitter design. *IEEE Trans. Circuits Syst. Regul. Pap.* 51 (7): 1269–1280.

86 Boahen, K.A. (2004). A burst-mode word-serial address-event Link-II: receiver design. *IEEE Trans. Circuits Syst. Regul. Pap.* 51 (7): 1281–1291.

87 Boahen, K.A. (2004). A burst-mode word-serial address-event Link-III: analysis and test results. *IEEE Trans. Circuits Syst. Regul. Pap.* 51 (7): 1292–1300.

88 Mitra, S., Fusi, S., and Indiveri, G. (2009). Real-time classification of complex patterns using spike-based learning in neuromorphic VLSI. *IEEE Trans. Biomed. Circuits Syst.* 3 (1): 32–42.

89 Qiao, N., Mostafa, H., McQuinn, E. et al. (2015). A reconfigurable on-line learning spiking neuromorphic processor comprising 256 neurons and 128 K synapses. *Front. Neurosci.* 9.

90 Hsieh, H. and Tang, K. (2012). VLSI implementation of a bio-inspired olfactory spiking neural network. *IEEE Trans. Neural Networks Learn. Syst.* 23 (7): 1065–1073.

91 Sun, Q., Schwartz, F., Michel, J. et al. (2011). Implementation study of an analog spiking neural network for assisting cardiac delay prediction in a cardiac resynchronization therapy device. *IEEE Trans. Neural Networks* 22 (6): 858–869.

92 Brader, J.M., Senn, W., and Fusi, S. (2007). Learning real-world stimuli in a neural network with spike-driven synaptic dynamics. *Neural Comput.* 19 (11): 2881–2912.

93 Burr, G.W., Shelby, R.M., Sebastian, A. et al. (2017). Neuromorphic computing using non-volatile memory. *Adv. Phys. X* 2 (1): 89–124.

94 Sheridan, P.M., Cai, F., Du, C. et al. (2017). Sparse coding with memristor networks. *Nat. Nanotechnol.* 12 (8): 784.

95 Basu, A., Acharya, J., Karnik, T. et al. (2018). Low-power, adaptive neuromorphic systems: recent progress and future directions. *IEEE J. Emerging Sel. Top. Circuits Syst.* 8 (1): 6–27.

96 Narasimman, G., Roy, S., Fong, X. et al. (2016). A low-voltage, low power STDP synapse implementation using domain-wall magnets for spiking neural networks. In: *2016 IEEE International Symposium on Circuits and Systems (ISCAS)*, 914–917.

97 Zhang, D., Zeng, L., Cao, K. et al. (2016). All spin artificial neural networks based on compound spintronic synapse and neuron. *IEEE Trans. Biomed. Circuits Syst.* 10 (4): 828–836.

98 Srinivasan, G., Sengupta, A., and Roy, K. (2016). Magnetic tunnel junction based long-term short-term stochastic synapse for a spiking neural network with on-chip STDP learning. *Sci. Rep.* 6: 29545.

99 Sengupta, A., Banerjee, A., and Roy, K. (2016). Hybrid spintronic-CMOS spiking neural network with on-chip learning: devices, circuits, and systems. *Phys. Rev. Appl.* 6 (6): 64003.

100 Sengupta, A., Parsa, M., Han, B., and Roy, K. (2016). Probabilistic deep spiking neural systems enabled by magnetic tunnel junction. *IEEE Trans. Electron Devices* 63 (7): 2963–2970.

101 Sengupta, A., Ankit, A., and Roy, K. (2017). Performance analysis and benchmarking of all-spin spiking neural networks (Special session paper). In: *Neural Networks (IJCNN), 2017 International Joint Conference on*, 4557–4563.

102 Vincent, A.F., Larroque, J., Locatelli, N. et al. (2015). Spin-transfer torque magnetic memory as a stochastic memristive synapse for neuromorphic systems. *IEEE Trans. Biomed. Circuits Syst.* 9 (2): 166–174.

103 Sengupta, A. and Roy, K. (2016). A vision for all-spin neural networks: a device to system perspective. *IEEE Trans. Circuits Syst. Regul. Pap.* 63 (12): 2267–2277.

104 Sengupta, A., Han, B., and Roy, K. (2016). Toward a spintronic deep learning spiking neural processor. In: *Biomedical Circuits and Systems Conference (BioCAS), 2016 IEEE*, 544–547.

105 Diehl, P.U. and Cook, M. (2015). Unsupervised learning of digit recognition using spike-timing-dependent plasticity. *Front. Comput. Neurosci.* 9: 99.

106 Diehl, P.U., Neil, D., Binas, J. et al. (2015). Fast-classifying, high-accuracy spiking deep networks through weight and threshold balancing. In: *2015 International Joint Conference on Neural Networks (IJCNN)*, 1–8.

107 Kvatinsky, S., Ramadan, M., Friedman, E.G., and Kolodny, A. (2015). VTEAM: a general model for voltage-controlled memristors. *IEEE Trans. Circuits Syst. II Express Briefs* 62 (8): 786–790.

108 Serrano-Gotarredona, T., Prodromakis, T., and Linares-Barranco, B. (2013). A proposal for hybrid memristor-CMOS spiking neuromorphic learning systems. *IEEE Circuits Syst. Mag.* 13 (2): 74–88.

109 Querlioz, D., Bichler, O., Dollfus, P., and Gamrat, C. (2013). Immunity to device variations in a spiking neural network with memristive nanodevices. *IEEE Trans. Nanotechnol.* 12 (3): 288–295.

110 Kuzum, D., Jeyasingh, R.G.D., Lee, B., and Wong, H.-S.P. (2011). Nanoelectronic programmable synapses based on phase change materials for brain-inspired computing. *Nano Lett.* 12 (5): 2179–2186.

111 Jo, S.H., Chang, T., Ebong, I. et al. (2010). Nanoscale memristor device as synapse in neuromorphic systems. *Nano Lett.* 10 (4): 1297–1301.

112 Kim, S., Ishii, M., Lewis, S. et al. (2015). NVM neuromorphic core with 64k-cell (256-by-256) phase change memory synaptic array with on-chip neuron circuits for continuous in-situ learning. In: *Electron Devices Meeting (IEDM), 2015 IEEE International*, 11–17.

113 Linares-Barranco, B., Serrano-Gotarredona, T., Camuñas-Mesa, L.A. et al. (2011). On spike-timing-dependent-plasticity, memristive devices, and building a self-learning visual cortex. *Front. Neurosci.* 5: 26.

114 Serrano-Gotarredona, T., Masquelier, T., Prodromakis, T. et al. (2013). STDP and STDP variations with memristors for spiking neuromorphic learning systems. *Front. Neurosci.* 7 (2).

115 Wong, H.-S.P., Raoux, S., Kim, S. et al. (2010). Phase change memory. *Proc. IEEE* 98 (12): 2201–2227.

116 Eryilmaz, S.B., Kuzum, D., Jeyasingh, R. et al. (2014). Brain-like associative learning using a nanoscale non-volatile phase change synaptic device array. *Front. Neurosci.* 8.

117 Lee, J.H. and Likharev, K.K. (May 2007). Defect-tolerant nanoelectronic pattern classifiers. *Int. J. Circuit Theory Appl.* 35 (3): 239–264.

118 Suri, M., Bichler, O., Querlioz, D. et al. (2012). CBRAM devices as binary synapses for low-power stochastic neuromorphic systems: Auditory (Cochlea) and visual (Retina) cognitive processing applications. In: *2012 International Electron Devices Meeting*, 10.3.1–10.3.4.

119 Du, C., Ma, W., Chang, T. et al. (2015). Biorealistic implementation of synaptic functions with oxide memristors through internal ionic dynamics. *Adv. Funct. Mater.* 25 (27): 4290–4299.

120 Kim, S., Du, C., Sheridan, P. et al. (2015). Experimental demonstration of a second-order memristor and its ability to biorealistically implement synaptic plasticity. *Nano Lett.* 15 (3): 2203–2211.

121 Wang, Z., Joshi, S., Savel'ev, S.E. et al. (2017). Memristors with diffusive dynamics as synaptic emulators for neuromorphic computing. *Nat. Mater.* 16 (1): 101–108.

122 Hu, M., Chen, Y., Yang, J.J. et al. (2017). A compact memristor-based dynamic synapse for spiking neural networks. *IEEE Trans. Comput. Aided Des. Integr. Circuits Syst.* 36 (8): 1353–1366.

123 Knag, P., Lu, W., and Zhang, Z. (2014). A native stochastic computing architecture enabled by memristors. *IEEE Trans. Nanotechnol.* 13 (2): 283–293.

124 Jo, S.H., Kim, K.-H., and Lu, W. (2008). Programmable resistance switching in nanoscale two-terminal devices. *Nano Lett.* 9 (1): 496–500.

125 Stata, R. (1967). Operational integrators. *Analog Dialogue* 1: 1–9.

6

Conclusions

By learning you will teach; by teaching you will learn.

–Latin Proverb

6.1 Outlooks

The looming end of Moore's law drives researchers to look for promising alternatives to conventional ways of computing. As neuromorphic computing started gaining popularity in recent years, this book aims at developing energy-efficient neural network hardware for energy-constrained applications. There is still a long way to go before achieving brain-like computing in hardware. The problem becomes even more complicated when the power consumption is also a major design consideration. A few outlooks for the future research directions in this field are presented in this section.

6.1.1 Brain-Inspired Computing

Even though the development of artificial intelligence (AI) does not necessarily follow the principle of how the brain functions, the research in the AI community and the neuroscience community has kept inspiring each other in the past. The amazing power of brains have inspired many algorithms and methods that are widely used in the AI community. At the same time, many algorithms and models developed for AI also helped in hypothesizing how the brain might actually work.

Hebbian rules and the spike-timing-dependent plasticity (STDP) that are discussed in Chapter 4 are two well-observed biological phenomena, which have long been believed to be the underlying mechanism of how the brain learns. In contrast to that, the backpropagation-based gradient descent learning algorithm that is widely adopted by the AI community has long been criticized to be biologically infeasible. The backpropagation originated from mathematical modeling. It emerged as an elegant way of solving the optimization problems in artificial neural networks (ANNs). Its birth had little, if any, to do with the learning in biological neural networks. Even though it has been long argued that for brain-like computing, one does not have to mimic exactly what a biological brain does, Hinton showed in [1] that the STDP protocol might be a way of conducting gradient descent optimization and backpropagation in the brain. Later on, this hypothesis was further developed in both the AI community [2–4] and the neuroscience community [5–7].

Learning in Energy-Efficient Neuromorphic Computing: Algorithm and Architecture Co-Design,
First Edition. Nan Zheng and Pinaki Mazumder.
© 2020 John Wiley & Sons Ltd. Published 2020 by John Wiley & Sons Ltd.

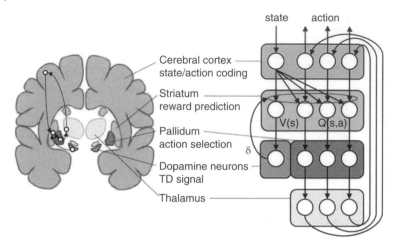

Figure 6.1 Implementation of reinforcement learning in the cortico-basal ganglia circuit. Different neurons in the brain are in charge of each stage needed in a TD-based learning [9]. Reproduced with permission of Taylor & Francis.

Reinforcement learning, which is discussed in Chapter 2, is a well-studied subject in the field of AI. It was found that the brain utilized a learning mechanism similar to that used in the actor–critic-based reinforcement learning framework. Conventionally, reinforcement learning is a reward-based method, which means that all decisions, actions, and evaluations are based on the reward received. On the other hand, animals rely more on a correlation-based learning. Wörgötter and Porr argued that these two methods were actually similar, if not identical [8]. In a classical reinforcement-learning task, the temporal difference (TD) error is often used as the cost function, which needs to be minimized in the learning, as demonstrated in Section 2.2.4. It was hypothesized and later on observed in biological experiments that dopamine might serve as the TD signal in the brain [9]. It was found in biological neural networks that a few types of neurons might be responsible for an actor–critic-based learning. As shown in Figure 6.1, different parts in the brain are in charge of the work that need to be done by the major blocks found in a classic reinforcement-learning framework [9]. There have been constant efforts in mapping components in reinforcement learning to the brain. Nevertheless, many discrepancies still exist between how the real brain functions and how reinforcement learning is formulated. Therefore, more experiments and theories are needed to provide a unified theory [10].

Another example of mutual inspiration comes from the widely adopted convolutional neural networks (CNNs). As discussed in previous sections, the CNN is one of the most popular and powerful neural networks in conducting image recognition tasks. The CNN was inspired by the simple cells and complex cells in biological visual systems [11, 12], which was discovered by Hubel and Wiesel in the 1960s [13]. The receptive field of a cell in the visual system can be defined as the visual field over which the firing of that cell can be influenced [13]. The convolutional layer in a CNN behaves similarly to a receptive field. Furthermore, the overall architecture of a CNN resembles the LGN-V1-V2-V4-IT hierarchy in the visual cortex [14]. In [15], Cadieu et al. compared the neural features measured from the inferior temporal cortex of a monkey with the features derived from deep ANNs. It was reported that the latest deep neural network was able to achieve an equivalent performance compared to the inferior cortex.

It is expected that, in the future, the two communities, the neuroscience community and the AI community, will continue to inspire each other to definitively understand the mechanisms disclosing how the brain works and how to build better machine-learning models. On one hand, the brain is by far our best roadmap for chartering the path of emergent research in AI. Therefore, we expect to discover and develop better brain-inspired algorithms. On the other hand, even though the long history of evolution might have shaped the brain into a piece of optimized "machine," it does not necessarily mean that duplicating a brain is the optimum solution, especially with different materials, e.g. silicon. In the future, much effort is needed from both the AI community and the circuit and device community to explore the best way to construct intelligent machines under the constraints of technologies, materials, and a reasonable power budget.

6.1.2 Emerging Nanotechnologies

In the previous chapters, the possibilities of utilizing many emerging nanotechnologies as the synapses in neuromorphic systems have been discussed. It is expected that many emerging non-volatile memories (NVMs) can provide a higher integration density, lower power consumption, and non-volatile data storage. Rajendran et al. conducted a study to compare the system implemented with conventional CMOS digital circuits and the system implemented with emerging nanoscale devices [16]. A network that contained 1 million neurons was chosen for comparison to evaluate the performance of large-scale learning systems. The comparison was based on a projected 10-nm technology scaled from a 22-nm CMOS technology. The comparison of the areas taken by these two types of systems is shown in Figure 6.2a. The area of the analog implementation with NVM devices is around 14× smaller than its digital counterpart. Such a great advantage in density is achieved by the compact sizes of both the nanoscale synapses and the analog neurons. As mentioned in previous chapters, a high integration density

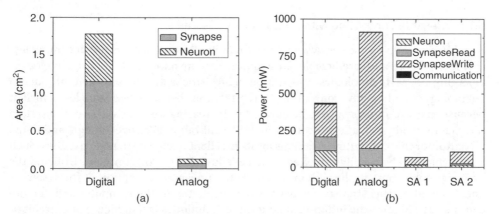

Figure 6.2 Comparison of (a) the area and (b) the power consumption of analog and digital implementations of SNNs [16]. The analog design is around 14× smaller than the digital counterpart thanks to the more compact analog neurons and synapses. The power consumption of the analog neural network varies significantly with the amplitude and duration of the current pulse needed to program the synapses. The "Analog," "SA1," and "SA2" configurations in the figure assume the amplitude of the current to be 1, 0.1, and 1 µA and the duration of the current to be 1, 1, and 0.1 µs, respectively. Reproduced with permission of IEEE.

is critical to large-scale neuromorphic systems because they normally contain an enormous number of synapses and neurons.

Despite savings in silicon area, the power consumption of the analog implementation is approximately twice as high as that consumed by the digital baseline. The high power consumption in the analog implementation is caused by the large programming current (1 μA) and the long programming duration (1 μs) assumed in [16]. It is also shown in Figure 6.2b that by reducing either the magnitude of the programming current or the programming duration, the power efficiency of the analog implementation can be readily boosted. In the figure, a programming current of 0.1 μA and a duration of 1 μs are assumed in the SA1 scheme, whereas a programming current of 1 μA and a duration of 0.1 μs are assumed in the SA2 scheme. It was further claimed in [16] that a nanoscale device that could be programmed by a current less than 100 nA at a voltage less than 0.5 V with a programming duration less than 100 ns could potentially improve both the density and energy efficiency of an analog neuromorphic system by at least a factor a 10 compared to its digital counterpart. Inspired by these results, it is expected that more and more researchers in the future will focus on optimizing existing devices and building new nanoscale NVM devices that can be operated and programmed with a lower power consumption.

In addition to reducing the power consumption of the NVM devices, it is paramount to improve the programming accuracy. In contrast to many digital systems where the synaptic weights can be arbitrarily manipulated, the weight stored in an analog synapse might not be controlled perfectly. This problem can be tackled from both the device level and the algorithm level. From the device level, it is expected that better devices will be engineered to exhibit less spatial and temporal variations. For example, in [17], Prezioso et al. have improved the fabrication process in order to produce memristor devices that are less prone to variations. From the algorithm level, algorithms that are intrinsically resilient to unreliable and defective devices will be explored. The learning examples with memristor crossbars illustrated in Section 5.3.5 fall into this category.

6.1.3 Reliable Computing with Neuromorphic Systems

As CMOS technology scales down to a few nanometers, transistors are becoming less and less reliable and predictable. In addition to conventional CMOS technologies, emerging NVM technologies also exhibit many stochastic behaviors. With such a technology trend, many conventional architectures become less and less efficient because many resources have to be devoted to increasing the robustness of the system through methods such as error correction [18], calibration [19], over-design, and so on.

Neuromorphic computing is famous for its excellent resilience to non-idealities. Such a resilience can be readily exploited to achieve low power consumption. Utilizing the excessive robustness in the system to reduce the power consumption of the system is not a new idea. This strategy has been widely exploited in the computer architecture community [20, 21]. The intuition behind these techniques is that failure of a circuit or a system only occurs rarely. Instead of over-designing the system so that it burns more power for an excessive reliability, one can eliminate this energy waste by allowing the system to fail. The circuit failure, however, can be captured and recovered with the help of specialized designed modules.

To exploit the excessive resilience in neuromorphic computing, many designs have been demonstrated in the past [22–24]. For example, voltage over-scaling, a technique that is widely used in approximate computing [25], has been adopted in neural networks. By aggressively reducing the supply voltage, the power consumption of the circuit can be quadratically reduced. Such a reduced voltage, however, might cause certain parts in the circuit to fail. The failure might be because the timing is not met or the memory cell loses its state. Unlike in conventional computing where even a single bit error might introduce catastrophic consequences, neuromorphic computing is more robust against errors. Such a feature can be readily exploited to increase the energy efficiency of the design [23, 24].

In addition to exploiting the robustness of the neural network hardware itself, it is worth exploring how to use neuromorphic computing to improve the reliability of conventional computing systems. For example, it has been shown that computing errors resulted from hardware faults or non-idealities can be mitigated through embedded machine learning [26, 27]. The intuition behind this approach is that through learning from the data, the system can acquire the knowledge on how to compensate for defects that exist in the system. Such an increased resilience can be leveraged to relax the requirement on the hardware, which indirectly reduces the power consumption of the system. Another possibility of utilizing a neural network to enhance the reliability of the hardware was demonstrated in [28] and [29]. Hopfield networks were employed to conduct memory self-repair. The problem of replacing faulty cells with spare ones was formulated as a problem of minimizing certain cost functions. A Hopfield network was then constructed according to the cost function. By exploiting the property of a Hopfield network, the cost function could be minimized and a repair scheme could be generated. This method of utilizing neural networks for memory repair is treated in more detail in the Appendix.

With all the robustness provided by the neural network, it is expected that, in the future, more and more researchers will devote to developing neuromorphic systems that can tolerate defects and errors in the hardware. Such an error-tolerance can be leveraged to achieve a lower power consumption or can be useful for systems deployed in extreme conditions or hazardous environments, such as high temperature, high pressure, etc., where the hardware failure rate is high. Furthermore, it is also expected that neuromorphic systems will be employed to enhance the reliability of conventional computing systems through leveraging their learning capability.

6.1.4 Blending of ANNs and SNNs

Conventionally, ANNs and spiking neural networks (SNNs) were two very different types of neural network. The development of these two neural networks was also largely separated. Many researchers whose main interests are ANNs may have little or no knowledge about SNNs. Indeed, as outlined in previous chapters, most successes to date have been achieved by the ANNs, partially thanks to the better support from the hardware side. However, as more researchers started entering the field of SNNs and more hardware platforms are specially designed for SNNs, it is expected that the development of SNNs will accelerate soon.

Regardless of the differences existing in these two kinds of neural networks, a trend in recent years is that the boundary between these two types of neural networks becomes less and less clear. It is sometimes advantageous to blend these two kinds of neural networks in certain applications. There are at least two ways to combine SNNs with ANNs. The first one is to directly connect these two types of neural networks to form a new network. Conventionally, how to bridge an SNN with an ANN is not clear as they operate in two distinct modes. With the learning algorithm presented in Section 4.3.5, this combination becomes feasible. With the proposed learning algorithm, an SNN can be trained in a similar way in which an ANN is trained. It can be foreseen that these two types of networks can be naturally connected and trained together. In the ANN domain, learning is carried out according to the conventional backpropagation method, with the gradients being derived analytically. For the SNN domain, the weight-dependent STDP modulation scheme can be readily employed to further backpropagate the error signal. By doing so, these two different neural networks can cooperate and exploit their advantages. For example, SNNs can be used as the frontend to handle the vast and highly sparse input signals by leveraging its better scalability and energy efficiency when dealing with sparse signals. ANNs can be employed as the backend to leverage its strength in dealing with dense and high-precision data.

The second way of blending ANNs and SNNs is to build SNN-inspired ANNs or ANN-inspired SNNs. For example, the learning algorithm presented in Section 4.3.5 is largely inspired by the learning algorithm for ANNs. By doing so, many well-developed theories and techniques can be inherited, which helps train SNNs better. On the other hand, many ANN algorithms and hardware developed in recent years have features that are inspired by SNNs. For example, the binarized network [30, 31] resembles an SNN in the sense that it uses a binary number to carry information, which can significantly improve the energy efficiency of the hardware. In addition, many newly developed hardware architectures [32, 33] break the activations of neurons in an ANN into binary streams in order to take advantage of the sparsity of the data and the variable precision needed by the application.

It can be foreseen that, in the future, these two types of networks will keep blending with each other. The boundary between them might become vaguer and vaguer as well. By combining the merits of these two types of networks, it is expected that faster and more energy-efficient hardware can be built.

6.2 Conclusions

In this book, the latest developments and trends in energy-efficient neuromorphic computing are discussed. A holistic approach across algorithm, architecture, and circuit is adopted. The objective of the book is to develop a methodology for designing energy-efficient neural network hardware for new forms of computing that is beyond the conventional von Neumann architectures. Based on the current status of the development of the algorithms and hardware architectures for neural-network-based computing, Chapters 2 and 3 are devoted to rate-based ANNs and Chapters 4 and 5 focus on spike-based SNNs. Starting from the basic operational principles of neural networks, various network types and topologies are presented. As the learning capability of neural networks is the fundamental reason that the neuromorphic approach can outperform conventional rule-based computing and even many other data-driven machine-learning tools in various real-life applications, a significant portion of this book is devoted to discussing various hardware-friendly learning algorithms and learning-friendly hardware architectures. Numerous state-of-the-art learning algorithms, hardware implementations, and low-power design techniques are discussed. Several case studies are also presented in great depth to help understand the details of the algorithms and architectures.

Neuromorphic computing is not a new topic, yet it keeps bringing up new findings and record-breaking performance in recent years. A new round of innovative research has just started. More and more researchers have embraced AI, machine learning, and neural networks as powerful tools to solve complex problems. We hope that this book will catalyze a good many future innovations to enrich the burgeoning field of neuromorphic engineering and computing.

References

1 Hinton, G. (2007). How to do backpropagation in a brain. In: *Invited talk at the NIPS'2007 Deep Learning Workshop.*

2 Bengio, Y., Lee, D.-H., Bornschein, J., et al., "Towards biologically plausible deep learning," *arXiv Prepr. arXiv1502.04156,* 2015.

3 Bengio, Y., Mesnard, T., Fischer, A., et al., "An objective function for STDP," *arXiv Prepr. arXiv1509.05936,* 2015.

4 Bengio, Y., and Fischer, A., "Early inference in energy-based models approximates back-propagation," *arXiv Prepr. arXiv1510.02777,* 2015.

5 Potjans, W., Diesmann, M., and Morrison, A. (2011). An imperfect dopaminergic error signal can drive temporal-difference learning. *PLoS Comput. Biol.* 7 (5): e1001133.

6 Potjans, W., Morrison, A., and Diesmann, M. (2009). A spiking neural network model of an actor-critic learning agent. *Neural Comput.* 21 (2): 301–339.

7 Frémaux, N., Sprekeler, H., and Gerstner, W. (2013). Reinforcement learning using a continuous time actor-critic framework with spiking neurons. *PLoS Comput. Biol.* 9 (4).

8 Wörgötter, F. and Porr, B. (2005). Temporal sequence learning, prediction, and control: a review of different models and their relation to biological mechanisms. *Neural Comput.* 17 (2): 245–319.

9 Doya, K. (2007). Reinforcement learning: computational theory and biological mechanisms. *HFSP J.* 1 (1): 30.

10 Dayan, P. and Niv, Y. (2008). Reinforcement learning: the good, the bad and the ugly. *Curr. Opin. Neurobiol.* 18 (2): 185–196.

11 LeCun, Y., Bengio, Y., and Hinton, G. (2015). Deep learning. *Nature* 521 (7553): 436–444.

12 LeCun, Y., Bottou, L., Bengio, Y., and Haffner, P. (1998). Gradient-based learning applied to document recognition. *Proc. IEEE* 86 (11): 2278–2323.

13 Hubel, D.H. and Wiesel, T.N. (1962). Receptive fields, binocular interaction and functional architecture in the cat's visual cortex. *J. Physiol.* 160 (1): 106–154.

14 Felleman, D.J. and Van Essen, D.C. (1991). Distributed hierarchical processing in the primate cerebral cortex. *Cereb. Cortex* 1 (1): 1–47.

15 Cadieu, C.F., Hong, H., Yamins, D.L. et al. (2014). Deep neural networks rival the representation of primate IT cortex for core visual object recognition. *PLoS Comput. Biol.* 10 (12): e1003963.

16 Rajendran, B., Liu, Y., Seo, J.S. et al. (2013). Specifications of nanoscale devices and circuits for neuromorphic computational systems. *IEEE Trans. Electron Devices* 60 (1): 246–253.

17 Prezioso, M., Merrikh-Bayat, F., Hoskins, B.D. et al. (2015). Training and operation of an integrated neuromorphic network based on metal-oxide memristors. *Nature* 521 (7550): 61–64.

18 Zheng, N. and Mazumder, P. (2017). An efficient eligible error locator polynomial searching algorithm and hardware architecture for one-pass chase decoding of BCH codes. *IEEE Trans. Circuits Syst. II Express Briefs* 64 (5): 580–584.

19 Zheng, N. and Mazumder, P. (2017). Modeling and mitigation of static noise margin variation in subthreshold SRAM cells. *IEEE Trans. Circuits Syst. I Regul. Pap.* 64 (10): 2726–2736.

20 Austin, T., Bertacco, V., Blaauw, D., and Mudge, T. (2005). Opportunities and challenges for better than worst-case design. In: *Proceedings of the 2005 Asia and South Pacific Design Automation Conference*, 2–7.

21 Ernst, D., Kim, N.S., Das, S. et al. (2003). Razor: a low-power pipeline based on circuit-level timing speculation. In: *Microarchitecture, 2003. MICRO-36. Proceedings. 36th Annual IEEE/ACM International Symposium on*, 7–18.

22 Kung, J., Kim, D., and Mukhopadhyay, S. (2015). A power-aware digital feedforward neural network platform with backpropagation driven approximate synapses. In: *Low Power Electronics and Design (ISLPED), 2015 IEEE/ACM International Symposium on*, 85–90.

23 Knag, P., Kim, J.K., Chen, T., and Zhang, Z. (2015). A sparse coding neural network ASIC with on-chip learning for feature extraction and encoding. *IEEE J. Solid-State Circuits* 50 (4): 1070–1079.

24 Reagen, B., Whatmough, P., Adolf, R. et al. (2016). Minerva: enabling low-power, highly-accurate deep neural network accelerators. In: *Proceedings of the 43rd International Symposium on Computer Architecture*, 267–278.

25 Han, J. and Orshansky, M. (2013). Approximate computing: an emerging paradigm for energy-efficient design. In: *2013 18th IEEE European Test Symposium (ETS)*, 1–6.

26 Zhang, J., Huang, L., Wang, Z., and Verma, N. (2015). A seizure-detection IC employing machine learning to overcome data-conversion and analog-processing non-idealities. In: *Custom Integrated Circuits Conference (CICC), 2015 IEEE*, 1–4.

27 Wang, Z., Lee, K.H., and Verma, N. (2015). Overcoming computational errors in sensing platforms through embedded machine-learning kernels. *IEEE Trans. Very Large Scale Integr. Syst.* 23 (8): 1459–1470.

28 Mazumder, P. and Jih, Y.-S. (1993). A new built-in self-repair approach to VLSI memory yield enhancement by using neural-type circuits. *IEEE Trans. Comput. Des. Integr. Circuits Syst.* 12 (1): 124–136.

29 Smith, M.D. and Mazumder, P. (1996). Generation of minimal vertex covers for row/column allocation in self-repairable arrays. *IEEE Trans. Comput.* 45 (1): 109–115.

30 Hubara, I., Courbariaux, M., Soudry, D. et al. (2016). Binarized neural networks. In: *Advances in Neural Information Processing Systems*, 4107–4115. Curran Associates, Inc.

31 Rastegari, M., Ordonez, V., Redmon, J., and Farhadi, A. (2016). XNOR-Net: ImageNet classification using binary convolutional neural networks. In: *European Conference on Computer Vision*, 525–542.

32 Judd, P., Albericio, J., Hetherington, T. et al. (2016). Stripes: bit-serial deep neural network computing. In: *2016 49th Annual IEEE/ACM International Symposium on Microarchitecture (MICRO)*, vol. 16, no. 1, 1–12.

33 Albericio, J., Judd, P., Delmás, A. et al. (2016). Bit-pragmatic deep neural network Computing. In: *Proceedings of the 50th Annual IEEE/ACM International Symposium on Microarchitecture*, 382–394.

Appendix A

As geometrical sizes of the devices used in a memory chip become smaller and con-comitantly the number of devices in a memory chip quadruples due to continual scaling of devices, the yield of memory chips reduces significantly due to the increase of manu-facturing defects. In order to improve the yield of memory chips, spare rows and spare columns are often incorporated in a memory chip so that they can be used to replace rows and columns containing faulty cells. To accomplish the repair, certain algorithmic methodologies are warranted to guide the replacement and repairing methods. In [1], the problem of minimizing the cost function in a memory repair problem is reformulated as a problem of minimizing the energy function in a lateral neural network, which can be solved elegantly by using the property of the Hopfield network.

A.1 Hopfield Network

The Hopfield network is a type of recurrent neural network with binary threshold units [2, 3]. The output of the binary neuron is either 0 or 1, similar to the perceptrons. Inter-actions between neurons occur through synapses. The transition function of a neuron can be defined as

$$
s_i' = \begin{cases} 0, & \text{if } \sum_j w_{ij}s_j < \theta_i \\ 1, & \text{if } \sum_j w_{ij}s_j > \theta_i \\ s_i, & \text{otherwise} \end{cases} \tag{A.1}
$$

where s_i and s_i' denote the current and the next state value of neuron i, respectively, and w_{ij} is the synaptic weight associated with the synapse connecting neuron i and neuron j. Conveniently, one can use a neuron with a threshold of zero and supply bias b_i to that neuron in order to achieve an effective non-zero threshold of $\theta_i = b_i$.

Learning in Energy-Efficient Neuromorphic Computing: Algorithm and Architecture Co-Design,
First Edition. Nan Zheng and Pinaki Mazumder.
© 2020 John Wiley & Sons Ltd. Published 2020 by John Wiley & Sons Ltd.

The energy function of a Hopfield network can be shown to follow the form

$$E^{NN} = -\frac{1}{2} \sum_i \sum_j w_{ij} s_i s_j - \sum_i s_i b_i \tag{A.2}$$

The Hopfield network has the inherent optimization property in the sense that when the network starts from a random initial state comprising arbitrary binary outputs of neurons in the entire network, ultimately the network will operate in such a way that the overall energy of the network monotonically decreases. The network eventually converges to a stable state corresponding to a minimum (possibly a local one) of the energy function shown in Eq. (A.2).

A.2 Memory Self-Repair with Hopfield Network

To utilize the Hopfield network for memory repair, the first step is to formulate a proper energy function for the problem [1]. Let us consider an $N \times N$ memory array with p spare rows and q spare columns. Suppose the memory has defects and the faulty cells are located at m distinct rows and n distinct columns. For the convenience of analysis, let us extract the $m \times n$ compacted subarray from the original $N \times N$ memory array, as illustrated in Figure A.1. The subarray contains all the defective cells in the memory. The $m \times n$ subarray can be conveniently represented by an $m \times n$ matrix D. The element d_{ij} in the matrix D is 1 if the corresponding cell is faulty, otherwise it is 0.

We can use a neural network with a size of $M = m + n$ to solve the problem of replacing faulty cells. We denote the first m neurons as s_{1i} and the remaining n neurons as s_{2j}; s_{1i}

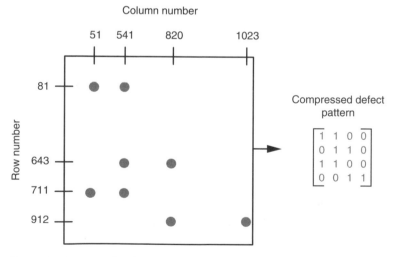

Figure A.1 An example of generating a compressed defect pattern from a memory array. Only rows and columns that contain faulty cells are extracted to form the compressed defect pattern. Defective cells are denoted as "1" in the array. Source: adapted from [1].

and s_{2j} are employed to determine whether a certain row or column needs to be replaced. More specifically, the row neuron s_{1i} is 1 when row i is selected for replacement, and it remains 0 otherwise. Similarly, the column neuron s_{2j} is 1 when column j is selected for replacement, and is 0 otherwise. With such a denotation, the cost functions of the repairing problem can be formulated as

$$C_1 = \frac{A}{2}\left[\left(\sum_{i=1}^{m} s_{1i}\right) - p\right]^2 + \frac{A}{2}\left[\left(\sum_{j=1}^{n} s_{2j}\right) - q\right]^2 \tag{A.3}$$

$$C_2 = B\left[\sum_{i=1}^{m}\sum_{j=1}^{n} d_{ij}(1 - s_{1i})(1 - s_{2j})\right] \tag{A.4}$$

The overall cost function is the sum of the above two cost functions. The first cost function deals with non-feasible repairing. It is worth noting that C_1 encourages to use up all the spare rows and columns. The cost function grows quadratically when the used spare rows and columns deviate from p and q. Nevertheless, the optimum usage of spare rows and columns can be experimentally decided, if desired. Alternatively, the first cost function can be defined as

$$C_1' = \frac{A}{2}\left(\sum_{i=1}^{m} s_{1i}\right)^2 + \frac{A}{2}\left(\sum_{j=1}^{n} s_{2j}\right)^2 \tag{A.5}$$

Clearly, such a cost function encourages to use as fewer spares as possible. The second cost function C_2 deals with the incomplete coverage. If all the faulty cells are covered by spares, this cost function becomes zero.

The next step is to determine the parameters of the neural network such that its energy function is equivalent to the cost function defined in Eqs. (A.3) and (A.4). Through comparing the energy function of the Hopfield network and the formulated cost function, the following parameters for the neural network can be obtained:

$$w_{1i.1j} = w_{2i.2j} = -A(1 - \delta_{ij}) \tag{A.6}$$

$$w_{1i.2j} = -Bd_{ij} \tag{A.7}$$

$$w_{2i.1j} = -Bd_{ji} \tag{A.8}$$

$$b_{1i} = \left(p - \frac{1}{2}\right)A + B\sum_{j} d_{ij} \tag{A.9}$$

$$b_{2j} = \left(q - \frac{1}{2}\right)A + B\sum_{i} d_{ij} \tag{A.10}$$

In the above equations, in order to set the main diagonal of the neural network's synaptic weight matrix to zero, the terms $A/2 \cdot \sum s_{1i}^2$ and $A/2 \cdot \sum s_{2j}^2$ are recast as $A/2 \cdot \sum s_{1i}$ and $A/2 \cdot \sum s_{2j}$. Nevertheless, the rewritten expressions are equivalent to the original forms, considering s_{1i} and s_{2j} are binary variables. Figure A.2 supplies an example of translating the defect pattern of a memory array into the connections in a neural network

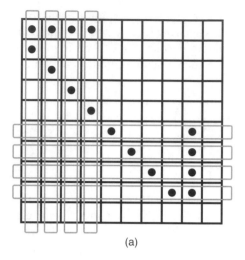

(a)

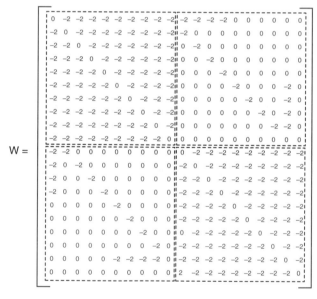

$$b = \begin{bmatrix} 15 & 9 & 9 & 9 & 9 & 11 & 11 & 11 & 11 & 7 & 11 & 11 & 11 & 11 & 9 & 9 & 9 & 9 & 15 & 7 \end{bmatrix}$$

(b)

Figure A.2 (a) An example of a faulty pattern. (b) The corresponding synaptic weights and biases for the faulty pattern. Source: adapted from [1].

used for obtaining the replacement solution. Sixteen faulty cells are assumed to exist in the memory array. Such a faulty pattern can be repaired with four spare rows and four spare columns, as illustrated in Figure A.2a. If one set A and B to be 2, then the weight matrix and bias vector shown in Figure A.2b can be obtained.

Table A.1 Averaged successful repairs achieved by RM and GD-zero.

	10 × 10 array			20 × 20 array		
		% of success			% of success	
% of Defects	Spares	RM	GD-zero	Spares	GD-zero	GD-zero
10	(3, 3)	36	70	(9, 9)	25	84
15	(3, 4)	33	57	(9, 12)	19	78
20	(4, 5)	30	72	(12, 12)	15	81

Source: data are from [1]. Reproduced with permission of IEEE.

The Hopfield network-based memory-repairing scheme was compared with the repair most (RM) algorithm that was widely used in memory repair [4]. The RM algorithm is a greedy algorithm that attempts to assign a spare row or spare column to replace the row or column that currently has the maximum uncovered faulty cells until all the defects are covered. The comparison results are shown in Table A.1. In the table, the results labeled with "GD-zero" correspond to those obtained from the neural network with an initial condition where the outputs from all neurons are zero. Such a setting was employed to provide an equal starting setup. Two array sizes, 10 × 10 and 20 × 20, are compared in Table A.1. It can be observed from the table that the gradient descent method outperforms the RM algorithm. Furthermore, as the defect pattern becomes large, the advantage of the gradient descent method is more obvious.

Similar to any other optimization method based on gradient descent, the process of minimizing the cost function may be stuck at the local minima. To circumvent this problem, a technique called hill-climbing (HC) approach was proposed in [1]. With this technique, the synaptic weights and biases in the neural network can be set as

$$w_{1i.1j} = w_{2i.2j} = -A \tag{A.11}$$

$$w_{1i.2j} = -Bd_{ij} \tag{A.12}$$

$$w_{2i.1j} = -Bd_{ji} \tag{A.13}$$

$$b_{1i} = pA + B \sum_j d_{ij} \tag{A.14}$$

$$b_{2j} = qA + B \sum_i d_{ij} \tag{A.15}$$

With these parameters, the diagonals in the weight matrix are no longer zero. In other words, self-feedback is allowed. The intuition behind this arrangement is that when the gradient descent is stuck at a local minimum, we can force the neural network to move by turning a neuron in the "Off" state into the "On" state (i.e. forcing it to output 1). This increases the energy of the system. The system is then expected to turn off some neurons (by setting their output to 0) to enter into a new lower energy state. Through

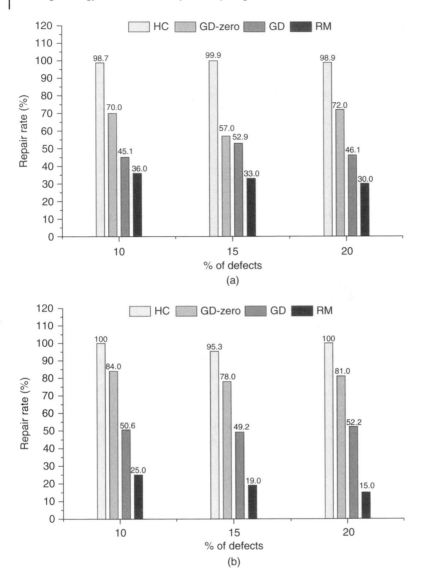

Figure A.3 (a) Performance summary of the four repair methods when the size of array is (a) 10×10 and (b) 20×20. All gradient-descent-based methods outperform the conventional RM algorithm. The HC technique can result in a nearly perfect coverage. Source: adapted from [1].

these operations, it is empirically established that the neural network can often escape from the local minima and can be guided to achieving the global optimization.

Four repairing schemes, HC, GD, GD-zero, and the baseline RM, are compared in Figure A.3. It can be observed that all of the Hopfield network-based methods proposed in [1] outperform the conventional RM algorithm. In addition, as the size of the

Table A.2 Comparison of the performances between GD and HC.

% of Defects	Spares		% of success		Avg number of steps		σ	
			GD	HC	GD	HC	GD	HC
10×10 array	10	(3, 3)	45.1	98.7	7.0	14.4	2.0	11.5
	15	(3, 4)	52.9	99.9	6.6	14.0	2.0	10.4
	20	(4, 5)	46.1	98.9	6.6	22.6	1.9	21.8
20×20 array	10	(9, 9)	50.6	100.0	13.0	22.6	2.6	13.1
	15	(9, 12)	49.2	95.3	12.6	40.2	2.7	25.3
	20	(12, 12)	52.2	100.0	12.7	20.4	2.6	16.8

Source: data are from [1]. Reproduced with permission of IEEE.

array increases, the advantage of the neural network approach becomes more obvious. The HC method with a global search capability can help achieve a better cover rate. Nearly perfect successful repairs can be achieved by this algorithm. To show the trade-off of employing the HC technique, simulations were conducted on networks with and without the HC algorithm and the obtained results are compared in Table A.2. The HC approach can almost always find a successful replacement strategy that can cover all the defects, thanks to its capability of escaping from local minima. In contrast, the gradient descent method is only able to achieve a successful repair rate of approximately half. Nevertheless, the number of steps needed by the GD method is considerably fewer in comparison to the HC approach. In addition, the variation in the run-time of the GD algorithm is much smaller compared to that of the HC approach.

The self-repair scheme based on the Hopfield network was implemented as an asynchronous mixed-signal circuit in [1]. The schematic of the circuit is shown in Figure A.4. It may be noted that all synaptic weights are negative for the memory repair problem. When a neuron in the figure fires, it turns on a transistor that sinks current from the input of the postsynaptic neuron. The interconnection matrix can be divided into four equal parts if we assume that the number of neurons represent spare rows and columns are equal. In the self-repair algorithm, only the upper-right and the lower-left quadrants need to be programmed according to the defect patterns existing in the memory.

How the circuit shown in Figure A.4 can be employed in a memory chip for self-repair is illustrated in Figure A.5. It is assumed that a built-in tester is also included to provide the fault pattern. The tester checks the status of each row and column in the memory and this information is recorded by the faulty-row-indicator shifter-register (FRISR) and the faulty-column-indicator shift-register (FCISR). The defect pattern can then be fed to the row-defect-pattern shift-register (RDPSR) row by row. This faulty information is used to obtain a compressed array. The bits in the RDPSR corresponding to the non-zero bits of the FCISR are shifted into the compressed-row-defect-pattern shit-register (CRDPSR). The CRDPSR then helps program a row or a column in the first and the third quadrants of the neural network. After a replacement pattern is obtained from the neural network, the information is expanded in the reverse order and fed into the actual reconfiguration circuits.

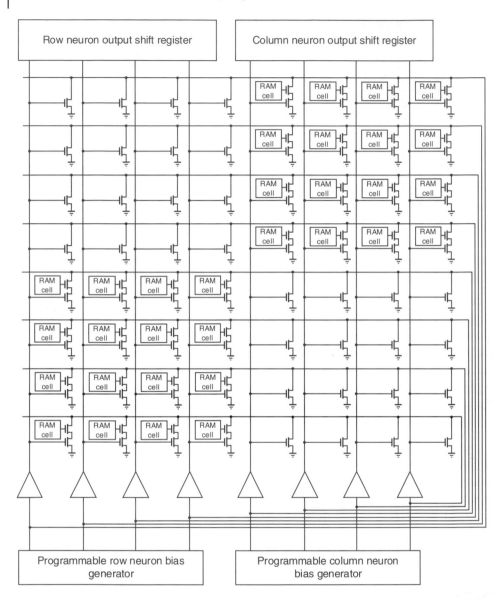

Figure A.4 Schematic of the neural network for memory repair. The weight matrix can be partitioned into four parts, as shown in the figure. When a neuron fires, the NMOS transistor is turned on and sinks current. Source: adapted from [1].

The asynchronous implementation in [1] worked when the size of the defect array was small. As the size of the array increases, the network tends to behave chaotically, considering the ever-growing uncertainties due to process, supply voltage, and temperature (PVT) variations. In [5], it was shown that the problem of spare row and column

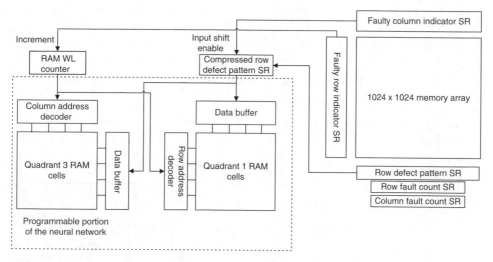

Figure A.5 Schematic of a neural network for embedded memory. A built-in memory tester is used to generate the fault pattern for the memory that needs to be repaired. The compressed row and column defect pattern are then extracted and used to program the neural network for repairing the faulty cells. Source: adapted from [1].

allocation could be treated as a special case of the generalized vertex cover (GVC) problem, which could be solved by a laterally connected neural network. Through constructing a new cost function, a more compact neural network was developed. The new neural network was implemented in a synchronous style in [5]. Compared to its asynchronous counterpart, the synchronous version of the network behaved more reliably and predictably. It was estimated in [5] that the neural network that could correct up to 32 by 32 faulty cells induced only a 0.29% area overhead for a 1-Mb DRAM.

As another example of employing a Hopfield network for memory self-repairing, a memristor-based neural network can be employed to repair defects in memristor-based memory arrays. One of the most promising applications of the emerging memristor technology is the non-volatile memory. Many attempts have been made to develop memristor-based memory [6, 7]. However, one of the biggest problems with memristor-based memory is its poor reliability. Therefore, the self-repair scheme can be readily employed in a memristor array to replace faulty cells with spare ones. The schematic of the self-repair circuit for a memristor memory array is illustrated in Figure A.6. The schematic resembles the purely-CMOS version shown in Figure A.4. The main difference here is that synaptic weights of the neural network are held by memristors instead of RAM cells. To program the faulty pattern into this array, the interface circuit needs to be changed accordingly so that each memristor can be written individually. Other operational schemes are similar to the CMOS version of the self-repair circuit shown in Figure A.4.

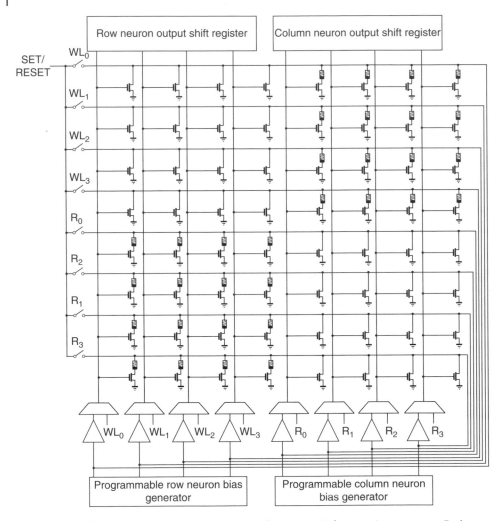

Figure A.6 Configuration of the memristor-based self-repair circuit for memristor memory. Faulty patterns are stored in memristors. An interface circuit is added to conveniently program each memristor cell.

References

1 Mazumder, P. and Jih, Y.-S. (1993). A new built-in self-repair approach to VLSI memory yield enhancement by using neural-type circuits. *IEEE Trans. Comput. Des. Integr. Circuits Syst.* 12 (1): 124–136.

2 Hopfield, J.J. and Tank, D.W. (1985). 'Neural' computation of decisions in optimization problems. *Biol. Cybern.* 52 (3): 141–152.

3 Hopfield, J.J. (1987). Neural networks and physical systems with emergent collective computational abilities. In: *Spin Glass Theory and Beyond: An Introduction to the Replica Method and Its Applications*, 411–415. World Scientific.

4 Tarr, M., Boudreau, D., and Murphy, R. (1984). Defect analysis system speeds test and repair of redundant memories. *Electronics* 57 (1): 175–179.

5 Smith, M.D. and Mazumder, P. (1996). Generation of minimal vertex covers for row/column allocation in self-repairable arrays. *IEEE Trans. Comput.* 45 (1): 109–115.

6 Ebong, I.E. and Mazumder, P. (2011). Self-controlled writing and erasing in a memristor crossbar memory. *IEEE Trans. Nanotechnol.* 10 (6): 1454–1463.

7 Ho, Y., Huang, G.M., and Li, P. (2009). Nonvolatile memristor memory: device characteristics and design implications. In: *Proceedings of the 2009 International Conference on Computer-Aided Design*, 485–490.

Index

Learning in Energy-Efficient Neuromorphic Computing: Algorithm and Architecture Co-Design,
First Edition. Nan Zheng and Pinaki Mazumder.
© 2020 John Wiley & Sons Ltd. Published 2020 by John Wiley & Sons Ltd.